D0849855

Parameter Estimation
and Inverse Problems

This is Volume 90 in the
INTERNATIONAL GEOPHYSICS SERIES

A series of monographs and textbooks
Edited by RENATA DMOWSKA, JAMES R. HOLTON, and H. THOMAS ROSSBY

A complete list of books in this series appears at the end of this volume.

Parameter Estimation
and Inverse Problems

Richard C. Aster, Brian Borchers, and Clifford H. Thurber

ELSEVIER

ACADEMIC
PRESS

Amsterdam • Boston • Heidelberg • London • New York • Oxford
Paris • San Diego • San Francisco • Singapore • Sydney • Tokyo

Acquisition Editor	Frank Cynar
Project Manager	Kyle Sarofeen
Editorial Coordinator	Jennifer Helé
Marketing Manager	Linda Beattie
Cover Design	Suzanne Rogers
Composition	Cepha Imaging Private Ltd.
Cover Printer	Phoenix Color Corporation
Interior Printer	Maple-Vail Book Manufacturing Group

Elsevier Academic Press
30 Corporate Drive, Suite 400, Burlington, MA 01803, USA
525 B Street, Suite 1900, San Diego, CA 92101-4495, USA
84 Theobald's Road, London WC1X 8RR, UK

This book is printed on acid-free paper. ∞

Copyright © 2005, Elsevier Inc. All rights reserved.

No part of this publication may be reproduced or transmitted in any form or by any means, electronic or mechanical, including photocopy, recording, or any information storage and retrieval system, without permission in writing from the publisher.

Permissions may be sought directly from Elsevier's Science & Technology Rights Department in Oxford, UK: phone: (+44) 1865 843830, fax: (+44) 1865 853333, e-mail: permissions@elsevier.com.uk. You may also complete your request online via the Elsevier homepage (http://elsevier.com), by selecting "Customer Support" and then "Obtaining Permissions."

Library of Congress Cataloging-in-Publication Data
Application submitted

British Library Cataloguing in Publication Data
A catalogue record for this book is available from the British Library

ISBN: 0-12-065604-3

For information on all Elsevier Academic Press Publications
visit our Web site at www.books.elsevier.com

Printed in the United States of America

04 05 06 07 08 09 9 8 7 6 5 4 3 2 1

Contents

11 BAYESIAN METHODS   201

A REVIEW OF LINEAR ALGEBRA   219

B REVIEW OF PROBABILITY AND STATISTICS   251

Preface

This textbook evolved from a course in geophysical inverse methods taught during the past decade at New Mexico Tech, first by Rick Aster and, for the past 5 years, jointly between Rick Aster and Brian Borchers. The audience for the course has included a broad range of first- or second-year graduate students (and occasionally advanced undergraduates) from geophysics, hydrology, mathematics, astronomy, and other disciplines. Cliff Thurber joined this collaboration during the past 3 years and has taught a similar course at the University of Wisconsin-Madison.

Our principal goal for this text is to promote fundamental understanding of parameter estimation and inverse problem philosophy and methodology, specifically regarding such key issues as uncertainty, ill–posedness, regularization, bias, and resolution. We emphasize theoretical points with illustrative examples, and MATLAB codes that implement these examples are provided on a companion CD. Throughout the examples and exercises, a CD icon indicates that there is additional material on the CD. Exercises include a mix of programming and theoretical problems. Please refer to the very last page of the book for more information on the contents of the CD as well as a link to a companion Web site.

This book has necessarily had to distill a tremendous body of mathematics and science going back to (at least) Newton and Gauss. We hope that it will find a broad audience of students and professionals interested in the general problem of estimating physical models from data. Because this is an introductory text surveying a very broad field, we have not been able to go into great depth. However, each chapter has a "Notes and Further Reading" section to help guide the reader to further exploration of specific topics. Where appropriate, we have also directly referenced research contributions to the field.

Some advanced topics have been deliberately omitted from the book because of space limitations and/or because we expect that many readers would not be sufficiently familiar with the required mathematics. For example, readers with a strong mathematical background may be surprised that we consider inverse problems with discrete data and discretized models. By doing this we avoid much of the technical complexity of functional analysis. Some advanced applications and topics that we have omitted include inverse scattering problems, seismic diffraction tomography, wavelets, data assimilation, and expectation maximization (EM) methods.

We expect that readers of this book will have prior familiarity with calculus, differential equations, linear algebra, probability, and statistics at the undergraduate level. In our experience, many students are in need of at least a review of these topics, and we typically spend the first 2 to 3 weeks of the course reviewing this material from Appendices A, B, and C.

Chapters 1 through 5 form the heart of the book and should be covered in sequence. Chapters 6, 7, and 8 are independent of each other, but depend strongly on the material in Chapters 1 through 5. As such, they may be covered in any order. Chapters 9 and 10 are independent of Chapters 6 through 8, but are most appropriately covered in sequence. Chapter 11 is independent of the specifics of Chapters 6 through 10 and provides an alternate view on, and summary of, key statistical and inverse theory issues.

If significant time is allotted for review of linear algebra, vector calculus, probability, and statistics in the appendices, there will probably not be time to cover the entire book in one semester. However, it should be possible for teachers to cover the majority of the material in the chapters following Chapter 5.

We especially wish to acknowledge our own professors and mentors in this field, including Kei Aki, Robert Parker, and Peter Shearer. We thank our many colleagues, including our own students, who provided sustained encouragement and feedback, particularly Kent Anderson, James Beck, Elena Resmerita, Charlotte Rowe, Tyson Strand, and Suzan van der Lee. Stuart Anderson, Greg Beroza, Ken Creager, Ken Dueker, Eliza Michalopoulou, Paul Segall, Anne Sheehan, and Kristy Tiampo deserve special mention for their classroom testing of early versions of this text. Robert Nowack, Gary Pavlis, Randall Richardson, and Steve Roecker provided thorough reviews that substantially improved the final manuscript. We offer special thanks to Per Christian Hansen of the Technical University of Denmark for collaboration in the incorporation of his Regularization Tools, which we highly recommend as an adjunct to this text. We also thank the editorial staff at Academic Press, especially Frank Cynar, Kyle Sarofeen, and Jennifer Helé, for essential advice and direction. Suzanne Borchers and Susan Delap provided valuable proofreading and graphics expertise. Brian Borchers was a visiting fellow at the Institute for Pure and Applied Mathematics (IPAM) at UCLA, and Rick Aster was partially supported by the New Mexico Tech Geophysical Research Center during the preparation of the text.

Rick Aster, Brian Borchers, and Cliff Thurber
December 2004

1

INTRODUCTION

Synopsis: General issues associated with parameter estimation and inverse problems are introduced through the concepts of the forward problem and its inverse solution. Scaling and superposition properties that characterize linear systems are given, and common situations leading to linear and nonlinear mathematical models are discussed. Examples of discrete and continuous linear and nonlinear parameter estimation problems to be revisited in more detail in later chapters are shown. Mathematical demonstrations highlighting the key issues of solution existence, uniqueness, and instability are presented and discussed.

1.1 CLASSIFICATION OF INVERSE PROBLEMS

Scientists and engineers frequently wish to relate physical parameters characterizing a **model**, m, to collected observations making up some set of **data**, d. We will commonly assume that the fundamental physics are adequately understood, so a function, G, may be specified relating m and d

$$G(m) = d. \tag{1.1}$$

In practice, d may be a function of time and/or space, or may be a collection of discrete observations. An important issue is that actual observations always contain some amount of noise. Two common ways that noise may arise are unmodeled influences on instrument readings or numerical round-off. We can thus envision data as generally consisting of noiseless observations from a "perfect" experiment, d_{true}, plus a noise component η,

$$d = G(m_{\text{true}}) + \eta \tag{1.2}$$

$$= d_{\text{true}} + \eta \tag{1.3}$$

where d_{true} exactly satisfies (1.1) for m equal to the true model, m_{true}, and we assume that the forward modeling is exact. We will see that it is commonly mathematically possible, although practically undesirable, to also fit all or part of η by (1.1). It may seem remarkable that it is often the case that a solution for m that is influenced by even a small noise amplitude η can have little or no correspondence to m_{true}. Another key issue that may seem astounding at first is that there are commonly an infinite number of models aside from m_{true} that fit the perfect data, d_{true}.

When m and d are functions, we typically refer to G as an **operator**. G will be called a function when m and d are vectors. The operator G can take on many forms. In some cases G is an ordinary differential equation (ODE) or partial differential equation (PDE). In other cases, G is a linear or nonlinear system of algebraic equations.

Note that there is some inconsistency between mathematicians and other scientists in modeling terminology. Applied mathematicians usually refer to $G(m) = d$ as the "mathematical model" and to m as the "parameters." On the other hand, scientists often refer to G as the "forward operator" and to m as the "model." We will adopt the scientific parlance and refer to m as the "model" while referring to the equation $G(m) = d$ as the "mathematical model."

The **forward problem** is to find d given m. Computing $G(m)$ might involve solving an ODE or PDE, evaluating an integral, or applying an algorithm for which there is no explicit analytical formula for $G(m)$. Our focus in this text is on the **inverse problem** of finding m given d. A third problem, not addressed here, is the **system identification problem** of determining G given examples of m and d.

In many cases, we will want to determine a finite number of parameters, n, to define a model. The parameters may define a physical entity directly (e.g., density, voltage, seismic velocity), or may be coefficients or other constants in a functional relationship that describes a physical process. In this case, we can express the model parameters as an n-element vector **m**. Similarly, if there are a finite number of data points then we can express the data as an m-element vector **d** (note that the use of the integer m here for the number of data points is easily distinguishable from the model m by its context). Such problems are called **discrete inverse problems** or **parameter estimation problems**. A general parameter estimation problem can be written as a system of equations

$$G(\mathbf{m}) = \mathbf{d}. \tag{1.4}$$

In other cases, where the model and data are functions of time and space, the associated task of estimating m from d is called a **continuous inverse problem**. A central theme of this book is that continuous inverse problems can often be well approximated by discrete inverse problems. The process of **discretizing** a continuous inverse problem is discussed in Chapter 3.

We will generally refer to problems with small numbers of parameters as "parameter estimation problems." Problems with a larger number of parameters will be referred to as "inverse problems." A key aspect of many inverse problems is that they are ill conditioned in a sense that will be discussed later in this chapter. In both parameter estimation and inverse problems we solve for a set of parameters that characterize a model, and a key point of this text is that the treatment of all such problems can be sufficiently generalized so that the distinction is largely irrelevant. In practice, the distinction that is important is the distinction between ill-conditioned and well-conditioned parameter estimation problems.

A class of mathematical models for which many useful results exist are **linear systems**. Linear systems obey superposition

$$G(m_1 + m_2) = G(m_1) + G(m_2) \tag{1.5}$$

and scaling

$$G(\alpha m) = \alpha G(m). \tag{1.6}$$

In the case of a discrete linear inverse problem, (1.4) can always be written in the form of a linear system of algebraic equations. See Exercise 1.1.

$$G(\mathbf{m}) = \mathbf{Gm} = \mathbf{d}. \tag{1.7}$$

In a continuous linear inverse problem, G can often be expressed as a linear integral operator, where (1.1) has the form

$$\int_a^b g(s, x) \, m(x) \, dx = d(s) \tag{1.8}$$

and the function $g(s, x)$ is called the **kernel**. The linearity of (1.8) is easily seen because

$$\int_a^b g(s, x)(m_1(x) + m_2(x)) \, dx = \int_a^b g(s, x) \, m_1(x) \, dx + \int_a^b g(s, x) m_2(x) \, dx \tag{1.9}$$

and

$$\int_a^b g(s, x) \, \alpha m(x) \, dx = \alpha \int_a^b g(s, x) \, m(x) \, dx. \tag{1.10}$$

Equations in the form of (1.8), where $m(x)$ is the unknown, are called **Fredholm integral equations of the first kind (IFKs)**. IFKs arise in a surprisingly large number of inverse problems. A key property of these equations is that they have mathematical properties that make it difficult to obtain useful solutions by straightforward methods.

In many cases the kernel in (1.8) can be written to depend explicitly on $s - x$, producing a **convolution equation**

$$\int_{-\infty}^{\infty} g(s - x) \, m(x) \, dx = d(s). \tag{1.11}$$

Here we have written the interval of integration as extending from minus infinity to plus infinity, but other intervals can easily be accommodated by having $g(s - x) = 0$ outside of the interval of interest. When a forward problem has the form of (1.11), determining $d(s)$ from $m(x)$ is called **convolution**, and the inverse problem of determining $m(x)$ from $d(s)$ is called **deconvolution**.

Another IFK arises in the problem of inverting a **Fourier transform**

$$\Phi(f) = \int_{-\infty}^{\infty} e^{-i2\pi f x} \phi(x) \, dx \tag{1.12}$$

to obtain $\phi(x)$. Although there are many tables and analytic methods of obtaining Fourier transforms and their inverses, numerical estimates of $\phi(x)$ may be of interest, such as where there is no analytic inverse, or where we wish to estimate $\phi(x)$ from spectral data collected at discrete frequencies.

It is an intriguing question as to why linearity appears in many interesting geophysical problems. One answer is that many physical systems encountered in practice are accompanied by only small departures from equilibrium. An important example is seismic wave propagation, where the stresses associated with elastic wave fields are often very small relative to the elastic moduli that characterize the medium. This situation leads to small strains and to a very nearly linear stress–strain relationship. Because of this, seismic wave-field problems in many useful circumstances obey superposition and scaling. Other fields such as gravity and magnetism, at the strengths typically encountered in geophysics, also show effectively linear physics.

Because many important inverse problems are linear, and because linear theory assists in solving nonlinear problems, Chapters 2 through 8 of this book cover theory and methods for the solution of linear inverse problems. Nonlinear mathematical models arise when the parameters of interest have an inherently nonlinear relationship to the observables. This situation commonly occurs, for example, in electromagnetic field problems where we wish to relate geometric model parameters such as layer thicknesses to observed field properties. We discuss methods for nonlinear parameter estimation and inverse problems in Chapters 9 and 10, respectively.

1.2 EXAMPLES OF PARAMETER ESTIMATION PROBLEMS

■ **Example 1.1** A canonical parameter estimation problem is the fitting of a function, defined by a collection of parameters, to a data set. In cases where this function fitting procedure can be cast as a linear inverse problem, the procedure is referred to as **linear regression**. An ancient example of linear regression is the characterization of a ballistic trajectory. In a basic take on this problem, the data, $\mathbf{y}$, are altitude observations of a ballistic body at a set of times $\mathbf{t}$. See Figure 1.1. We wish to solve for a model, $\mathbf{m}$, that contains the initial altitude (m_1), initial vertical velocity (m_2), and effective gravitational acceleration (m_3) experienced by the body during its trajectory. This and related problems are naturally of practical interest in rocketry and warfare, but are also of fundamental geophysical interest, for example, in absolute gravity meters capable of estimating g from the acceleration of a falling object in a vacuum to accuracies approaching one part in 10^9 [86].

The mathematical model is a quadratic function in the (t, y) plane

$$y(t) = m_1 + m_2 t - (1/2)m_3 t^2 \tag{1.13}$$

that we expect to apply at all times of interest not just at the times t_i when we happen to have observations. The data will consist of m observations of the height of the body y_i at corresponding times t_i. Assuming that the t_i are measured precisely, and applying (1.13) to each observation, we obtain a system of equations with m rows and $n = 3$ columns that

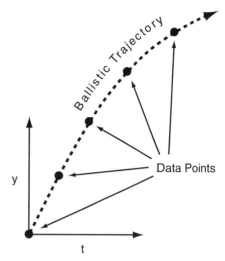

Figure 1.1 The parabolic trajectory problem.

relates the data y_i to the model parameters, m_j:

$$
\begin{bmatrix}
1 & t_1 & -\frac{1}{2}t_1^2 \\
1 & t_2 & -\frac{1}{2}t_2^2 \\
1 & t_3 & -\frac{1}{2}t_3^2 \\
\vdots & \vdots & \vdots \\
1 & t_m & -\frac{1}{2}t_m^2
\end{bmatrix}
\begin{bmatrix}
m_1 \\
m_2 \\
m_3
\end{bmatrix}
=
\begin{bmatrix}
y_1 \\
y_2 \\
y_3 \\
\vdots \\
y_m
\end{bmatrix}.
\tag{1.14}
$$

Although the mathematical model of (1.13) is quadratic, the equations for the three parameters m_i in (1.14) are linear, so solving for **m** is a linear parameter estimation problem.

If there are more data points than model parameters in (1.14) ($m > n$), then the m constraint equations in (1.14) will likely be inconsistent. In this case, it will typically be impossible to find a model **m** that satisfies every equation exactly. Geometrically, the nonexistence of a model that exactly satisfies the observations can be interpreted as being because there is no parabola that goes through all of the observed (t_i, y_i) points. See Exercise 1.2.

Such a situation could arise in practice because of noise in the determinations of the y_i, and because the forward model of (1.13) is approximate (e.g., we have neglected the physics of atmospheric drag, so the true trajectory is not exactly parabolic). In elementary linear algebra, where an exact solution is expected, we might throw up our hands and state that no solution exists. However, useful solutions to such systems may be found by solving for model parameters that satisfy the data in an approximate "best fit" sense. ■

A reasonable approach to finding the "best" approximate solution to an inconsistent system of linear equations is to find an **m** that produces a minimum misfit, or **residual**, between the data and the theoretical predictions of the forward problem. A traditional strategy, least squares, is to find the model that minimizes the **2-norm** (or Euclidean length) of the residual

$$\|\mathbf{y} - \mathbf{Gm}\|_2 = \sqrt{\sum_{i=1}^{m}(y_i - (\mathbf{Gm})_i)^2}. \tag{1.15}$$

However, (1.15) is not the only misfit measure that can be applied in solving such systems of equations. An alternative misfit measure that is better in many situations is the 1-norm

$$\|\mathbf{y} - \mathbf{Gm}\|_1 = \sum_{i=1}^{m}|y_i - (\mathbf{Gm})_i|. \tag{1.16}$$

We shall see in Chapter 2 that a solution minimizing (1.16) is less sensitive to data points that are wildly discordant with the mathematical model than a solution minimizing (1.15). Solution techniques that are resistant to such **outliers** are called **robust estimation procedures**.

■ **Example 1.2** A classic nonlinear parameter estimation problem in geophysics is determining an earthquake hypocenter in space and time specified by the 4-vector

$$\mathbf{m} = \begin{bmatrix} \mathbf{x} \\ \tau \end{bmatrix} \tag{1.17}$$

where **x** is the three-dimensional earthquake location and τ is the earthquake origin time. See Figure 1.2. The hypocenter model sought is that which best fits a vector of seismic phase **arrival times**, **t**, observed at an m-station seismic network. The mathematical model is

$$G(\mathbf{m}) = \mathbf{t} \tag{1.18}$$

where G models the physics of seismic wave propagation to map a hypocenter into a vector of predicted seismic arrival times at m stations. G depends on the seismic velocity structure, $v(\mathbf{x})$, which we assume is known.

The earthquake location problem is nonlinear even if $v(\mathbf{x})$ is a constant, c. In this case, all of the ray paths in Figure 1.2 would be straight, and the arrival time of the seismic signal at station i would be

$$t_i = \frac{\|\mathbf{S}_{\cdot,i} - \mathbf{x}\|_2}{c} + \tau \tag{1.19}$$

where the ith column of the matrix **S**, $\mathbf{S}_{\cdot,i}$ specifies the coordinates for station i. Equation (1.19) is nonlinear with respect to the spatial parameters x_i in **m**, and thus the problem cannot be expressed as a linear system of equations.   ■

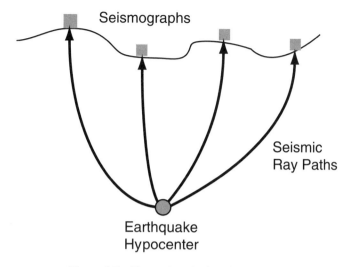

Figure 1.2 The earthquake location problem.

In special cases, a change of variables can convert a nonlinear problem to a linear one. Nonlinear parameter estimation problems can often be solved by choosing a starting model and then iteratively improving it until a good solution is obtained. General methods for solving nonlinear parameter estimation problems are discussed in Chapter 9.

1.3 EXAMPLES OF INVERSE PROBLEMS

■ **Example 1.3** In **vertical seismic profiling** we wish to know the vertical seismic velocity of the material surrounding a borehole. A downward-propagating seismic wavefront is generated at the surface by a source, and seismic waves are sensed by a string of seismometers in the borehole. See Figure 1.3.

The arrival time of the seismic wavefront at each instrument is measured from the recorded seismograms. The problem is nonlinear if expressed in terms of velocity parameters. However, we can linearize it by parameterizing it in terms of **slowness**, s, the reciprocal of the velocity v. The observed travel time at depth z is the definite integral of the vertical slowness, s, from the surface to z,

$$t(z) = \int_0^z s(\xi)\, d\xi \tag{1.20}$$

$$= \int_0^\infty s(\xi) H(z - \xi)\, d\xi, \tag{1.21}$$

where the kernel function H is the **Heaviside step function**, which is equal to one when its argument is nonnegative and zero when its argument is negative. The explicit dependence of the kernel on $z - \xi$ shows that (1.21) is a convolution.

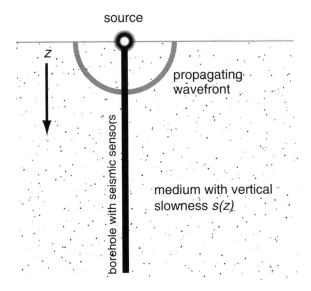

Figure 1.3   The vertical seismic profiling problem.

In theory, we can solve (1.21) quite easily because, by the fundamental theorem of calculus,

$$t'(z) = s(z). \tag{1.22}$$

In practice, there will be noise present in the data, $t(z)$, and simply differentiating the observations may result in a wildly noisy solution.   ∎

∎ **Example 1.4** A further instructive linear continuous inverse problem is the inversion of a vertical gravity anomaly, $d(s)$ observed at some height h to estimate an unknown buried line mass density distribution, $m(x) = \Delta\rho(x)$. See Figure 1.4. The mathematical model for this problem can be written as an IFK, because the data are a superposition of the vertical

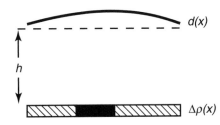

Figure 1.4 A linear inverse problem: determine the density of a buried line mass, $\Delta\rho(x)$, from gravity anomaly observations, $d(x)$.

gravity contributions from the differential elements comprising the line mass,

$$d(s) = \Gamma \int_{-\infty}^{\infty} \frac{h}{\left((x - s)^2 + h^2\right)^{3/2}} \, m(x) \, dx \tag{1.23}$$

$$= \int_{-\infty}^{\infty} g(x - s)m(x) \, dx, \tag{1.24}$$

where Γ is Newton's gravitational constant. Note that the kernel has the form $g(x - s)$, and (1.24) is thus another example of convolution. Because the kernel is a smooth function, $d(s)$ will be a smoothed transformation of $m(x)$. Conversely, solutions for $m(x)$ will be a roughened transformation of $d(s)$. For this reason we again need to be wary of the possibly severe deleterious effects of noise in the data. ∎

∎ **Example 1.5** Consider a variation on Example 1.4, where the depth of the line density perturbation varies, rather than the density contrast. The gravity anomaly is now attributed to a model describing variation in the burial depth, $m(x) = h(x)$, of a fixed line density perturbation, $\Delta\rho$. See Figure 1.5. The physics is the same as in Example 1.4, so the data are still given by the superposition of density perturbation contributions to the gravitational anomaly field, but the mathematical model now has the form

$$d(s) = \Gamma \int_{-\infty}^{\infty} \frac{m(x)}{\left((x - s)^2 + m^2(x)\right)^{3/2}} \, \Delta\rho \, dx. \tag{1.25}$$

This equation is nonlinear in $m(x)$ because (1.25) does not follow the superposition and scaling rules (1.5) and (1.6). ∎

Nonlinear inverse problems are generally far more difficult to solve than linear ones. In special instances they may be solvable by coordinate transformations that linearize the problem or other clever special-case methods. In other cases the problem cannot be linearized so nonlinear optimization techniques must be applied. The essential differences in the treatment of linear and nonlinear problems arise because, as we shall see in Chapters 2 through 4, all linear problems can be generalized to be the "same" in an important sense. In contrast, nonlinear problems tend to be nonlinear in mathematically different ways.

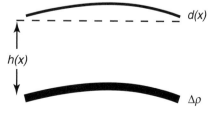

Figure 1.5   A nonlinear inverse problem: determine the depth $h(x)$ to a buried line mass density anomaly from observed gravity anomaly observations $d(x)$.

■ **Example 1.6** An important and instructive inverse problem is **tomography**, from the Greek roots *tomos*, "to section" or "to cut" (the ancient concept of an *atom* was that of an irreducible, uncuttable object), and *graphein*, "to write." Tomography is the general technique of determining a model from path-integrated properties such as attenuation (e.g., X-ray, radar, seismic), travel time (e.g., electromagnetic, seismic, or acoustic), or source intensity (e.g., positron emission). Although tomography problems originally involved models that were two-dimensional slices of three-dimensional objects, the term is now commonly used in situations where the model is two- or three-dimensional. Tomography has many applications in medicine, engineering, acoustics, and Earth science. One important geophysical example is cross-well tomography, where the sources are in a borehole, and the signals are received by sensors in another borehole. Another example is joint earthquake location/velocity structure inversion carried out on scales ranging from a fraction of a cubic kilometer to global [162, 163].

The physical model for tomography in its most basic form assumes that geometric ray theory (essentially the high-frequency limiting case of the wave equation) is valid, so that wave energy traveling between a source and receiver can be considered to be propagating along narrow ray paths. The density of ray path coverage in a tomographic problem may vary significantly throughout the study region, and may thus provide much better constraints on physical properties in some densely sampled regions than in other sparsely sampled ones.

If the slowness at a point $\mathbf{x}$ is $s(\mathbf{x})$, and the ray path ℓ is known, then the travel time along that ray path is given by the line integral along ℓ,

$$t = \int_\ell s(\mathbf{x}(l)) \, dl. \tag{1.26}$$

In general, ray paths can change direction because of refraction and/or reflection. In the simplest case where such effects are negligible, ray paths can be approximated as straight lines and the inverse problem can be cast in a linear form. If the ray paths depend on the model parameters (i.e., slowness), the inverse problem will be nonlinear.

A common way of discretizing the model in a tomographic problem is as uniform blocks. In this parameterization, the elements of $\mathbf{G}$ are just the lengths of the ray paths within the individual blocks. Consider the example of Figure 1.6, where nine homogeneous blocks with sides of unit length and unknown slowness are intersected by eight ray paths. For straight ray paths, the constraint equations in the mathematical model are

$$\mathbf{Gm} = \begin{bmatrix} 1 & 0 & 0 & 1 & 0 & 0 & 1 & 0 & 0 \\ 0 & 1 & 0 & 0 & 1 & 0 & 0 & 1 & 0 \\ 0 & 0 & 1 & 0 & 0 & 1 & 0 & 0 & 1 \\ 1 & 1 & 1 & 0 & 0 & 0 & 0 & 0 & 0 \\ 0 & 0 & 0 & 1 & 1 & 1 & 0 & 0 & 0 \\ 0 & 0 & 0 & 0 & 0 & 0 & 1 & 1 & 1 \\ \sqrt{2} & 0 & 0 & 0 & \sqrt{2} & 0 & 0 & 0 & \sqrt{2} \\ 0 & 0 & 0 & 0 & 0 & 0 & 0 & 0 & \sqrt{2} \end{bmatrix} \begin{bmatrix} s_{1,1} \\ s_{1,2} \\ s_{1,3} \\ s_{2,1} \\ s_{2,2} \\ s_{2,3} \\ s_{3,1} \\ s_{3,2} \\ s_{3,3} \end{bmatrix} = \begin{bmatrix} t_1 \\ t_2 \\ t_3 \\ t_4 \\ t_5 \\ t_6 \\ t_7 \\ t_8 \end{bmatrix}. \tag{1.27}$$

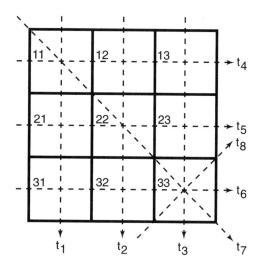

Figure 1.6   A simple tomography example.

Because there are nine unknown parameters $s_{i,j}$ in the model, but only eight constraints, the **G** matrix is clearly rank deficient. In fact, rank(**G**) is only seven. In addition, there is clearly redundant information in (1.27), in that the slowness $s_{3,3}$ is completely determined by t_8, yet $s_{3,3}$ also influences the observations t_3, t_6, and t_7.   ∎

1.4 WHY INVERSE PROBLEMS ARE HARD

Scientists and engineers need to be concerned with far more than simply finding mathematically acceptable answers to parameter estimation and inverse problems. One reason is that there may be many models that adequately fit the data. It is essential to characterize just what solution has been obtained, how "good" it is in terms of physical plausibility and fit to the data, and perhaps how consistent it is with other constraints. Essential issues that must be considered include **solution existence**, **solution uniqueness**, and **instability of the solution process**.

1. *Existence.* There may be no model that exactly fits the data. This can occur in practice because our mathematical model of the system's physics is approximate or because the data contain noise.
2. *Uniqueness.* If exact solutions do exist, they may not be unique, even for an infinite number of exact data points. That is, there may be other solutions besides m_{true} that exactly satisfy $G(m) = d_{\text{true}}$. This situation commonly occurs in potential field problems. A classic example is the external gravitational field from a spherically symmetric mass distribution, which depends only on the total mass, and not on the radial density distribution.

   Nonuniqueness is a characteristic of rank-deficient discrete linear inverse problems because the matrix **G** has a nontrivial **null space**. In linear inverse problems, models, $\mathbf{m}_0$, that lie in the null space of **G** are solutions to $\mathbf{Gm_0} = \mathbf{0}$. By superposition, any linear combination of these **null space models** can be added to a particular model that satisfies (1.7) and not change the fit to the data. There are thus an infinite number of mathematically acceptable models in such cases. In practical terms, suppose that there exists a nonzero model $\mathbf{m_0}$ which results in an instrument reading of zero. We cannot discriminate in such an experiment from the situation where $\mathbf{m_0}$ is truly zero.

An important and thorny issue with problems that have nonunique solutions is that an estimated model may be significantly smoothed or otherwise biased relative to the true model. Characterizing such bias is essential to interpreting models in terms of their possible correspondence to reality. This issue falls under the general topic of **model resolution analysis**.

3. *Instability*. The process of computing an inverse solution can be, and often is, extremely unstable in that a small change in measurement [e.g., a small η in (1.3)] can lead to an enormous change in the estimated model. Inverse problems where this situation arises are referred to as **ill-posed** in the case of continuous systems, or **ill-conditioned** in the case of discrete linear systems. A key point is that it is commonly possible to stabilize the inversion process by imposing additional constraints that bias the solution, a process that is generally referred to as **regularization**. Regularization is frequently essential to producing a usable solution to an otherwise intractable ill-posed or ill-conditioned inverse problem.

To examine existence, uniqueness, and instability issues, let us consider some simple mathematical examples using an IFK

$$\int_0^1 g(s, x)\, m(x)\, dx = y(s).$$ (1.28)

First, consider the trivial case of

$$g(s, x) = 1$$ (1.29)

that produces the integral equation

$$\int_0^1 m(x)\, dx = y(s).$$ (1.30)

Because the left-hand side of (1.30) is independent of s, this system has no solution unless $y(s)$ is a constant. Thus, there are an infinite number of mathematically conceivable data sets $y(s)$ that are not constant and for which no exact solution exists. This is an existence issue.

Where a solution to (1.30) does exist, the solution is nonunique because there are an infinite number of functions that, when integrated over the unit interval, produce the same constant and thus satisfy the IFK exactly. This is a uniqueness issue.

A more subtle example of nonuniqueness can be seen by letting

$$g(s, x) = s \sin(\pi x)$$ (1.31)

in (1.28), so that the IFK becomes

$$\int_0^1 s \cdot \sin(\pi x)\, m(x)\, dx = y(s). \tag{1.32}$$

The functions $\sin(k\pi x)$ are orthogonal in the sense that

$$\int_0^1 \sin(k\pi x) \sin(l\pi x)\, dx = -\frac{1}{2} \int_0^1 \cos(\pi(k+l)x) - \cos(\pi(k-l)x)\, dx,$$

$$= -\frac{1}{2\pi} \left(\frac{\sin(\pi(k+l))}{k+l} d - \frac{\sin(\pi(k-l))}{k-l} \right),$$

$$= 0 \qquad (k \neq \pm l; \quad k, l \neq 0). \tag{1.33}$$

Thus in (1.32) for $k = \pm 2, \pm 3, \ldots$ we have

$$\int_0^1 g(s, x) \sin(k\pi x)\, dx = 0. \tag{1.34}$$

Because (1.32) is a linear system, we can add any function of the form

$$m_0(x) = \sum_{k=2}^{\infty} \alpha_k \sin(k\pi x) \tag{1.35}$$

to a solution, $m(x)$, and obtain a new model that fits the data equally well.

$$\int_0^1 s \cdot \sin(\pi x)(m(x) + m_0(x))\, dx = \int_0^1 s \cdot \sin(\pi x)\, m(x)\, dx + \int_0^1 s \cdot \sin(\pi x)\, m_0(x)\, dx, \tag{1.36}$$

$$= \int_0^1 s \cdot \sin(\pi x)\, m(x)\, dx + 0. \tag{1.37}$$

There are thus an infinite number of very different solutions that fit the data equally well.

Even if we do not encounter existence or uniqueness issues, instability is a fundamental feature of IFKs. In the limit as k goes to infinity,

$$\lim_{k \to \infty} \int_{-\infty}^{\infty} g(s, t) \sin(k\pi t)\, dt = 0 \tag{1.38}$$

for *all* square-integrable functions $g(s, t)$. This result is known as the **Riemann–Lebesgue lemma** [135].

Without proving (1.38) rigorously, we can still understand why this occurs. The "wiggliness" of the sine function is smoothed by integration with the kernel $g(s, t)$. For large enough k, the postive and negative values of the sine effectively average out to 0. The inverse problem has this situation reversed; an inferred model can be very sensitive to small changes in the data, including random noise that has nothing to do with the physical system that we are trying to study.

The unstable character of IFK solutions is similar to the situation encountered in solving linear systems of equations where the condition number of the matrix is very large, or equivalently, where the matrix is nearly singular. In both cases, the difficulty lies in the mathematical model itself, and not in the particular algorithm used to solve the problem. Ill-posed behavior is a fundamental feature of many inverse problems because of the smoothing that occurs in most forward problems and the corresponding roughening that occurs in solving them.

1.5 EXERCISES

1.1 Consider a mathematical model of the form $G(\mathbf{m}) = \mathbf{d}$, where $\mathbf{m}$ is a vector of length n, and $\mathbf{d}$ is a vector of length m. Suppose that the model obeys the superposition and scaling laws and is thus linear. Show that $G(\mathbf{m})$ can be written in the form

$$G(\mathbf{m}) = \mathbf{\Gamma m} \tag{1.39}$$

where $\mathbf{\Gamma}$ is an m by n matrix. What are the elements of $\mathbf{\Gamma}$? Hint: Consider the standard basis, and write $\mathbf{m}$ as a linear combination of the vectors in the standard basis. Apply the superposition and scaling laws. Finally, recall the definition of matrix–vector multiplication.

1.2 Can (1.14) be inconsistent, even with only $m = 3$ data points? How about just $m = 2$ data points? If the system can be inconsistent, give an example. If not, explain why not.

1.3 Find a journal article that discusses the solution of an inverse problem in a discipline of special interest to you. What are the data? Are the data discrete or continuous? Have the authors discussed possible sources of noise in the data? What is the model? Is the model continuous or discrete? What physical laws determine the forward operator G? Is G linear or nonlinear? Do the authors discuss any issues associated with existence, uniqueness, or instability of solutions?

1.6 NOTES AND FURTHER READING

Some important references on inverse problems in geophysics and remote sensing include [21, 95, 122, 168, 54]. Instructive examples of ill-posed problems and their solution can be found in the book edited by Tikhonov and Goncharsky [165]. More mathematically oriented references on inverse problems include [9, 38, 53, 59, 85, 90, 104, 100, 164, 158].

Tomography, and in particular tomography in medical imaging, is a very large field. Some general references on tomography are [64, 78, 94, 96, 109, 117, 74].

2

LINEAR REGRESSION

Synopsis: Linear regression is introduced as a parameter estimation problem, and least squares solutions are derived. Maximum likelihood is defined, and its association with least squares solutions under normally distributed data errors is demonstrated. Statistical tests based on χ^2 that provide insight into least squares solutions are discussed. The mapping of data errors into model errors in least squares is described. The determination of confidence intervals using the model covariance matrix and the interpretation of model parameter correlations is discussed. The problems of estimating unknown data standard deviations and recognizing proportional data errors are addressed. The issue of data outliers and the concept of robust estimation are described and 1-norm minimization is introduced as a robust estimation technique. General propagation of errors between data and model using Monte Carlo methods is discussed in the context of the iteratively reweighted least squares 1-norm minimization algorithm.

2.1 INTRODUCTION TO LINEAR REGRESSION

The problem of finding a parameterized curve that approximately fits a set of data is referred to as **regression**. When the regression model is linear in the fitted parameters, then we have a **linear regression** problem. In this chapter linear regression problems are analyzed as discrete linear inverse problems.

Consider a discrete linear inverse problem. We begin with a data vector, $\mathbf{d}$, of m observations, and a vector $\mathbf{m}$ of n model parameters that we wish to determine. As we have seen, the inverse problem can be written as a linear system of equations

$$\mathbf{Gm} = \mathbf{d}. \tag{2.1}$$

Recall that if $\text{rank}(\mathbf{G}) = n$, then the matrix has full column rank. In this chapter we will assume that the matrix $\mathbf{G}$ has full column rank. In Chapter 4 we will consider rank-deficient problems.

For a full column rank matrix, it is frequently the case that no solution $\mathbf{m}$ satisfies (2.1) exactly. This happens because the dimension of the range of $\mathbf{G}$ is smaller than m and a noisy data vector can easily lie outside of the range of $\mathbf{G}$.

A useful approximate solution may still be found by finding a particular model $\mathbf{m}$ that minimizes some measure of the misfit between the actual data and $\mathbf{Gm}$. The **residual** vector, with elements that are frequently simply referred to as the **residuals**, is

$$\mathbf{r} = \mathbf{d} - \mathbf{Gm}. \tag{2.2}$$

One commonly used measure of the misfit is the 2-norm of the residuals. A model that minimizes this 2-norm is called a **least squares solution**. The least squares or 2-norm solution is of special interest both because it is very amenable to analysis and geometric intuition, and because it turns out to be statistically the most likely solution if data errors are normally distributed (see Section 2.2).

The least squares solution is, from the normal equations (A.73),

$$\mathbf{m}_{L_2} = (\mathbf{G}^T\mathbf{G})^{-1}\mathbf{G}^T\mathbf{d}. \tag{2.3}$$

It can be shown that if $\mathbf{G}$ is of full column rank then $(\mathbf{G}^T\mathbf{G})^{-1}$ exists. See Exercise A.17f.

A common linear regression problem is finding parameters m_1 and m_2 for a line, $y = m_1 + m_2 x$, that best fits a set of $m > 2$ data points. The system of equations in this problem is

$$\mathbf{Gm} = \begin{bmatrix} 1 & x_1 \\ 1 & x_2 \\ \vdots & \vdots \\ 1 & x_m \end{bmatrix} \begin{bmatrix} m_1 \\ m_2 \end{bmatrix} = \begin{bmatrix} d_1 \\ d_2 \\ \vdots \\ d_m \end{bmatrix} = \mathbf{d}. \tag{2.4}$$

Applying (2.3) to find a least squares solution gives

$$\mathbf{m}_{L_2} = (\mathbf{G}^T\mathbf{G})^{-1}\mathbf{G}^T\mathbf{d} = \left(\begin{bmatrix} 1 & \cdots & 1 \\ x_1 & \cdots & x_m \end{bmatrix} \begin{bmatrix} 1 & x_1 \\ \vdots & \vdots \\ 1 & x_m \end{bmatrix} \right)^{-1} \cdot \begin{bmatrix} 1 & \cdots & 1 \\ x_1 & \cdots & x_m \end{bmatrix} \begin{bmatrix} d_1 \\ d_2 \\ \vdots \\ d_m \end{bmatrix}$$

$$= \begin{bmatrix} m & \sum_{i=1}^{m} x_i \\ \sum_{i=1}^{m} x_i & \sum_{i=1}^{m} x_i^2 \end{bmatrix}^{-1} \begin{bmatrix} \sum_{i=1}^{m} d_i \\ \sum_{i=1}^{m} x_i d_i \end{bmatrix}$$

$$= \frac{1}{m \sum_{i=1}^{m} x_i^2 - \left(\sum_{i=1}^{m} x_i \right)^2} \begin{bmatrix} -\sum_{i=1}^{m} x_i^2 & -\sum_{i=1}^{m} x_i \\ -\sum_{i=1}^{m} x_i & -m \end{bmatrix} \begin{bmatrix} \sum_{i=1}^{m} d_i \\ \sum_{i=1}^{m} x_i d_i \end{bmatrix}. \tag{2.5}$$

2.2 STATISTICAL ASPECTS OF LEAST SQUARES

If we consider data points to be imperfect measurements that include random errors, then we are faced with the problem of finding the solution which is best from a statistical point of view. **Maximum likelihood estimation** considers the question from the following perspective. Given that we observed a particular data set, that we know the statistical characteristics of these observations, and that we have a mathematical model for the forward problem, what is the model from which these observations would most likely arise?

Maximum likelihood estimation can be applied to any estimation problem where a joint probability density function (B.26) can be assigned to the observations. The essential problem is to find the most likely model, as characterized by the elements of the parameter vector **m**, for the set of observations contained in the vector **d**. We will assume that the observations are independent so that we can use the product form of the joint probability density function (B.28).

Given a model **m**, we have a probability density function $f_i(d_i|\mathbf{m})$ for each of the i observations. In general, these probability density functions will vary depending on **m**. The joint probability density for a vector of independent observations **d** will be

$$f(\mathbf{d}|\mathbf{m}) = f_1(d_1|\mathbf{m}) \cdot f_2(d_2|\mathbf{m}) \cdots f_m(d_m|\mathbf{m}). \tag{2.6}$$

Note that the $f(\mathbf{d}|\mathbf{m})$ are probability densities, not probabilities. We can only compute the probability of observing data in some range for a given model **m** by integrating $f(\mathbf{d}|\mathbf{m})$ over that range. In fact, the probability of getting any particular set of data exactly is precisely zero! This conceptual conundrum can be avoided by considering the probability of getting a data set that lies within a small m-dimensional box around a particular data set **d**. This probability will be nearly proportional to the probability density $f(\mathbf{d}|\mathbf{m})$.

In practice, we measure a particular data vector and wish to find the "best" model to match it in the maximum likelihood sense. That is, **d** will be a fixed set of observations, and **m** will be a vector of parameters to be estimated. The **likelihood function** is

$$L(\mathbf{m}|\mathbf{d}) = f(\mathbf{d}|\mathbf{m}) \tag{2.7}$$

$$= f_1(d_1|\mathbf{m}) \cdot f_2(d_2|\mathbf{m}) \cdots f_m(d_m|\mathbf{m}). \tag{2.8}$$

For many possible models **m** the likelihood (2.8) will be extremely close to zero. Such models would be very unlikely to produce the observed data set **d**. The likelihood might be much larger for other models, and these would be relatively likely to produce the observed data.

According to the **maximum likelihood principle** we should select the model **m** that maximizes the likelihood function, (2.8). Model estimates obtained in this manner have many desirable statistical properties [22, 33].

For the remainder of this section we consider the important case of the least squares solution when the data contain normally distributed errors. It is particularly interesting that when we have a discrete linear inverse problem and the data errors are independent and normally distributed, then the maximum likelihood principle solution is the least squares solution. To show this, assume that the data have independent random errors that are normally distributed with expected value zero, and where the standard deviation of the ith observation, d_i is σ_i. The probability density for d_i then takes the form of (B.6)

$$f_i(d_i|\mathbf{m}) = \frac{1}{(2\pi)^{1/2}\sigma_i} e^{-(d_i-(\mathbf{Gm})_i)^2/2\sigma_i^2}. \tag{2.9}$$

The likelihood function for the complete data set is the product of the individual likelihoods (2.8).

$$L(\mathbf{m}|\mathbf{d}) = \frac{1}{(2\pi)^{m/2}\prod_{i=1}^{m}\sigma_i} \Pi_{i=1}^{m} e^{-(d_i-(\mathbf{Gm})_i)^2/2\sigma_i^2}. \tag{2.10}$$

The constant factor does not affect the maximization of L, so we can solve

$$\max \ \Pi_{i=1}^{m} e^{-(d_i-(\mathbf{Gm})_i)^2/2\sigma_i^2}. \tag{2.11}$$

The logarithm is a monotonically increasing function, so we can equivalently solve

$$\max \ \log \Pi_{i=1}^{m} e^{-(d_i-(\mathbf{Gm})_i)^2/2\sigma_i^2} = \max \ -\sum_{i=1}^{m} \frac{(d_i-(\mathbf{Gm})_i)^2}{2\sigma_i^2}. \tag{2.12}$$

Finally, if we turn the maximization into a minimization by changing sign and ignore the constant factor of $1/2$, the problem becomes

$$\min \ \sum_{i=1}^{m} \frac{(d_i-(\mathbf{Gm})_i)^2}{\sigma_i^2}. \tag{2.13}$$

Aside from the distinct $1/\sigma_i^2$ factors in each term, this is identical to solving (2.3), the least squares problem for $\mathbf{Gm} = \mathbf{d}$.

To incorporate the data standard deviations into a solution, we scale the system of equations. Let a diagonal weighting matrix be

$$\mathbf{W} = \text{diag}(1/\sigma_1, 1/\sigma_2, \dots, 1/\sigma_m). \tag{2.14}$$

Then let

$$\mathbf{G}_w = \mathbf{WG} \tag{2.15}$$

and

$$\mathbf{d}_w = \mathbf{Wd}. \tag{2.16}$$

The weighted system of equations is then

$$\mathbf{G}_w\mathbf{m} = \mathbf{d}_w. \tag{2.17}$$

The normal equations (A.73) solution to (2.17) is

$$\mathbf{m}_{L_2} = (\mathbf{G}_w^T \mathbf{G}_w)^{-1} \mathbf{G}_w^T \mathbf{d}_w. \tag{2.18}$$

Now,

$$\|\mathbf{d}_w - \mathbf{G}_w \mathbf{m}_w\|_2^2 = \sum_{i=1}^m (d_i - (\mathbf{Gm}_{L_2})_i)^2 / \sigma_i^2. \tag{2.19}$$

Thus the least squares solution to $\mathbf{G}_w \mathbf{m} = \mathbf{d}_w$, (2.18), is the maximum likelihood solution.

The sum of the squares of the residuals also provides useful statistical information about the quality of model estimates obtained with least squares. The **chi-square statistic** is

$$\chi_{\text{obs}}^2 = \sum_{i=1}^m (d_i - (\mathbf{Gm}_{L_2})_i)^2 / \sigma_i^2. \tag{2.20}$$

Since χ_{obs}^2 depends on the random measurement errors in $\mathbf{d}$, it is itself a random variable. It can be shown that under our assumptions, χ_{obs}^2 has a χ^2 distribution with $\nu = m - n$ degrees of freedom [22, 33].

The probability density function for the χ^2 distribution is

$$f_{\chi^2}(x) = \frac{1}{2^{\nu/2} \Gamma(\nu/2)} x^{\frac{1}{2}\nu - 1} e^{-x/2}. \tag{2.21}$$

See Figure B.5. The χ^2 **test** provides a statistical assessment of the assumptions that we used in finding the least squares solution. In this test, we compute χ_{obs}^2 and compare it to the theoretical χ^2 distribution with $\nu = m - n$ degrees of freedom.

The probability of obtaining a χ^2 value as large or larger than the observed value is

$$p = \int_{\chi_{\text{obs}}^2}^\infty f_{\chi^2}(x) \, dx. \tag{2.22}$$

This is called the *p-value* of the test. When data errors are independent and normally distributed, and the mathematical model is correct, it can be shown that the *p*-value will be uniformly distributed between zero and one. See Exercise 2.4. In practice, particular *p*-values that are very close to either end of the unit interval indicate that one or more of these assumptions are incorrect.

There are three general cases.

1. The *p*-value is not too small and not too large. Our least squares solution produces an acceptable data fit and our statistical assumptions of data errors are consistent. Practically, p does not actually have to be very large to be deemed marginally "acceptable" in many cases (e.g., $p \approx 10^{-2}$), as truly "wrong" models will typically produce extraordinarily small *p*-values (e.g., 10^{-12}) because of the short-tailed nature of the normal distribution.

Because the p-value will be uniformly distributed when we have a correct mathematical model and our statistical data assumptions are valid, it is inappropriate to conclude anything based on the differences between p-values in this range. For example, one should not conclude that a p-value of 0.7 is "better" than a p-value of 0.2.

2. The p-value is very small. We are faced with three nonexclusive possibilities, but something is clearly wrong.

    (a) The data truly represent an extremely unlikely realization. This is easy to rule out for p-values very close to zero. For example, suppose an experiment produced a data realization where the probability of a worse fit was 10^{-9}. If the model was correct, then we would have to perform on the order of a billion experiments to get a comparably poor fit to the data. It is far more likely that something else is wrong.

    (b) The mathematical model $\mathbf{Gm} = \mathbf{d}$ is incorrect. Most often this happens because we are using an inappropriate mathematical model.

    (c) The data errors are underestimated (in particular, we may have underestimated the σ_i), or are not normally distributed.

3. The p-value is very large (very close to one). The fit of the model predictions to the data is almost exact. We should investigate the possibility that we have overestimated the data errors. A rare possibility is that a very high p-value is indicative of data fraud, such as might happen if data were cooked up ahead of time to fit a particular model.

A rule of thumb for problems with a large number of degrees of freedom, ν, is that the expected value of χ^2 approaches ν. This arises because, by the central limit theorem (see Section B.6), the χ^2 random variable, which is itself a sum of random variables, will become nearly normally distributed as the number of terms in the sum becomes large. The mean of the resulting distribution will approach ν and the standard deviation will approach $(2\nu)^{1/2}$.

In addition to examining χ^2_{obs}, it is important to examine the residuals corresponding to a model. They should be roughly normally distributed with standard deviation one and should show no obvious patterns. In some cases where an incorrect model has been fitted to the data, the residuals will reveal the nature of the modeling error. For example, in linear regression to a line, it might be that all of the residuals are negative for small and large values of the independent variable x but positive for intermediate values of x. This would suggest that perhaps an additional quadratic term is required in the regression model.

Parameter estimates obtained via linear regression are linear combinations of the data. See (2.18). If the data errors are normally distributed, then the parameter estimates will also be normally distributed because a linear combination of normally distributed random variables is normally distributed [5, 22]. To derive the mapping between data and model covariances, consider the covariance of a data vector, $\mathbf{d}$, of normally distributed, independent random variables, operated on by a general linear transformation specified by a matrix, $\mathbf{A}$. Recall from (B.65) in Appendix B that the appropriate covariance mapping is

$$\text{Cov}(\mathbf{Ad}) = \mathbf{A}\text{Cov}(\mathbf{d})\mathbf{A}^T. \tag{2.23}$$

The least squares solution (2.18) has $\mathbf{A} = (\mathbf{G}_w^T \mathbf{G}_w)^{-1} \mathbf{G}_w^T$. Since the weighted data have an identity covariance matrix, the covariance for the model parameters is

$$\text{Cov}(\mathbf{m}_{L_2}) = (\mathbf{G}_w^T \mathbf{G}_w)^{-1} \mathbf{G}_w^T \mathbf{I}_m \mathbf{G}_w (\mathbf{G}_w^T \mathbf{G}_w)^{-1} = (\mathbf{G}_w^T \mathbf{G}_w)^{-1}. \tag{2.24}$$

In the case of independent and identically distributed normal data errors, the data covariance matrix $\text{Cov}(\mathbf{d})$ is simply the variance σ^2 times the m by m identity matrix, $\mathbf{I}_m$, and (2.24) simplifies to

$$\text{Cov}(\mathbf{m}_{L_2}) = \sigma^2 (\mathbf{G}^T \mathbf{G})^{-1}. \tag{2.25}$$

Note that the covariance matrix of the model parameters is typically not diagonal indicating that the model parameters are correlated. Because elements of least squares models are each constructed from linear combinations of the data vector elements, this statistical dependence between the elements of $\mathbf{m}$ should not be surprising.

The expected value of the least squares solution to $\mathbf{G}_w \mathbf{m} = \mathbf{d}_w$ is

$$E[\mathbf{m}_{L_2}] = (\mathbf{G}_w^T \mathbf{G}_w)^{-1} \mathbf{G}_w^T E[\mathbf{d}_w]. \tag{2.26}$$

Because $E[\mathbf{d}_w] = \mathbf{d}_{\text{true}_w}$, and $\mathbf{G}_w \mathbf{m}_{\text{true}} = \mathbf{d}_{\text{true}_w}$, we have

$$\mathbf{G}_w^T \mathbf{G}_w \mathbf{m}_{\text{true}} = \mathbf{G}_w^T \mathbf{d}_{\text{true}_w}. \tag{2.27}$$

Thus

$$E[\mathbf{m}_{L_2}] = (\mathbf{G}_w^T \mathbf{G}_w)^{-1} \mathbf{G}_w^T \mathbf{G}_w \mathbf{m}_{\text{true}} \tag{2.28}$$

$$= \mathbf{m}_{\text{true}} \tag{2.29}$$

In statistical terms, the least squares solution is said to be **unbiased**.

We can compute 95% confidence intervals for individual model parameters using the fact that each model parameter m_i has a normal distribution with mean given by the corresponding element of $\mathbf{m}_{\text{true}}$ and variance $\text{Cov}(\mathbf{m}_{L_2})_{i,i}$. The 95% confidence intervals are given by

$$\mathbf{m}_{L_2} \pm 1.96 \cdot \text{diag}(\text{Cov}(\mathbf{m}_{L_2}))^{1/2} \tag{2.30}$$

where the 1.96 factor arises from

$$\frac{1}{\sigma \sqrt{2\pi}} \int_{-1.96\sigma}^{1.96\sigma} e^{-\frac{x^2}{2\sigma^2}} \, dx \approx 0.95. \tag{2.31}$$

■ **Example 2.1** Let us recall Example 1.1 of linear regression of ballistic observations to a quadratic regression model

$$y(t) = m_1 + m_2 t - (1/2) m_3 t^2. \tag{2.32}$$

Here y is measured in the upward direction, and the minus sign is applied to the third term because gravitational acceleration is downward. Consider a synthetic data set with $m = 10$ observations and independent normal data errors ($\sigma = 8$ m), generated using

$$\mathbf{m}_{\text{true}} = [10 \text{ m}, 100 \text{ m/s}, 9.8 \text{ m/s}^2]^T. \tag{2.33}$$

See Table 2.1.

To obtain the least squares solution, we construct the $\mathbf{G}$ matrix. The ith row of $\mathbf{G}$ is

$$\mathbf{G}_{i,\cdot} = [1, t_i, -(1/2)t_i^2] \tag{2.34}$$

so that

$$\mathbf{G} = \begin{bmatrix} 1 & 1 & -0.5 \\ 1 & 2 & -2.0 \\ 1 & 3 & -4.5 \\ 1 & 4 & -8.0 \\ 1 & 5 & -12.5 \\ 1 & 6 & -18.0 \\ 1 & 7 & -24.5 \\ 1 & 8 & -32.0 \\ 1 & 9 & -40.5 \\ 1 & 10 & -50.0 \end{bmatrix}. \tag{2.35}$$

We solve for the parameters using the weighted normal equations, (2.18), to obtain a model estimate

$$\mathbf{m}_{L_2} = [16.4 \text{ m}, 97.0 \text{ m/s}, 9.4 \text{ m/s}^2]^T. \tag{2.36}$$

Figure 2.1 shows the observed data and the fitted curve. The model covariance matrix associated with $\mathbf{m}_{L_2}$ is

$$\text{Cov}(\mathbf{m}_{L_2}) = \begin{bmatrix} 88.53 & -33.60 & -5.33 \\ -33.60 & 15.44 & 2.67 \\ -5.33 & 2.67 & 0.48 \end{bmatrix}. \tag{2.37}$$

Table 2.1 Data for the ballistics example.

t (s)	1	2	3	4	5
y (m)	109.4	187.5	267.5	331.9	386.1
t (s)	6	7	8	9	10
y (m)	428.4	452.2	498.1	512.3	513.0

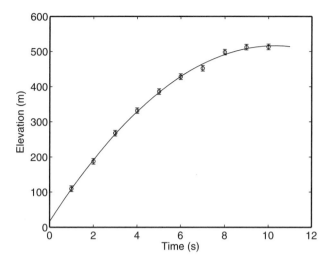

Figure 2.1   Data and model predictions for the ballistics example.

Equation (2.30) gives

$$\mathbf{m}_{L_2} = [16.4 \pm 18.4 \text{ m}, 97.0 \pm 7.7 \text{ m/s}, 9.4 \pm 1.4 \text{ m/s}^2]^T. \tag{2.38}$$

The χ^2 value for this regression is about 4.2, and the number of degrees of freedom is $v = m - n = 10 - 3 = 7$, so the p-value, (2.22), is

$$p = \int_{4.2}^{\infty} \frac{1}{2^{7/2}\Gamma(7/2)} x^{\frac{5}{2}} e^{-\frac{x}{2}} \, dx \approx 0.76 \tag{2.39}$$

which is in the realm of plausibility. This means that the fitted model is consistent with the modeling and data uncertainty assumptions.   ■

If we consider combinations of model parameters, the interpretation of model uncertainty becomes more complex. To characterize model uncertainty more effectively, we can examine 95% **confidence regions** for pairs or larger sets of parameters. When joint parameter confidence regions are projected onto the coordinate axes, m_i, we obtain intervals for parameters that may be significantly larger than we would estimate when considering parameters individually, as in (2.38).

For a vector of estimated model parameters characterized by an n-dimensional multivariate normal distribution with mean $\mathbf{m}_{\text{true}}$ and covariance matrix $\mathbf{C}$, the random variable

$$(\mathbf{m} - \mathbf{m}_{L_2})^T \mathbf{C}^{-1}(\mathbf{m} - \mathbf{m}_{L_2}) \tag{2.40}$$

can be shown to have a χ^2 distribution with n degrees of freedom [84]. Thus if Δ^2 is the 95th percentile of the χ^2 distribution with n degrees of freedom, the 95% confidence region is defined by the inequality

$$(\mathbf{m} - \mathbf{m}_{L_2})^T \mathbf{C}^{-1} (\mathbf{m} - \mathbf{m}_{L_2}) \leq \Delta^2. \tag{2.41}$$

The confidence region defined by this inequality is an n-dimensional ellipsoid.

If we wish to find an error ellipsoid for a lower dimensional subset of the model parameters, we can project the n-dimensional error ellipsoid onto the lower dimensional space by taking only those rows and columns of $\mathbf{C}$ and elements of $\mathbf{m}$ that correspond to the dimensions that we want to keep [2]. In this case, the number of degrees of freedom in the associated χ^2 calculation should also be reduced to match the number of model parameters in the projected error ellipsoid.

Since the covariance matrix and its inverse are symmetric and positive definite, we can diagonalize $\mathbf{C}^{-1}$ using (A.79) as

$$\mathbf{C}^{-1} = \mathbf{P}^T \Lambda \mathbf{P} \tag{2.42}$$

where Λ is a diagonal matrix of positive eigenvalues and the columns of $\mathbf{P}$ are orthonormal eigenvectors. The semiaxes defined by the columns of $\mathbf{P}$ are referred to as error ellipsoid **principal axes**, where the ith semimajor error ellipsoid axis direction is defined by $\mathbf{P}_{.,i}$ and has length $\Delta/\sqrt{\Lambda_{i,i}}$.

Because the model covariance matrix is typically not diagonal, the principal axes are typically not aligned in the m_i axis directions. However, we can project the appropriate confidence ellipsoid onto the m_i axes to obtain a "box" which includes the entire 95% error ellipsoid, along with some additional external volume. Such a box provides a conservative confidence interval for a joint collection of model parameters.

Correlations for parameter pairs (m_i, m_j) are measures of the inclination of the error ellipsoid with respect to the parameter axes. A correlation approaching $+1$ means the projection is highly eccentric with its long principal axis having a positive slope, a zero correlation means that the projection has principal axes that are aligned with the axes of the (m_i, m_j) plane, and a correlation approaching -1 means that the projection is highly eccentric with its long principal axis having a negative slope.

■ **Example 2.2** The parameter correlations for Example 2.1 are

$$\rho_{m_i, m_j} = \frac{\text{Cov}(m_i, m_j)}{\sqrt{\text{Var}(m_i) \cdot \text{Var}(m_j)}} \tag{2.43}$$

which give

$$\rho_{m_1, m_2} = -0.91 \tag{2.44}$$

$$\rho_{m_1, m_3} = -0.81 \tag{2.45}$$

$$\rho_{m_2, m_3} = 0.97. \tag{2.46}$$

The three model parameters are highly statistically dependent, and the error ellipsoid is thus inclined and eccentric. Figure 2.2 shows the 95% confidence ellipsoid.

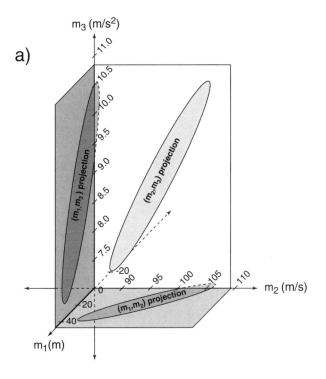

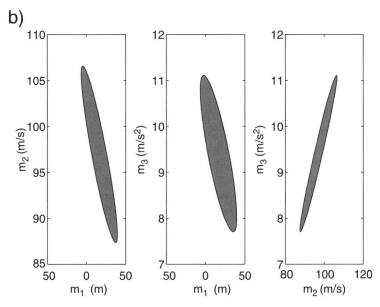

Figure 2.2 Projections of the 95% error ellipsoid onto model axes. (a) Projections in perspective; (b) projections onto the parameter axis planes.

Diagonalization of (2.37) shows that the directions of the semiaxes for the error ellipsoid are

$$\mathbf{P} = [\mathbf{P}_{\cdot,1}, \mathbf{P}_{\cdot,2}, \mathbf{P}_{\cdot,3}] \approx \begin{bmatrix} -0.03 & -0.93 & 0.36 \\ -0.23 & 0.36 & 0.90 \\ 0.97 & 0.06 & 0.23 \end{bmatrix} \tag{2.47}$$

with corresponding eigenvalues

$$[\lambda_1, \lambda_2, \lambda_3] \approx [104.7, 0.0098, 0.4046]. \tag{2.48}$$

The corresponding 95% confidence ellipsoid semiaxis lengths are

$$\sqrt{F_{\chi^2,3}^{-1}(0.95)} \left[1/\sqrt{\lambda_1}, 1/\sqrt{\lambda_2}, 1/\sqrt{\lambda_3} \right] \approx [0.24, 24.72, 3.85] \tag{2.49}$$

where $F_{\chi^2,3}^{-1}(0.95) \approx 2.80$ is the 95th percentile of the χ^2 distribution with three degrees of freedom.

Projecting the 95% confidence ellipsoid into the (m_1, m_2, m_3) coordinate system we obtain 95% confidence intervals for the parameters considered jointly,

$$[m_1, m_2, m_3] = [16.42 \pm 23.03 \text{ m}, 96.97 \pm 9.62 \text{ m/s}, 9.41 \pm 1.70 \text{ m/s}^2], \tag{2.50}$$

that are about 40% broader than the single parameter confidence estimates obtained using only the diagonal covariance matrix terms in (2.38). Note that (2.50) is a conservative estimate because there is actually a greater than 95% probability that the box defined by (2.50) will include the true values of the parameters. The reason is that these intervals, considered together as a region, include many points that lie outside of the 95% confidence ellipsoid. ∎

It is insightful to note that the covariance matrix (2.25) contains information only about where and how often we made measurements, and on what the standard deviations of those measurements were. Covariance is thus exclusively a characteristic of experimental design that reflects how much influence the noise in a *general* data set will have on a model estimate. Conversely, it does not depend upon *particular* data values from an individual experiment. This is why it is essential to evaluate the *p*-value or some other "goodness-of-fit" measure for an estimated model. Examining the solution parameters and the covariance matrix alone does *not* reveal whether we are fitting the data adequately.

2.3 UNKNOWN MEASUREMENT STANDARD DEVIATIONS

Suppose that we do not know the standard deviations of the measurement errors *a priori*. In this case, if we assume that the measurement errors are independent and normally distributed with expected value 0 and standard deviation σ, then we can perform the linear regression and estimate σ from the residuals.

First, we find the least squares solution to the unweighted problem $\mathbf{Gm} = \mathbf{d}$, and let

$$\mathbf{r} = \mathbf{d} - \mathbf{Gm}_{L_2}. \tag{2.51}$$

To estimate the standard deviation from the residuals, let

$$s = \sqrt{\frac{1}{m-n} \sum_{i=1}^{m} r_i^2}. \tag{2.52}$$

As you might expect, there is a statistical cost associated with not knowing the true standard deviation. If the data standard deviations are known ahead of time, then the model errors

$$m_i' = \frac{m_i - m_{\text{true}_i}}{\sigma} \tag{2.53}$$

have the standard normal distribution. If instead of a known σ, we have an estimate of σ, s, obtained from (2.52), then the model errors

$$m_i' = \frac{m_i - m_{\text{true}_i}}{s} \tag{2.54}$$

have a Student's t distribution (B.12) with $\nu = m - n$ degrees of freedom. For smaller degrees of freedom this produces appreciably broader confidence intervals, but as ν becomes large, s becomes an increasingly better estimate of σ. Confidence ellipsoids for these problems can also be computed, but the formula is somewhat more complicated than in the case of known standard deviations [33].

A problem arises in that we can no longer use the χ^2 test of goodness-of-fit in this case. The χ^2 test was based on the assumption that the data errors were normally distributed with known standard deviations σ_i. If the actual residuals were too large relative to the σ_i, then χ^2 would be large, and we would reject the linear regression fit based on a very small p-value. However, if we substitute the estimate (2.52) into (2.20), we find that $\chi^2_{\text{obs}} = \nu$, so such a model will always pass the χ^2 test.

■ **Example 2.3** Consider the analysis of a linear regression problem in which the measurement errors are assumed to be independent and normally distributed, with equal but unknown standard deviations, σ. We are given a set of x and y data that appear to follow a linear relationship.

In this case,

$$\mathbf{G} = \begin{bmatrix} 1 & x_1 \\ 1 & x_2 \\ \vdots & \vdots \\ 1 & x_n \end{bmatrix}. \tag{2.55}$$

The least squares solution to

$$\mathbf{Gm} = \mathbf{y} \tag{2.56}$$

has

$$y = -1.03 + 10.09x. \tag{2.57}$$

Figure 2.3 shows the data and the linear regression line. Our estimate of the standard deviation of the measurement errors using (2.52) is $s = 30.74$. The estimated covariance matrix for the fitted parameters is

$$\mathbf{C} = s^2 (\mathbf{G}^T \mathbf{G})^{-1}$$

$$= \begin{bmatrix} 338.24 & -4.93 \\ -4.93 & 0.08 \end{bmatrix}. \tag{2.58}$$

Confidence intervals, evaluated for each parameter separately, are

$$m_1 = -1.03 \pm \sqrt{338.24} t_{n-2,0.975}$$

$$= -1.03 \pm 38.05 \tag{2.59}$$

and

$$m_2 = 10.09 \pm \sqrt{0.08} t_{n-2,0.975}$$

$$= 10.09 \pm 0.59. \tag{2.60}$$

Since the actual standard deviation of the measurement errors is unknown, we cannot perform a χ^2 test of goodness-of-fit. However, we can still examine the residuals. Figure 2.4 shows the residuals. It is clear that although they appear to be random, the standard deviation seems to increase as x and y increase. This is a common phenomenon in linear regression, called a **proportional effect**. One possible way that such an effect might occur is if the size of measurement errors were proportional to the measurement magnitude due to characteristics of the instrument used.

We will address the proportional effect by assuming that the standard deviation is proportional to y. We then rescale the system of equations (2.56) by dividing each equation by the corresponding y_i, to obtain

$$\mathbf{G}_w \mathbf{m} = \mathbf{y}_w. \tag{2.61}$$

For this weighted system, we obtain a revised least squares estimate of

$$y = -12.24 + 10.25x \tag{2.62}$$

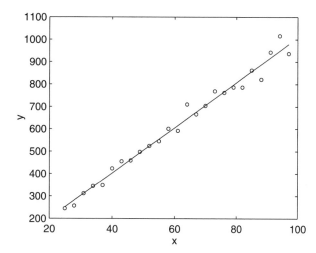

Figure 2.3 Data and linear regression line.

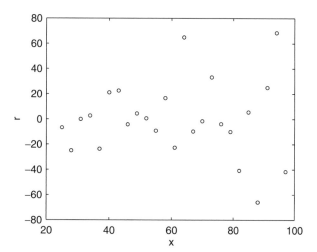

Figure 2.4 Residuals.

with 95% parameter confidence intervals, evaluated in the same manner as (2.59) and (2.60), of

$$m_1 = -12.24 \pm 22.39 \tag{2.63}$$

and

$$m_2 = 10.25 \pm 0.47. \tag{2.64}$$

Figure 2.5 shows the data and least squares fit. Figure 2.6 shows the scaled residuals. Note that there is now no obvious trend in the magnitude of the residuals as x and y increase, as there was in Figure 2.4. The estimated standard deviation is 0.045, or 4.5% of the y value.

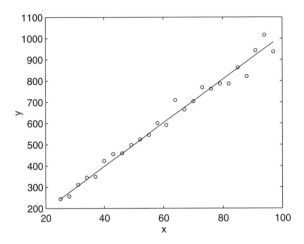

Figure 2.5   Data and linear regression line, weighted.

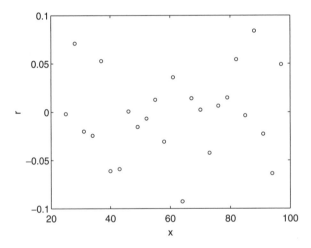

Figure 2.6   Residuals for the weighted problem.

In fact, these data were generated according to the true model $y = 10x + 0$, using standard deviations for the measurement errors that were 5% of the y value.   ■

2.4 L_1 REGRESSION

Least squares solutions are highly susceptible to even small numbers of discordant observations, or **outliers**. Outliers are data points that are highly inconsistent with the other data. Outliers may arise from procedural measurement error, for example from incorrectly recording the position of a decimal point in a floating-point number. Outliers should be investigated carefully, since the data may actually be showing us that the form of the mathematical model that we are trying to fit is incorrect. However, if we conclude that there are only a small number of outliers in the data due to incorrect measurements, we need to analyze the data in a way which minimizes their effect on the estimated model.

We can readily appreciate the strong effect of outliers on least squares solutions from a maximum likelihood perspective by noting the very rapid fall-off of the tails of the normal distribution. For example, for a normally distributed error, the probability of a single data point occurring more than 5 standard deviations away from its expected value is less than one in a million:

$$P(|X - E[X]| \geq 5\sigma) = \frac{2}{\sqrt{2\pi}} \int_5^\infty e^{-\frac{1}{2}x^2} \, dx \approx 6 \times 10^{-7}. \tag{2.65}$$

If an outlier occurs in the data set because of a nonnormal error process, the least squares solution will go to great lengths to accommodate it to prevent its contribution to the total likelihood (2.10) from being vanishingly small.

As an alternative to least squares, consider the solution that minimizes the 1-norm of the residual vector,

$$\mu^{(1)} = \sum_{i=1}^m \frac{|d_i - (\mathbf{Gm})_i|}{\sigma_i}$$

$$= \|\mathbf{d}_w - \mathbf{G}_w \mathbf{m}\|_1. \tag{2.66}$$

The **1-norm solution**, $\mathbf{m}_{L_1}$, will be more outlier resistant, or **robust**, than the least squares solution, $\mathbf{m}_{L_2}$, because (2.66) does not square each of the terms in the misfit measure, as (2.13) does. The 1-norm solution $\mathbf{m}_{L_1}$ also has a maximum likelihood interpretation; it is the maximum likelihood estimator for data with errors distributed according to a double-sided exponential distribution

$$f(x) = \frac{1}{2\sigma} e^{-|x-\mu|/\sigma}. \tag{2.67}$$

Data sets distributed as (2.67) are unusual. Nevertheless, it may be worthwhile to consider a solution where (2.66) is minimized rather than (2.13), even if most of the measurement errors are normally distributed, if there is reason to suspect the presence of outliers. This solution strategy may be useful if the data outliers occur for reasons that do not undercut our belief that the mathematical model is otherwise correct.

■ **Example 2.4** We can demonstrate the advantages of 1-norm minimization using the quadratic regression example discussed in Examples 1.1 and 2.1. Figure 2.7 shows the original sequence of independent data points with 8 m standard deviations. One of the points (d_4) is now an outlier for a mathematical model of the form (2.32), specifically the original data point with 200 m subtracted from it. The least-squares model for this data set is

$$\mathbf{m}_{L_2} = [26.4 \text{ m}, 75.6 \text{ m/s}, 4.9 \text{ m/s}^2]^T. \tag{2.68}$$

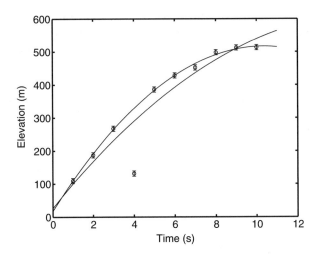

Figure 2.7 L_1 (upper) and L_2 (lower) solutions for a parabolic data set with an outlier at $t = 4$ s.

The least squares solution is skewed away from the majority of data points in trying to accommodate the outlier and is a poor estimate of the true model. We can also see that (2.68) fails to fit these data acceptably because of its huge χ^2 value ($\approx$1109). This is clearly astronomically out of bounds for a problem with 7 degrees of freedom, where the χ^2 value should not be far from 7. The corresponding p-value for $\chi^2 = 1109$ is effectively zero.

The upper curve in Figure 2.7, defined by the parameters

$$\mathbf{m}_{L_1} = [17.6 \text{ m}, 96.4 \text{ m/s}, 9.3 \text{ m/s}^2]^T, \tag{2.69}$$

is obtained using the 1-norm solution that minimizes (2.66). The data prediction from (2.69) faithfully fits the quadratic trend for the majority of the data points and ignores the outlier at $t = 4$. It is also much closer than (2.68) to the true model (2.33) and to the least squares model for the data set without the outlier (2.36).   ■

In examining the differences between 2- and 1-norm models, it is instructive to consider the almost trivial regression problem of estimating the value of a single parameter from repeated measurements. The system of equations $\mathbf{G}\mathbf{m} = \mathbf{d}$ is

$$\begin{bmatrix} 1 \\ 1 \\ 1 \\ \vdots \\ 1 \end{bmatrix} \mathbf{m} = \begin{bmatrix} d_1 \\ d_2 \\ d_3 \\ \vdots \\ d_m \end{bmatrix}. \tag{2.70}$$

The least squares solution to (2.70) can be seen from the normal equations (A.73) to be simply the observational average

$$\mathbf{m}_{L_2} = (\mathbf{G}^T\mathbf{G})^{-1}\mathbf{G}^T\mathbf{d} = m^{-1}\sum_{i=1}^{m} d_i. \qquad (2.71)$$

Finding the 1-norm solution is more complicated. The bad news is that the 1-norm of the residual vector

$$f(\mathbf{m}) = \|\mathbf{d} - \mathbf{Gm}\|_1 = \sum_{i=1}^{m} |d_i - \mathbf{m}| \qquad (2.72)$$

is a nondifferentiable function of $\mathbf{m}$ at each point where $\mathbf{m} = d_i$. The good news is that $f(\mathbf{m})$ is a convex function of $\mathbf{m}$. Thus any local minimum point is also a global minimum point. We can proceed by finding $f'(\mathbf{m})$ at those points where it is defined, and then separately consider the points at which the derivative is not defined. Every minimum point must either have $f'(\mathbf{m})$ undefined or $f'(\mathbf{m}) = 0$.

At those points where $f'(\mathbf{m})$ is defined, it is given by

$$f'(\mathbf{m}) = \sum_{i=1}^{n} \text{sgn}(d_i - \mathbf{m}), \qquad (2.73)$$

where the **signum function**, $\text{sgn}(x)$, is -1 if its argument is negative, 1 if its argument is positive, and 0 if its argument is zero. The derivative (2.73) is zero when exactly half of the data are less than $\mathbf{m}$ and half of the data are greater than $\mathbf{m}$. Of course, this can only happen when the number of observations, m, is even. In this case, any value of $\mathbf{m}$ lying between the two middle observations is a 1-norm solution. When there are an odd number of data, the median data point is the unique 1-norm solution. Even an extreme outlier will not have a large effect on the median of an otherwise clustered set of observations. This illuminates the reason for the robustness of the 1-norm solution.

The general problem of finding solutions that minimize $\|\mathbf{d} - \mathbf{Gm}\|_1$ is complex. One practical way is **iteratively reweighted least squares**, or **IRLS** [141]. The IRLS algorithm solves a sequence of weighted least squares problems whose solutions converge to a 1-norm minimizing solution. Beginning with the residual vector

$$\mathbf{r} = \mathbf{d} - \mathbf{Gm} \qquad (2.74)$$

we want to minimize

$$f(\mathbf{m}) = \|\mathbf{r}\|_1 = \sum_{i=1}^{m} |r_i|. \qquad (2.75)$$

The function in (2.75), like the function in (2.72), is nondifferentiable at any point where one of the elements of $\mathbf{r}$ is zero. Ignoring this issue for a moment, we can go ahead and compute the derivatives of f at other points:

$$\frac{\partial f(\mathbf{m})}{\partial m_k} = \sum_{i=1}^{m} \frac{\partial |r_i|}{m_k} = \sum_{i=1}^{m} G_{i,k}\, \mathrm{sgn}(r_i). \tag{2.76}$$

Writing $\mathrm{sgn}(r_i)$ as $r_i/|r_i|$ gives

$$\frac{\partial f(\mathbf{m})}{\partial m_k} = \sum_{i=1}^{m} G_{i,k} \frac{1}{|r_i|} r_i. \tag{2.77}$$

The gradient of f is

$$\nabla f(\mathbf{m}) = \mathbf{G}^T \mathbf{R}\mathbf{r} = \mathbf{G}^T \mathbf{R}(\mathbf{d} - \mathbf{G}\mathbf{m}) \tag{2.78}$$

where $\mathbf{R}$ is a diagonal weighting matrix with diagonal elements that are just the absolute values of the reciprocals of the residuals:

$$R_{i,i} = 1/|r_i|. \tag{2.79}$$

To find the 1-norm minimizing solution, we solve $\nabla f(\mathbf{m}) = \mathbf{0}$, specifically

$$\mathbf{G}^T \mathbf{R}(\mathbf{d} - \mathbf{G}\mathbf{m}) = \mathbf{0} \tag{2.80}$$

or

$$\mathbf{G}^T \mathbf{R}\mathbf{G}\mathbf{m} = \mathbf{G}^T \mathbf{R}\mathbf{d}. \tag{2.81}$$

Since $\mathbf{R}$ depends on $\mathbf{m}$, (2.81) is a nonlinear system of equations that we cannot solve directly. IRLS is a simple iterative algorithm to find the appropriate weights in $\mathbf{R}$. The algorithm begins with the least squares solution $\mathbf{m}^0 = \mathbf{m}_{L_2}$. We calculate the corresponding residual vector $\mathbf{r}^0 = \mathbf{d} - \mathbf{G}\mathbf{m}^0$. We then solve (2.81) to obtain a new model $\mathbf{m}^1$ and associated residual vector $\mathbf{r}^1$. The process is repeated until the model and residual vectors converge. A typical rule is to stop the iteration when

$$\frac{\|\mathbf{m}^{k+1} - \mathbf{m}^k\|_2}{1 + \|\mathbf{m}^{k+1}\|_2} < \tau \tag{2.82}$$

for some tolerance τ.

The above procedure will fail if any element of the residual vector becomes zero. A simple modification to the algorithm deals with this problem. We select a tolerance ϵ below which we consider the residuals to be effectively zero. If $|r_i| < \epsilon$, then we set $R_{i,i} = 1/\epsilon$. With this modification it can be shown that this procedure will always converge to an approximate 1-norm minimizing solution.

As with the χ^2 misfit measure for least squares solutions, there is a corresponding p-value that can be used under the assumption of normal data errors, but for the assessment of 1-norm solutions [123]. Let

$$\mu_{\text{obs}}^{(1)} = \|\mathbf{Gm}_{L_1} - \mathbf{d}\| \tag{2.83}$$

be the observed 1-norm misfit. For a 1-norm misfit measure given by (2.83), the probability that a worse misfit could have occurred given independent and normally distributed data and ν degrees of freedom is approximately given by

$$p^{(1)}(y, \nu) = P(\mu^{(1)} > \mu_{\text{obs}}^{(1)}) = S(x) - \frac{\gamma Z^{(2)}(x)}{6} \tag{2.84}$$

where

$$S(x) = \frac{1}{\sigma_1 \sqrt{2\pi}} \int_{-\infty}^{x} e^{-\frac{\xi^2}{2\sigma_1^2}} \, d\xi \tag{2.85}$$

$$\sigma_1 = (1 - 2/\pi)\nu \tag{2.86}$$

$$\gamma = \frac{2 - \pi/2}{(\pi/2 - 1)^{3/2}} \nu^{\frac{1}{2}} \tag{2.87}$$

$$Z^{(2)}(x) = \frac{x^2 - 1}{\sqrt{2\pi}} e^{-\frac{x^2}{2}} \tag{2.88}$$

$$x = \frac{\mu^{(1)} - \sqrt{2/\pi} \, \nu}{\sigma_1}. \tag{2.89}$$

2.5 MONTE CARLO ERROR PROPAGATION

For solution techniques that are nonlinear and/or algorithmic, such as IRLS, there is typically no simple way to propagate uncertainties in the data to uncertainties in the estimated model parameters. In such cases, one can apply **Monte Carlo error propagation** techniques, in which we simulate a collection of noisy data vectors and then examine the statistics of the corresponding models. We can obtain an approximate covariance matrix by first forward-propagating the L_1 solution into an assumed noise-free baseline data vector

$$\mathbf{Gm}_{L_1} = \mathbf{d}_b. \tag{2.90}$$

We next re-solve the IRLS problem many times for 1-norm models corresponding to independent data realizations, obtaining a suite of q 1-norm solutions to

$$\mathbf{Gm}_{L_1,i} = \mathbf{d}_b + \boldsymbol{\eta}_i \tag{2.91}$$

where η_i is the ith noise vector realization. Let $\mathbf{A}$ be a q by m matrix where the ith row contains the difference between the ith model estimate and the average model

$$\mathbf{A}_{i,\cdot} = \mathbf{m}_{L_1,i}^T - \bar{\mathbf{m}}_{L_1}^T. \tag{2.92}$$

Then an empirical estimate of the covariance matrix is

$$\text{Cov}(\mathbf{m}_{L_1}) = \frac{\mathbf{A}^T \mathbf{A}}{q}. \tag{2.93}$$

■ **Example 2.5** Recall Example 2.1. An estimate of $\text{Cov}(\mathbf{m}_{L_1})$ using 10,000 iterations of the Monte Carlo procedure is

$$\text{Cov}(\mathbf{m}_{L_1}) = \begin{bmatrix} 122.52 & -46.50 & -7.37 \\ -46.50 & 21.49 & 3.72 \\ -7.37 & 3.72 & 0.68 \end{bmatrix}, \tag{2.94}$$

which is about 1.4 times as large as the covariance matrix elements found from the least squares solution (2.37). Although we have no reason to believe that the model parameters will be normally distributed given that this solution was obtained with the IRLS algorithm, we can compute approximate 95% confidence intervals for the parameters:

$$\mathbf{m}_{L_1} = [17.6 \pm 21.8 \text{ m}, 96.4 \pm 7.7 \text{ m/s}, 9.3 \pm 1.4 \text{ m/s}^2]^T. \tag{2.95}$$

■

2.6 EXERCISES

2.1 A seismic profiling experiment is performed where the first arrival times of seismic energy from a midcrustal refractor are observed at distances (in kilometers) of

$$\mathbf{x} = \begin{bmatrix} 6.0000 \\ 10.1333 \\ 14.2667 \\ 18.4000 \\ 22.5333 \\ 26.6667 \end{bmatrix} \tag{2.96}$$

from the source, and are found to be (in seconds after the source origin time)

$$\mathbf{t} = \begin{bmatrix} 3.4935 \\ 4.2853 \\ 5.1374 \\ 5.8181 \\ 6.8632 \\ 8.1841 \end{bmatrix}. \tag{2.97}$$

A two-layer flat Earth structure gives the mathematical model

$$t_i = t_0 + s_2 x_i \tag{2.98}$$

where the intercept time t_0 depends on the thickness and slowness of the upper layer, and s_2 is the slowness of the lower layer. The estimated noise in the first arrival time measurements is believed to be independent and normally distributed with expected value 0 and standard deviation $\sigma = 0.1$ s.

(a) Find the least squares solution for the model parameters t_0 and s_2. Plot the data, the fitted model, and the residuals.

(b) Calculate and comment on the parameter correlation matrix.

(c) Plot the error ellipsoid in the (t_0, s_2) plane and calculate conservative 95% confidence intervals for t_0 and s_2. How are the correlations reflected in the appearance of the error ellipsoid in (t_0, s_2) space? Hint: The following MATLAB code will plot a two-dimensional covariance ellipse, where **covm** is the covariance matrix and m is the 2-vector of model parameters.

```
%diagonalize the covariance matrix
[u,lam]=eig(inv(covm));
%generate a vector of angles from 0 to 2*pi
theta=(0:.01:2*pi)';
%calculate the x component of the ellipsoid for all angles
r(:,1)=(delta/sqrt(lam(1,1)))*u(1,1)*cos(theta)+...
        (delta/sqrt(lam(2,2)))*u(1,2)*sin(theta);
%calculate the y component of the ellipsoid for all angles
r(:,2)=(delta/sqrt(lam(1,1)))*u(2,1)*cos(theta)+...
        (delta/sqrt(lam(2,2)))*u(2,2)*sin(theta);
%plot(x,y), adding in the model parameters
plot(m(1)+r(:,1),m(2)+r(:,2))
```

(d) Evaluate the p-value for this model. You may find the MATLAB Statistics Toolbox function **chi2cdf** to be useful here.

(e) Evaluate the value of χ^2 for 1000 Monte Carlo simulations using the data prediction from your model perturbed by noise that is consistent with the data assumptions. Compare a histogram of these χ^2 values with the theoretical χ^2 distribution for the correct number of degrees of freedom. You may find the MATLAB Statistical Toolbox function **chi2pdf** to be useful here.

(f) Are your p-value and Monte Carlo χ^2 distribution consistent with the theoretical modeling and the data set? If not, explain what is wrong.

(g) Use IRLS to find 1-norm estimates for t_0 and s_2. Plot the data predictions from your model relative to the true data and compare with (a).

(h) Use Monte Carlo error propagation and IRLS to estimate symmetric 95% confidence intervals on the 1-norm solution for t_0 and s_2.

(i)   Examining the contributions from each of the data points to the 1-norm misfit measure, can you make a case that any of the data points are statistical outliers?

2.2   In this chapter we have assumed that the measurement errors are independent. Suppose instead that the measurement errors have an MVN distribution with expected value $\mathbf{0}$ and a known covariance matrix $\mathbf{C}_D$. It can be shown that the likelihood function is then

$$L(\mathbf{m}|\mathbf{d}) = \frac{1}{(2\pi)^{m/2}} \frac{1}{\sqrt{\det(\mathbf{C}_D)}} e^{-(\mathbf{Gm}-\mathbf{d})^T \mathbf{C}_D^{-1}(\mathbf{Gm}-\mathbf{d})/2}. \qquad (2.99)$$

(a)   Show that the maximum likelihood estimate can be obtained by solving the minimization problem

$$\min \ (\mathbf{Gm} - \mathbf{d})^T \mathbf{C}_D^{-1}(\mathbf{Gm} - \mathbf{d}). \qquad (2.100)$$

(b)   Show that (2.100) can be solved by the system of equations

$$(\mathbf{G}^T \mathbf{C}_D^{-1} \mathbf{G})\mathbf{m} = \mathbf{G}^T \mathbf{C}_D^{-1} \mathbf{d}. \qquad (2.101)$$

(c)   Show that (2.100) is equivalent to the linear least squares problem

$$\min \ \|\mathbf{C}_D^{-1/2}\mathbf{Gm} - \mathbf{C}_D^{-1/2}\mathbf{d}\|_2. \qquad (2.102)$$

2.3   Use MATLAB to generate 10,000 realizations of a data set of $m = 5$ points $\mathbf{d} = a + b\mathbf{x} + \boldsymbol{\eta}$, where $\mathbf{x} = [1, 2, 3, 4, 5]^T$, the $n = 2$ true model parameters are $a = b = 1$, and $\boldsymbol{\eta}$ is an m-element vector of independent $N(0, 1)$ noise.

(a)   Assuming that the noise standard deviation is known *a priori* to be 1, solve for the least squares parameters for your realizations and histogram them in 100 bins.
(b)   Calculate the parameter covariance matrix, $\mathbf{C} = \sigma^2(\mathbf{G}^T\mathbf{G})^{-1}$, assuming independent $N(0, 1)$ data errors, and give standard deviations, σ_a and σ_b, for your estimates of a and b.
(c)   Calculate the standardized parameter estimates

$$a' = \frac{a - 1}{\sqrt{C_{11}}} \qquad (2.103)$$

and

$$b' = \frac{b - 1}{\sqrt{C_{22}}} \qquad (2.104)$$

and demonstrate using a Q–Q plot that your estimates for a' and b' are distributed as $N(0, 1)$.

(d) Show using a Q–Q plot that the squared residual lengths

$$\|\mathbf{r}\|_2^2 = \|\mathbf{d} - \mathbf{Gm}\|_2^2 \tag{2.105}$$

for your solutions in (a) are distributed as χ^2 with $m - n = \nu = 3$ degrees of freedom.

(e) Assume that the noise standard deviation for the synthetic data set is not known, and estimate it for each realization as

$$s = \sqrt{\frac{1}{m-n} \sum_{i=1}^{m} r_i^2}. \tag{2.106}$$

Histogram your standardized solutions

$$a' = \frac{a-1}{\sqrt{C_{11}}} \tag{2.107}$$

and

$$b' = \frac{b-1}{\sqrt{C_{22}}} \tag{2.108}$$

where $C = s^2 (\mathbf{G}^T \mathbf{G})^{-1}$.

(f) Demonstrate using a Q–Q plot that your estimates for a' and b' are distributed as the Student's t distribution (see Section B7) with $\nu = 3$ degrees of freedom.

2.4 Suppose that we analyze a large number of data sets $\mathbf{d}$ in a linear regression problem and compute p-values for each data set. The χ^2_{obs} values should be distributed according to a χ^2 distribution with $m - n$ degrees of freedom. Show that the corresponding p-values will be uniformly distributed between 0 and 1.

2.5 Use linear regression to fit a polynomial of the form

$$y = a_0 + a_1 x + a_2 x^2 + \ldots + a_{19} x^{19} \tag{2.109}$$

to the noise-free data points

$$(x_i, y_i) = (-0.95, -0.95), (-0.85, -0.85), \ldots, (0.95, 0.95). \tag{2.110}$$

Use the normal equations to solve the least squares problem.

Plot the data and your fitted model, and list the parameters, a_i obtained in your regression. Clearly, the correct solution has $a_1 = 1$, and all other $a_i = 0$. Explain why your answer differs.

2.7 NOTES AND FURTHER READING

Linear regression is a major subfield within statistics, and there are literally hundreds of associated textbooks. Many of these references focus on applications of linear regression in the social sciences. In such applications, the primary focus is often on determining which variables have an effect on response variables of interest (rather than on estimating parameter values for a predetermined model). In this context it is important to compare alternative regression models and to test the hypothesis that a predictor variable has a nonzero coefficient in the regression model. Since we normally know which predictor variables are important in the physical sciences, the approach commonly differs. Useful linear regression references from the standpoint of estimating parameters in the context considered here include [33, 106].

Robust statistical methods are an important topic. Huber discusses a variety of robust statistical procedures [72]. The computational problem of computing a 1-norm solution has been extensively researched. Techniques for 1-norm minimization include methods based on the simplex method for linear programming, interior point methods, and iteratively reweighted least squares [7, 25, 126, 141]. The IRLS method is the simplest to implement, but interior point methods can be the most efficient approaches for very large problems. Watson reviews the history of methods for finding p-norm solutions including the 1-norm case [177].

We have assumed that $\mathbf{G}$ is known exactly. In some cases elements of this matrix might be subject to measurement error. This problem has been studied as the **total least squares problem** [73]. An alternative approach to least squares problems with uncertainties in $\mathbf{G}$ that has recently received considerable attention is called **robust least squares** [11, 46].

3

DISCRETIZING CONTINUOUS
INVERSE PROBLEMS

Synopsis: Techniques for discretizing continuous inverse problems characterized by Fredholm integral equations of the first kind are discussed. Discretization based on quadrature formulas for numerically approximating integrals are introduced, and examples are given. Alternative methods of discretization based on expanding the model as a linear combination of basis functions are presented. The method of Backus and Gilbert is introduced.

3.1 INTEGRAL EQUATIONS

Consider problems of the form

$$d(s) = \int_a^b g(s, x)m(x)\, dx. \tag{3.1}$$

Here $d(s)$ is a known function, typically representing observed data. The function $g(s, x)$ is considered to be prescribed and encodes the physics that relates the unknown model $m(x)$ to the observed $d(s)$. The interval $[a, b]$ may be finite or infinite. The function $d(s)$ might in theory be known over an entire interval but in practice we will only have measurements of $d(s)$ at a finite set of points.

We wish to solve for the unknown function $m(x)$. This type of linear equation, introduced in Chapter 1, is called a Fredholm integral equation of the first kind or IFK. For reasons that were introduced in Chapter 1, a surprisingly large number of inverse problems can be written as Fredholm integral equations of the first kind. Unfortunately, IFKs have properties that can make them difficult to solve.

3.2 QUADRATURE METHODS

To obtain useful numerical solutions to IFKs, we will frequently seek to discretize them into forms that are tractably solvable using linear algebra methods. We first assume that $d(s)$ is known at a finite number of points $s_1, s_2, \ldots, s_m$. For a finite number of data points we can write the inverse problem as

$$d_i = d(s_i)$$

$$= \int_a^b g(s_i, x)m(x)\, dx \qquad i = 1, 2, \ldots, m \qquad (3.2)$$

or as

$$d_i = \int_a^b g_i(x)m(x)\, dx \qquad i = 1, 2, \ldots, m \qquad (3.3)$$

where $g_i(x) = g(s_i, x)$. The functions $g_i(x)$ are referred to as **representers** or **data kernels**.

In the quadrature approach to discretizing an IFK, we use a **quadrature rule** or numerical integration scheme to numerically approximate (3.3). Note that, although quadrature methods are applied in this chapter to linear integral equations, they will also have utility in the discretization of nonlinear problems. The simplest quadrature rule is the **midpoint rule**, where we divide the interval $[a, b]$ into n subintervals and pick points $x_1, x_2, \ldots, x_n$ in the middle of each subinterval. The points are given by

$$x_j = a + \frac{\Delta x}{2} + (j - 1)\Delta x \qquad (3.4)$$

where

$$\Delta x = \frac{b - a}{n}. \qquad (3.5)$$

The integral (3.3) is then approximated by

$$d_i = \int_a^b g_i(x)m(x)\, d$$

$$x \approx \sum_{j=1}^{n} g_i(x_j)m(x_j)\, \Delta x \qquad i = 1, 2, \ldots, m. \qquad (3.6)$$

See Figure 3.1.

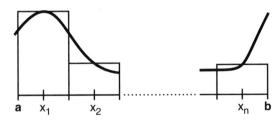

Figure 3.1 Grid for the midpoint rule.

If we let

$$G_{i,j} = g_i(x_j)\Delta x \quad \begin{pmatrix} i = 1, 2, \ldots, m \\ j = 1, 2, \ldots, n \end{pmatrix} \tag{3.7}$$

and

$$m_j = m(x_j) \quad j = 1, 2, \ldots, n \tag{3.8}$$

then we obtain a linear system of equations $\mathbf{Gm = d}$.

The approach of using the midpoint rule to approximate the integral is known as **simple collocation**. Of course, there are also more sophisticated quadrature rules for numerically approximating integrals (e.g., the trapezoidal rule, or Simpson's rule). In each case, we end up with a similar linear system of equations, but the formulas for evaluating the elements of $\mathbf{G}$ will be different.

■ **Example 3.1** Consider the vertical seismic profiling problem (Example 1.3) where we wish to estimate vertical seismic slowness using travel time measurements of downward-propagating seismic waves. See Figure 3.2. The data in this case are integrated values of the model parameters. We discretize the forward problem (1.21) for m observations taken at times t_i and at depths y_i that are equally spaced at intervals of Δy. The model is discretized at n model depths z_j that are equally spaced at intervals of Δz. The discretization is shown in Figure 3.2.

The discretized problem has

$$t_i = \sum_{j=1}^{n} H(y_i - z_j)s_j\,\Delta z, \tag{3.9}$$

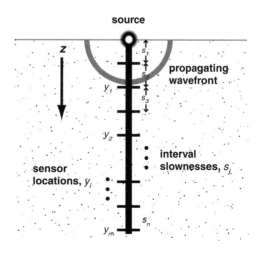

Figure 3.2 Discretization of the vertical seismic profiling problem ($n/m = 2$).

where $n/m = \Delta y/\Delta z$. The rows of the matrix $\mathbf{G}_{i,\cdot}$ each consist of $i \cdot n/m$ elements Δz on the left and $n - (i \cdot n/m)$ zeros on the right. For $n = m$, $\mathbf{G}$ is a lower triangular matrix with each nonzero entry equal to Δz.   ∎

■ **Example 3.2** Another instructive example is an optics experiment in which light passes through a thin slit. See Figure 3.3. This problem was studied by Shaw [144]. The data, $d(s)$, are measurements of diffracted light intensity as a function of outgoing angle $-\pi/2 \le s \le \pi/2$. Our goal is to find the intensity of the incident light on the slit $m(\theta)$, as a function of the incoming angle $-\pi/2 \le \theta \le \pi/2$.

The mathematical model relating d and m is

$$d(s) = \int_{-\pi/2}^{\pi/2} (\cos(s) + \cos(\theta))^2 \left(\frac{\sin(\pi(\sin(s) + \sin(\theta)))}{\pi(\sin(s) + \sin(\theta))} \right)^2 m(\theta) \, d\theta. \tag{3.10}$$

To discretize (3.10) we apply the method of simple collocation with $n = m$ equally sized intervals for the model and data functions, where n is even. For simplicity, we additionally define the model and data points at the same n angles

$$s_i = \theta_i = \frac{(i - 0.5)\pi}{n} - \frac{\pi}{2} \qquad i = 1, 2, \ldots, n. \tag{3.11}$$

Discretizing the data and model into n-length vectors

$$d_i = d(s_i) \qquad i = 1, 2, \ldots, n \tag{3.12}$$

and

$$m_j = m(\theta_j) \qquad j = 1, 2, \ldots, n \tag{3.13}$$

leads to a discrete linear system where

$$G_{i,j} = \Delta s (\cos(s_i) + \cos(\theta_j))^2 \left(\frac{\sin(\pi(\sin(s_i) + \sin(\theta_j)))}{\pi(\sin(s_i) + \sin(\theta_j))} \right)^2 \tag{3.14}$$

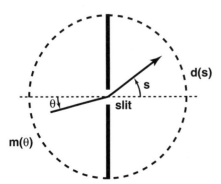

Figure 3.3   The Shaw problem (3.10), with example model and data plotted as functions of angles θ and s.

and

$$\Delta s = \frac{\pi}{n}. \tag{3.15}$$

The MATLAB Regularization Tools contain a routine **shaw** that computes the **G** matrix along with a sample model and data for this problem for $n = m$ [58].   ■

■ **Example 3.3** Consider the problem of recovering the history of groundwater pollution at a source site from later measurements of the contamination at downstream wells to which the contaminant plume has been transported by advection and diffusion. See Figure 3.4. This "source history reconstruction problem" has been considered by a number of authors [113, 147, 148, 179].

The mathematical model for contaminant transport is an advection–diffusion equation

$$\frac{\partial C}{\partial t} = D\frac{\partial^2 C}{\partial x^2} - v\frac{\partial C}{\partial x}$$
$$C(0, t) = C_{in}(t)$$
$$C(x, t) \rightarrow 0 \ \text{ as } \ x \rightarrow \infty$$
$$C(x, 0) = C_0(x) \tag{3.16}$$

where D is the diffusion coefficient, and v is the velocity of groundwater flow. The solution to (3.16) at time T is the convolution

$$C(x, T) = \int_0^T C_{in}(t) f(x, T - t) \, dt, \tag{3.17}$$

where $C_{in}(t)$ is the time history of contaminant injection at $x = 0$, and the kernel is

$$f(x, T - t) = \frac{x}{2\sqrt{\pi D(T - t)^3}} \exp\left(-\frac{[x - v(T - t)]^2}{4D(T - t)}\right). \tag{3.18}$$

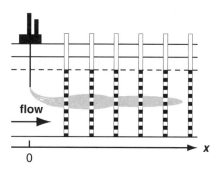

Figure 3.4 The contaminant plume source history reconstruction problem.

We assume that the parameters of (3.18) are known, and wish to estimate $C_{in}(t)$ from simultaneous observations at some later time T. The convolution (3.17) for $C(x, T)$ is discretized as

$$\mathbf{d} = \mathbf{Gm} \tag{3.19}$$

where $\mathbf{d}$ is a vector of sampled concentrations at different well locations, $\mathbf{x}$, at a time T, $\mathbf{m}$ is a vector of C_{in} values to be estimated, and

$$G_{i,j} = f(x_i, T - t_j)\Delta t \tag{3.20}$$

$$= \frac{x_i}{2\sqrt{\pi D(T - t_j)^3}} \exp\left(-\frac{[x_i - v(T - t_j)]^2}{4D(T - t_j)}\right) \Delta t. \tag{3.21}$$

$\blacksquare$

3.3 EXPANSION IN TERMS OF REPRESENTERS

In the **Gram matrix technique** for discretizing a linear inverse problem, a continuous model $m(x)$ is written as a linear combination of the m representers (3.3)

$$m(x) = \sum_{j=1}^{m} \alpha_j g_j(x), \tag{3.22}$$

where the α_j are coefficients to be determined. The representers form a basis for a subspace of the space of all functions on the interval (a, b). Substituting (3.22) into (3.3) gives

$$d(s_i) = \int_a^b g_i(x) \sum_{j=1}^{m} \alpha_j g_j(x)\, dx \tag{3.23}$$

$$= \sum_{j=1}^{m} \alpha_j \int_a^b g_i(x)g_j(x)\, dx \qquad i = 1, 2, \ldots, m. \tag{3.24}$$

Recall from Appendix A that an m by m matrix $\mathbf{\Gamma}$ with elements

$$\Gamma_{i,j} = \int_a^b g_i(x)g_j(x)\, dx \qquad i = 1, 2, \ldots, m \tag{3.25}$$

is called a Gram matrix. The IFK can thus be discretized as an m by m linear system of equations

$$\mathbf{\Gamma}\boldsymbol{\alpha} = \mathbf{d}. \tag{3.26}$$

Once (3.26) is solved for the vector of coefficients $\boldsymbol{\alpha}$, the corresponding model is given by (3.22). If the representers and Gram matrix are analytically expressible, then the Gram matrix formulation produces a continuous solution. Where only numerical representations for the representers exist, the method can still be applied, although the integrals in (3.25) must be obtained by numerical integration.

It can be shown that if the representers $g_j(x)$ are linearly independent, then the Gram matrix will be nonsingular. See Exercise 3.3. As we will see in Chapter 4, the Gram matrix tends to become very badly conditioned as m increases. On the other hand, we want to use as large as possible a value of m so as to increase the accuracy of the discretization. Thus there is a trade-off between the discretization error and ill-conditioning.

It can be shown that the model obtained by solving (3.26) minimizes $\|m(x)\|_2^2 = \int_a^b m(x)^2 dx$ for all models that match the data [97]. However, this result is of more theoretical than practical interest, since we typically do not want to match noisy data exactly.

3.4 EXPANSION IN TERMS OF ORTHONORMAL BASIS FUNCTIONS

In Section 3.3, we approximated $m(x)$ as a linear combination of m representers. Generalizing this approach, suppose we are given suitable functions $h_1(x), h_2(x), \ldots, h_n(x)$ that form a basis for a function space H. We could then approximate $m(x)$ by

$$m(x) = \sum_{j=1}^{n} \alpha_j h_j(x).$$

(3.27)

Substituting this approximation into (3.3) gives

$$d(s_i) = \int_a^b g_i(x) \sum_{j=1}^{n} \alpha_j h_j(x) \, dx$$

(3.28)

$$= \sum_{j=1}^{n} \alpha_j \int_a^b g_i(x) h_j(x) \, dx \qquad i = 1, 2, \ldots, m.$$

(3.29)

This leads to an m by n linear system

$$\mathbf{G}\boldsymbol{\alpha} = \mathbf{d}$$

(3.30)

where

$$G_{i,j} = \int_a^b g_i(x) h_j(x) \, dx.$$

(3.31)

If we define the dot product or **inner product** of two functions to be

$$f \cdot g = \int_a^b f(x) g(x) \, dx,$$

(3.32)

then the corresponding norm is

$$\|f\|_2 = \sqrt{\int_a^b f(x)^2 \, dx}. \tag{3.33}$$

If the basis functions $h_j(x)$ are orthonormal with respect to this inner product, then the projection of $g_i(x)$ onto the space H spanned by the basis is

$$\text{proj}_H \; g_i(x) = (g_i \cdot h_1)h_1(x) + (g_i \cdot h_2)h_2(x) + \cdots + (g_i \cdot h_n)h_n(x). \tag{3.34}$$

The elements in the **G** matrix are given by the same dot products

$$G_{i,j} = g_i \cdot h_j. \tag{3.35}$$

Thus we have effectively projected the original representers onto our function space H.

An important advantage of using an orthonormal basis is that it can be shown that $\|m(x)\|_2 = \|\boldsymbol{\alpha}\|_2$. See Exercise 3.4. In Chapter 5, we will regularize the solution of $\mathbf{G}\boldsymbol{\alpha} = \mathbf{d}$ by finding a vector $\boldsymbol{\alpha}$ that minimizes $\|\boldsymbol{\alpha}\|_2$ subject to a constraint on the misfit $\|\mathbf{G}\boldsymbol{\alpha} - \mathbf{d}\|_2$. By using an orthonormal basis, we can minimize $\|m(x)\|_2$ subject to the same constraint on the misfit.

3.5 THE METHOD OF BACKUS AND GILBERT

The method of Backus and Gilbert [3, 122] is applicable to continuous linear inverse problems of the form

$$d(s) = \int_a^b g(s, x)m(x) \, dx \tag{3.36}$$

where we have observations at points $s_1, s_2, \ldots, s_m$. Let

$$d_j = d(s_j) \qquad j = 1, 2, \ldots, m. \tag{3.37}$$

Using (3.36), we can write d_j as

$$d_j = \int_a^b g(s_j, x)m(x) \, dx \tag{3.38}$$

$$= \int_a^b g_j(x)m(x) \, dx. \tag{3.39}$$

We want to estimate $m(x)$ at some point $\hat{x}$ given the m data values d_j. Since the only data that we have are the d_j values, we will consider estimates of the form

$$m(\hat{x}) \approx \hat{m} = \sum_{j=1}^{m} c_j d_j \qquad (3.40)$$

where the c_j are coefficients to be determined.

Combining (3.39) and (3.40) gives

$$\hat{m} = \sum_{j=1}^{m} c_j \int_a^b g_j(x) m(x)\, dx \qquad (3.41)$$

$$= \int_a^b \left(\sum_{j=1}^{m} c_j g_j(x) \right) m(x)\, dx \qquad (3.42)$$

$$= \int_a^b A(x) m(x)\, dx \qquad (3.43)$$

where

$$A(x) = \sum_{j=1}^{m} c_j g_j(x). \qquad (3.44)$$

The function $A(x)$ is called an **averaging kernel**. Ideally, we would like the averaging kernel to closely approximate a delta function

$$A(x) = \delta(x - \hat{x}) \qquad (3.45)$$

because, assuming the data were exact, (3.43) would then produce exact agreement ($\hat{m} = m(\hat{x})$) between the estimated and the true model. Since this is not possible, we will instead select the coefficients so that the area under the averaging kernel is one, and so that the width of the averaging kernel around the $\hat{x}$ is as small as possible.

In terms of the coefficients $\mathbf{c}$, the unit area constraint can be written as

$$\int_a^b A(x)\, dx = 1 \qquad (3.46)$$

$$\int_a^b \sum_{j=1}^{m} c_j g_j(x)\, dx = 1 \qquad (3.47)$$

$$\sum_{j=1}^{m} c_j \left(\int_a^b g_j(x)\, dx \right) = 1. \qquad (3.48)$$

Letting

$$q_j = \int_a^b g_j(x)\,dx \tag{3.49}$$

the unit area constraint (3.46) can be written as

$$\mathbf{q}^T \mathbf{c} = 1. \tag{3.50}$$

Averaging kernel widths can be usefully characterized in a variety of ways [122]. The most commonly used measure is the second moment of $A(x)$ about $\hat{x}$:

$$w = \int_a^b A(x)^2 (x - \hat{x})^2\,dx. \tag{3.51}$$

In terms of the coefficients $\mathbf{c}$, this can be written as

$$w = \mathbf{c}^T \mathbf{H} \mathbf{c} \tag{3.52}$$

where

$$H_{j,k} = \int_a^b g_j(x) g_k(x)(x - \hat{x})^2\,dx. \tag{3.53}$$

Now, the problem of finding the optimal coefficients can be written as

$$\begin{aligned} \min\quad &\mathbf{c}^T \mathbf{H} \mathbf{c} \\ &\mathbf{c}^T \mathbf{q} = 1. \end{aligned} \tag{3.54}$$

This can be solved using the Lagrange multiplier technique (see Appendix C).

In practice, the data may be noisy, and the solution may be unstable because of this noise. For measurements with independent errors, the standard deviation of the estimate is given by

$$\mathrm{Var}(\hat{m}) = \sum_{j=1}^m c_j^2 \sigma_j^2 \tag{3.55}$$

where σ_j is the standard deviation of the jth observation.

The solution can be stabilized by adding a constraint on the variance to (3.54):

$$\begin{aligned} \min\quad &\mathbf{c}^T \mathbf{H} \mathbf{c} \\ &\mathbf{q}^T \mathbf{c} = 1 \\ &\sum_{j=1}^n c_j^2 \sigma_j^2 \le \Delta. \end{aligned} \tag{3.56}$$

Again, this problem can be solved by the method of Lagrange multipliers. Smaller values of Δ decrease the variance of the estimate but restrict the choice of coefficients so that the width of the averaging kernel increases. There is a trade-off between stability of the solution and the width of the averaging kernel.

The method of Backus and Gilbert produces an estimate of the model at a particular point $\hat{x}$. It is possible to use the method to compute estimates on a grid of points $x_1, x_2, \ldots, x_n$. However, since the averaging kernels at these points may not be well localized around the grid points and may overlap in complicated ways, this is not equivalent to the simple collocation method discussed earlier. Furthermore, this approach requires the computationally intensive solution of (3.56) for each point. For these reasons the method of Backus and Gilbert is not used as extensively as it once was.

■ **Example 3.4** For a spherically symmetric Earth model, the mass M_e and moment of inertia I_e are determined by the radial density $\rho(r)$, where

$$M_e = \int_0^{R_e} \left(4\pi r^2 \right) \rho(r)\, dr \tag{3.57}$$

and

$$I_e = \int_0^{R_e} \left(\frac{8}{3} \pi r^4 \right) \rho(r)\, dr. \tag{3.58}$$

Using $R_e = 6.3708 \times 10^6$ m as the radius of a spherical Earth, and supposing that from astronomical measurements we can infer that $M_e = 5.973 \pm 0.0005 \times 10^{24}$ kg and $I_e = 8.02 \pm 0.005 \times 10^{37}$ kg · m^2, we will estimate the density of the Earth in the lower mantle (e.g., at $r = 5000$ km), and core (e.g., at $r = 1000$ km).

Equations (3.57) and (3.58) include terms that span an enormous numerical range. Scaling so that

$$\hat{r} = r/R_e \qquad \hat{\rho} = \rho/1000 \qquad \hat{I}_e = I_e/10^{31} \qquad \hat{M}_e = M_e/10^{24}$$

gives

$$\hat{M}_e = 0.2586 \int_0^1 \left(4\pi \hat{r}^2 \right) \hat{\rho}(\hat{r})\, d\hat{r} \tag{3.59}$$

and

$$\hat{I}_e = 1.0492 \int_0^1 \left(\frac{8}{3} \pi \hat{r}^4 \right) \hat{\rho}(\hat{r})\, d\hat{r}. \tag{3.60}$$

Applying (3.54) for $r = 5000$ km gives the numerical coefficient values $\mathbf{c}^T = [1.1809, -0.1588]$ and a corresponding model density of 5.8 g/cm^3. This is not too bad of an estimate for this radius where standard Earth models estimated using seismological methods [92] have densities of approximately 5 g/cm^3. The associated standard deviation (3.55) is 0.001 g/cm^3, so there is very little sensitivity to data uncertainty.

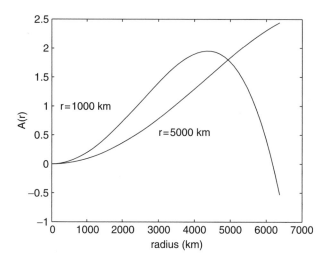

Figure 3.5   Averaging kernels for target radii of 1000 and 5000 km.

At $r = 1000$ km, we obtain the numerical coefficients $\mathbf{c}^T = [2.5537, -1.0047]$ and a corresponding density estimate of 7.2 g/cm^3. This is not a very accurate estimate for the density of the inner core, where standard Earth models have densities of around 13 g/cm^3. The corresponding standard deviation is just 0.005 g/cm^3, so this inaccuracy is not related to instability in the inverse problem.

Figure 3.5 shows the averaging kernels corresponding to these model element estimates and explains both the successful mantle and failed core-density estimates. In the mantle case, the averaging kernel has much of its area near the targeted radius of 5000 km. In the core case, however, the averaging kernel has most of its area at much greater r, and little area near the target radius of 1000 km. The fundamental reason for this situation is that both the mass and moment of inertia are insensitive to the density of the innermost Earth because of its relatively small mass and radius.   ∎

3.6 EXERCISES

3.1   Consider the data in Table 3.1. These data can also be found in the file **ifk.mat**.
The function $d(y)$, $0 \le y \le 1$ is related to an unknown function $m(x)$, $0 \le x \le 1$ by the mathematical model

$$d(y) = \int_0^1 xe^{-xy} m(x) \, dx. \tag{3.61}$$

(a)   Using the data provided, discretize the integral equation using simple collocation and solve the resulting system of equations.

Table 3.1 Data for Exercise 3.1.

y	0.0250	0.0750	0.1250	0.1750	0.2250
$d(y)$	0.2388	0.2319	0.2252	0.2188	0.2126
y	0.2750	0.3250	0.3750	0.4250	0.4750
$d(y)$	0.2066	0.2008	0.1952	0.1898	0.1846
y	0.5250	0.5750	0.6250	0.6750	0.7250
$d(y)$	0.1795	0.1746	0.1699	0.1654	0.1610
y	0.7750	0.8250	0.8750	0.9250	0.9750
$d(y)$	0.1567	0.1526	0.1486	0.1447	0.1410

(b) What is the condition number for this system of equations? Given that the data $d(y)$ are only accurate to about 4 digits, what does this tell you about the accuracy of your solution?

3.2 Use the Gram matrix technique to discretize the integral equation from Exercise 3.1.

(a) Solve the resulting linear system of equations, and plot the resulting model.

(b) What was the condition number of Γ? What does this tell you about the accuracy of your solution?

3.3 Show that if the representers $g_i(t)$ are linearly independent, then the Gram matrix Γ is nonsingular.

3.4 Show that if the basis functions in (3.27) are orthonormal, then $\|m(x)\|_2 = \|\alpha\|_2$. Hint: Expand $\|m(x)\|_2^2$ using (3.33), and then simplify using the orthogonality of the basis functions.

3.5 Recall the polynomial regression problem from Exercise 2.5. Instead of using the polynomials $1, x, \ldots, x^{19}$, we will use the basis of Legendre polynomials which are orthogonal on the interval $[-1, 1]$. These polynomials are generated by the recurrence relation

$$p_{n+1}(x) = \frac{(2n + 1)xp_n(x) - np_{n-1}(x)}{n + 1} \tag{3.62}$$

starting with

$$p_0(x) = 1 \tag{3.63}$$

and

$$p_1(x) = x. \tag{3.64}$$

The next two Legendre polynomials are $p_2(x) = (3x^2 - 1)/2$ and $p_3(x) = (5x^3 - 3x)/2$. This recurrence relation can be used both to compute coefficients of the Legendre polynomials and to compute values of the polynomials for particular values of x.

Use the first 20 Legendre polynomials to fit a polynomial of degree 19 to the data from Exercise 2.5. Express your solution as a linear combination of the Legendre polynomials and also as a regular polynomial. How well conditioned was this system of equations? Plot your solution and compare it with your solution to Exercise 2.5.

3.7 NOTES AND FURTHER READING

Techniques for discretizing integral equations are discussed in [38, 122, 168, 178]. A variety of basis functions have been used to discretize integral equations including sines and cosines, spherical harmonics, B-splines, and wavelets. In selecting the basis functions, it is important to select a basis that can reasonably represent likely models. The basis functions must be linearly independent, so that a function can be written in terms of the basis functions in exactly one way, and (3.27) is thus unique. As we have seen, the use of an orthonormal basis has the further advantage that $\|\boldsymbol{\alpha}\|_2 = \|m(x)\|_2$.

The selection of an appropriate basis for a particular problem is a fine art that requires detailed knowledge of the problem as well as of the behavior of the basis functions. Be aware that a poorly selected basis may not adequately approximate the solution, resulting in an estimated model $m(x)$ that is very wrong. The choice of basis can also have a very large effect on the condition number of the discretized problem, potentially making it very ill-conditioned.

An important theoretical question is whether the solutions to discretized versions of a continuous inverse problem with noise-free data will converge to a solution of the continuous inverse problem. Engl, Hanke, and Neubauer provide an explicit example showing that nonconvergence of a discretization scheme is possible [38]. They also provide conditions under which convergence is guaranteed. For Fredholm integral equations of the first kind, the Gram matrix scheme discussed in Section 3.3 can be shown to be convergent [38].

Using a coarser mesh in the discretization of an inverse problem typically results in a better-conditioned linear system of equations. By carefully selecting the grid size, it is possible to produce a well-conditioned linear system of equations. Of course, this also typically reduces the spatial or temporal resolution of the inverse solution. This process is known as regularization by discretization [38].

4

RANK DEFICIENCY AND
ILL-CONDITIONING

Synopsis: The characteristics of rank-deficient and ill-conditioned linear problems are explored using the singular value decomposition. The connection between model and data null spaces and solution uniqueness and ability to fit data is examined. Model and data resolution matrices are defined. The relationship between singular value size and singular vector roughness and its connection to solution stability is discussed in the context of the fundamental trade-off between model resolution and instability. Specific manifestations of these issues in rank-deficient and ill-conditioned discrete problems are shown in several examples.

4.1 THE SVD AND THE GENERALIZED INVERSE

A method of analyzing and solving least squares problems that is of particular interest in ill-conditioned and/or rank-deficient systems is the **singular value decomposition**, or **SVD**. In the SVD [49, 89, 155] an *m* by *n* matrix **G** is factored into

$$\mathbf{G} = \mathbf{U}\mathbf{S}\mathbf{V}^T \tag{4.1}$$

where

- **U** is an *m* by *m* orthogonal matrix with columns that are unit basis vectors spanning the **data space**, R^m.
- **V** is an *n* by *n* orthogonal matrix with columns that are basis vectors spanning the **model space**, R^n.
- **S** is an *m* by *n* diagonal matrix with nonnegative diagonal elements called **singular values**.

The SVD matrices can be computed in MATLAB with the **svd** command. It can be shown that every matrix has a singular value decomposition [49].

The singular values along the diagonal of **S** are customarily arranged in decreasing size, $s_1 \geq s_2 \geq \cdots \geq s_{\min(m,n)} \geq 0$. Note that some of the singular values may be zero. If only the first p singular values are nonzero, we can partition $\mathbf{S}$ as

$$\mathbf{S} = \begin{bmatrix} \mathbf{S}_p & \mathbf{0} \\ \mathbf{0} & \mathbf{0} \end{bmatrix} \tag{4.2}$$

where $\mathbf{S}_p$ is a p by p diagonal matrix composed of the positive singular values. Expanding the SVD representation of $\mathbf{G}$ in terms of the columns of $\mathbf{U}$ and $\mathbf{V}$ gives

$$\mathbf{G} = \begin{bmatrix} \mathbf{U}_{.,1}, \mathbf{U}_{.,2}, \ldots, \mathbf{U}_{.,m} \end{bmatrix} \begin{bmatrix} \mathbf{S}_p & \mathbf{0} \\ \mathbf{0} & \mathbf{0} \end{bmatrix} \begin{bmatrix} \mathbf{V}_{.,1}, \mathbf{V}_{.,2}, \ldots, \mathbf{V}_{.,n} \end{bmatrix}^T \tag{4.3}$$

$$= \begin{bmatrix} \mathbf{U}_p, \mathbf{U}_0 \end{bmatrix} \begin{bmatrix} \mathbf{S}_p & \mathbf{0} \\ \mathbf{0} & \mathbf{0} \end{bmatrix} \begin{bmatrix} \mathbf{V}_p, \mathbf{V}_0 \end{bmatrix}^T \tag{4.4}$$

where $\mathbf{U}_p$ denotes the first p columns of $\mathbf{U}$, $\mathbf{U}_0$ denotes the last $m - p$ columns of $\mathbf{U}$, $\mathbf{V}_p$ denotes the first p columns of $\mathbf{V}$, and $\mathbf{V}_0$ denotes the last $n - p$ columns of $\mathbf{V}$. Because the last $m - p$ columns of $\mathbf{U}$ and the last $n - p$ columns of $\mathbf{V}$ in (4.4) are multiplied by zeros in $\mathbf{S}$, we can simplify the SVD of $\mathbf{G}$ into its **compact form**

$$\mathbf{G} = \mathbf{U}_p \mathbf{S}_p \mathbf{V}_p^T. \tag{4.5}$$

For any vector $\mathbf{y}$ in the range of $\mathbf{G}$, applying (4.5) gives

$$\mathbf{y} = \mathbf{G}\mathbf{x} \tag{4.6}$$

$$= \mathbf{U}_p \left(\mathbf{S}_p \mathbf{V}_p^T \mathbf{x} \right). \tag{4.7}$$

Thus every vector in $R(\mathbf{G})$ can be written as $\mathbf{y} = \mathbf{U}_p \mathbf{z}$ where $\mathbf{z} = \mathbf{S}_p \mathbf{V}_p^T \mathbf{x}$. Writing out this matrix-vector multiplication, we see that any vector $\mathbf{y}$ in $R(\mathbf{G})$ can be written as a linear combination of the columns of $\mathbf{U}_p$:

$$\mathbf{y} = \sum_{i=1}^{p} z_i \mathbf{U}_{.,i}. \tag{4.8}$$

The columns of $\mathbf{U}_p$ span $R(\mathbf{G})$, are linearly independent, and form an orthonormal basis for $R(\mathbf{G})$. Because this orthonormal basis has p vectors, rank$(\mathbf{G}) = p$.

Since $\mathbf{U}$ is an orthogonal matrix, the columns of $\mathbf{U}$ form an orthonormal basis for R^m. We have already shown in (4.8) that the p columns of $\mathbf{U}_p$ form an orthonormal basis for $R(\mathbf{G})$. By Theorem A.5, $N(\mathbf{G}^T) + R(\mathbf{G}) = R^m$, so the remaining $m - p$ columns of $\mathbf{U}_0$ form an orthonormal basis for the null space of $\mathbf{G}^T$. We will sometimes refer to $N(\mathbf{G}^T)$ as the **data null space**. Similarly, because $\mathbf{G}^T = \mathbf{V}_p \mathbf{S}_p \mathbf{U}_p^T$, the columns of $\mathbf{V}_p$ form an orthonormal basis for $R(\mathbf{G}^T)$ and the columns of $\mathbf{V}_0$ form an orthonormal basis for $N(\mathbf{G})$. We will sometimes refer to $N(\mathbf{G})$ as the **model null space**.

Two other important SVD properties are similar to properties of eigenvalues and eigenvectors. See Section A.6. Because the columns of $\mathbf{V}$ are orthonormal,

$$\mathbf{V}^T\mathbf{V}_{.,i} = \mathbf{e}_i. \tag{4.9}$$

Thus

$$\mathbf{G}\mathbf{V}_{.,i} = \mathbf{U}\mathbf{S}\mathbf{V}^T\mathbf{V}_{.,i} \tag{4.10}$$

$$= \mathbf{U}\mathbf{S}\mathbf{e}_i \tag{4.11}$$

$$= s_i\mathbf{U}_{.,i} \tag{4.12}$$

and

$$\mathbf{G}^T\mathbf{U}_{.,i} = \mathbf{V}\mathbf{S}^T\mathbf{U}^T\mathbf{U}_{.,i} \tag{4.13}$$

$$= \mathbf{V}\mathbf{S}^T\mathbf{e}_i \tag{4.14}$$

$$= s_i\mathbf{V}_{.,i}. \tag{4.15}$$

There is also a connection between the singular values of $\mathbf{G}$ and the eigenvalues of $\mathbf{G}\mathbf{G}^T$ and $\mathbf{G}^T\mathbf{G}$:

$$\mathbf{G}\mathbf{G}^T\mathbf{U}_{.,i} = \mathbf{G}s_i\mathbf{V}_{.,i} \tag{4.16}$$

$$= s_i\mathbf{G}\mathbf{V}_{.,i} \tag{4.17}$$

$$= s_i^2\mathbf{U}_{.,i}. \tag{4.18}$$

Similarly,

$$\mathbf{G}^T\mathbf{G}\mathbf{V}_{.,i} = s_i^2\mathbf{V}_{.,i}. \tag{4.19}$$

These relations show that we could, in theory, compute the SVD by finding the eigenvalues and eigenvectors of $\mathbf{G}^T\mathbf{G}$ and $\mathbf{G}\mathbf{G}^T$. In practice, more efficient specialized algorithms are used [31, 49, 167].

The SVD can be used to compute a generalized inverse of $\mathbf{G}$, called the **Moore–Penrose pseudoinverse**, because it has desirable inverse properties originally identified by Moore and Penrose [102, 125]. The generalized inverse is

$$\mathbf{G}^\dagger = \mathbf{V}_p\mathbf{S}_p^{-1}\mathbf{U}_p^T. \tag{4.20}$$

MATLAB has a **pinv** command that generates $\mathbf{G}^\dagger$. This command allows the user to select a tolerance such that singular values smaller than the tolerance are not included in the computation.

Using (4.20), we define the pseudoinverse solution to be

$$\mathbf{m}_\dagger = \mathbf{G}^\dagger \mathbf{d} \tag{4.21}$$

$$= \mathbf{V}_p \mathbf{S}_p^{-1} \mathbf{U}_p^T \mathbf{d}. \tag{4.22}$$

Among the desirable properties of (4.22) is that $\mathbf{G}^\dagger$, and hence $\mathbf{m}_\dagger$, always exist. In contrast, the inverse of $\mathbf{G}^T\mathbf{G}$ that appears in the normal equations (2.3) does not exist when $\mathbf{G}$ is not of full column rank. We will shortly show that $\mathbf{m}_\dagger$ is a least squares solution.

To encapsulate what the SVD tells us about our linear matrix $\mathbf{G}$, and the corresponding generalized inverse matrix $\mathbf{G}^\dagger$, consider four cases:

1.   Both the model and data null spaces, $N(\mathbf{G})$ and $N(\mathbf{G}^T)$ are trivial (only include the zero vector). $\mathbf{U}_p = \mathbf{U}$ and $\mathbf{V}_p = \mathbf{V}$ are square orthogonal matrices, so that $\mathbf{U}_p^T = \mathbf{U}_p^{-1}$, and $\mathbf{V}_p^T = \mathbf{V}_p^{-1}$. Equation (4.22) gives

$$\mathbf{G}^\dagger = \mathbf{V}_p \mathbf{S}_p^{-1} \mathbf{U}_p^T \tag{4.23}$$

$$= (\mathbf{U}_p \mathbf{S}_p \mathbf{V}_p^T)^{-1} \tag{4.24}$$

$$= \mathbf{G}^{-1} \tag{4.25}$$

which is the matrix inverse for a square full rank matrix where $m = n = p$. The solution is unique, and the data are fit exactly.

2. $N(\mathbf{G})$ is nontrivial, but $N(\mathbf{G}^T)$ is trivial. $\mathbf{U}_p^T = \mathbf{U}_p^{-1}$ and $\mathbf{V}_p^T \mathbf{V}_p = \mathbf{I}_p$. $\mathbf{G}$ applied to the generalized inverse solution 4.21 gives

$$\mathbf{G}\mathbf{m}_\dagger = \mathbf{G}\mathbf{G}^\dagger \mathbf{d} \tag{4.26}$$

$$= \mathbf{U}_p \mathbf{S}_p \mathbf{V}_p^T \mathbf{V}_p \mathbf{S}_p^{-1} \mathbf{U}_p^T \mathbf{d} \tag{4.27}$$

$$= \mathbf{U}_p \mathbf{S}_p \mathbf{I}_p \mathbf{S}_p^{-1} \mathbf{U}_p^T \mathbf{d} \tag{4.28}$$

$$= \mathbf{d}. \tag{4.29}$$

The data are fit exactly but the solution is nonunique, because of the existence of the nontrivial model null space $N(\mathbf{G})$. Since $\mathbf{m}_\dagger$ is an exact solution to $\mathbf{Gm} = \mathbf{d}$, it is also a least-squares solution.

We need to characterize the least squares solutions to $\mathbf{Gm} = \mathbf{d}$. If $\mathbf{m}$ is any least squares solution, then it satisfies the normal equations. This is shown in Exercise C.5.

$$(\mathbf{G}^T\mathbf{G})\mathbf{m} = \mathbf{G}^T\mathbf{d}. \tag{4.30}$$

Since $\mathbf{m}_\dagger$ is a least squares solution, it also satisfies the normal equations.

$$(\mathbf{G}^T\mathbf{G})\mathbf{m}_\dagger = \mathbf{G}^T\mathbf{d}. \tag{4.31}$$

Subtracting (4.30) from (4.31), we find that

$$(\mathbf{G}^T\mathbf{G})(\mathbf{m}_\dagger - \mathbf{m}) = \mathbf{0}. \tag{4.32}$$

Thus $\mathbf{m}_\dagger - \mathbf{m}$ lies in $N(\mathbf{G}^T\mathbf{G})$. In Exercise A.17f it is shown that $N(\mathbf{G}^T\mathbf{G}) = N(\mathbf{G})$. This implies that $\mathbf{m}_\dagger - \mathbf{m}$ lies in $N(\mathbf{G})$.

The general solution is thus the sum of $\mathbf{m}_\dagger$ and an arbitrary vector in $N(\mathbf{G})$ that can be written as a linear combination of the basis vectors for $N(\mathbf{G})$:

$$\mathbf{m} = \mathbf{m}_\dagger + \mathbf{m}_0 \tag{4.33}$$

$$= \mathbf{m}_\dagger + \sum_{i=p+1}^{n} \alpha_i \mathbf{V}_{\cdot,i}. \tag{4.34}$$

Because the columns of $\mathbf{V}$ are orthonormal, the square of the 2-norm of a general solution $\mathbf{m}$ is

$$\|\mathbf{m}\|_2^2 = \|\mathbf{m}_\dagger\|_2^2 + \sum_{i=p+1}^{n} \alpha_i^2 \geq \|\mathbf{m}_\dagger\|_2^2 \tag{4.35}$$

where we have equality only if all of the model null space coefficients α_i are zero. The generalized inverse solution is thus a **minimum length solution**.

We can also write this solution in terms of $\mathbf{G}$ and $\mathbf{G}^T$.

$$\mathbf{m}_\dagger = \mathbf{V}_p \mathbf{S}_p^{-1} \mathbf{U}_p^T \mathbf{d} \tag{4.36}$$

$$= \mathbf{V}_p \mathbf{S}_p \mathbf{U}_p^T \mathbf{U}_p \mathbf{S}_p^{-2} \mathbf{U}_p^T \mathbf{d} \tag{4.37}$$

$$= \mathbf{G}^T (\mathbf{U}_p \mathbf{S}_p^{-2} \mathbf{U}_p^T) \mathbf{d} \tag{4.38}$$

$$= \mathbf{G}^T (\mathbf{G}\mathbf{G}^T)^{-1} \mathbf{d}. \tag{4.39}$$

In practice it is better to compute a solution using the SVD than to use (4.39) because of numerical accuracy issues.

3. $N(\mathbf{G})$ is trivial but $N(\mathbf{G}^T)$ is nontrivial and $R(\mathbf{G})$ is a strict subset of R^m. Here

$$\mathbf{G}\mathbf{m}_\dagger = \mathbf{U}_p \mathbf{S}_p \mathbf{V}_p^T \mathbf{V}_p \mathbf{S}_p^{-1} \mathbf{U}_p^T \mathbf{d} \tag{4.40}$$

$$= \mathbf{U}_p \mathbf{U}_p^T \mathbf{d}. \tag{4.41}$$

The product $\mathbf{U}_p \mathbf{U}_p^T \mathbf{d}$ gives the projection of $\mathbf{d}$ onto $R(\mathbf{G})$. Thus $\mathbf{G}\mathbf{m}_\dagger$ is the point in $R(\mathbf{G})$ that is closest to $\mathbf{d}$, and $\mathbf{m}_\dagger$ is a least squares solution to $\mathbf{G}\mathbf{m} = \mathbf{d}$. If $\mathbf{d}$ is actually in $R(\mathbf{G})$, then $\mathbf{m}_\dagger$ will be an exact solution to $\mathbf{G}\mathbf{m} = \mathbf{d}$.

We can see that this solution is exactly that obtained from the normal equations because

$$(\mathbf{G}^T\mathbf{G})^{-1} = (\mathbf{V}_p\mathbf{S}_p\mathbf{U}_p^T\mathbf{U}_p\mathbf{S}_p\mathbf{V}_p^T)^{-1} \tag{4.42}$$

$$= (\mathbf{V}_p\mathbf{S}_p^2\mathbf{V}_p^T)^{-1} \tag{4.43}$$

$$= \mathbf{V}_p\mathbf{S}_p^{-2}\mathbf{V}_p^T \tag{4.44}$$

and

$$\mathbf{m}_\dagger = \mathbf{G}^\dagger\mathbf{d} \tag{4.45}$$

$$= \mathbf{V}_p\mathbf{S}_p^{-1}\mathbf{U}_p^T\mathbf{d} \tag{4.46}$$

$$= \mathbf{V}_p\mathbf{S}_p^{-2}\mathbf{V}_p^T\mathbf{V}_p\mathbf{S}_p\mathbf{U}_p^T\mathbf{d} \tag{4.47}$$

$$= (\mathbf{G}^T\mathbf{G})^{-1}\mathbf{G}^T\mathbf{d}. \tag{4.48}$$

This solution is unique, but cannot fit general data exactly. As with (4.39), it is better in practice to use the generalized inverse solution than to use (4.48) because of numerical accuracy issues.

4. Both $N(\mathbf{G}^T)$ and $N(\mathbf{G})$ are nontrivial and p is less than both m and n. In this case, the generalized inverse solution encapsulates the behavior of both of the two previous cases, minimizing both $\|\mathbf{Gm} - \mathbf{d}\|_2$ and $\|\mathbf{m}\|_2$.
As in case 3,

$$\mathbf{Gm}_\dagger = \mathbf{U}_p\mathbf{S}_p\mathbf{V}_p^T\mathbf{V}_p\mathbf{S}_p^{-1}\mathbf{U}_p^T\mathbf{d} \tag{4.49}$$

$$= \mathbf{U}_p\mathbf{U}_p^T\mathbf{d} \tag{4.50}$$

$$= \mathrm{proj}_{R(\mathbf{G})}\mathbf{d}. \tag{4.51}$$

Thus $\mathbf{m}_\dagger$ is a least squares solution to $\mathbf{Gm} = \mathbf{d}$.
As in case 2 we can write the model and its norm using (4.34) and (4.35). Thus $\mathbf{m}_\dagger$ is the least squares solution of minimum length.

We have shown that the generalized inverse provides an inverse solution (4.22) that always exists, is both least squares and minimum length, and properly accommodates the rank and dimensions of $\mathbf{G}$. Relationships between the subspaces $R(\mathbf{G})$, $N(\mathbf{G}^T)$, $R(\mathbf{G}^T)$, $N(\mathbf{G})$ and the operators $\mathbf{G}$ and $\mathbf{G}^\dagger$ are schematically depicted in Figure 4.1. Table 4.1 summarizes the SVD and its properties.

The existence of a nontrivial model null space (one that includes more than just the zero vector) is at the heart of solution nonuniqueness. There are an infinite number of solutions that will fit the data equally well, because model components in $N(\mathbf{G})$ have no effect on data fit. To select a particular preferred solution from this infinite set thus requires more constraints (such as minimum length or smoothing constraints) than are encoded in the matrix $\mathbf{G}$.

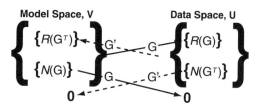

Figure 4.1   SVD model and data space mappings, where $\mathbf{G}^\dagger$ is the generalized inverse. $N(\mathbf{G}^T)$ and $N(\mathbf{G})$ are the data and model null spaces, respectively.

Table 4.1   Summary of the SVD and its associated scalars and matrices.

Object	Size	Properties
p	Scalar	$\mathrm{rank}(\mathbf{G}) = p$
m	Scalar	Dimension of the data space
n	Scalar	Dimension of the model space
$\mathbf{G}$	m by n	Forward problem matrix; $\mathbf{G} = \mathbf{U}\mathbf{S}\mathbf{V}^T = \mathbf{U}_p \mathbf{S}_p \mathbf{V}_p^T$
$\mathbf{U}$	m by m	Orthogonal matrix; $\mathbf{U} = [\mathbf{U}_p, \mathbf{U}_0]$
$\mathbf{S}$	m by n	Diagonal matrix of singular values; $\mathbf{S}_{i,i} = s_i$
$\mathbf{V}$	n by n	Orthogonal matrix $\mathbf{V} = [\mathbf{V}_p, \mathbf{V}_0]$
$\mathbf{U}_p$	m by p	Columns form an orthonormal basis for $R(\mathbf{G})$
$\mathbf{S}_p$	p by p	Diagonal matrix of nonzero singular values
$\mathbf{V}_p$	n by p	Columns form an orthonormal basis for $R(\mathbf{G}^T)$
$\mathbf{U}_0$	m by $m - p$	Columns form an orthonormal basis for $N(\mathbf{G}^T)$
$\mathbf{V}_0$	n by $n - p$	Columns form an orthonormal basis for $N(\mathbf{G})$
$\mathbf{U}_{\cdot,i}$	m by 1	Eigenvector of $\mathbf{G}\mathbf{G}^T$ with eigenvalue s_i^2
$\mathbf{V}_{\cdot,i}$	n by 1	Eigenvector of $\mathbf{G}^T\mathbf{G}$ with eigenvalue s_i^2
$\mathbf{G}^\dagger$	n by m	Pseudoinverse of $\mathbf{G}$; $\mathbf{G}^\dagger = \mathbf{V}_p \mathbf{S}_p^{-1} \mathbf{U}_p^T$
$\mathbf{m}_\dagger$	n by 1	Generalized inverse solution; $\mathbf{m}_\dagger = \mathbf{G}^\dagger \mathbf{d}$

To see the significance of the $N(\mathbf{G}^T)$ subspace, consider an arbitrary data vector, $\mathbf{d}_0$, which lies in $N(\mathbf{G}^T)$:

$$\mathbf{d}_0 = \sum_{i=p+1}^{m} \beta_i \mathbf{U}_{\cdot,i}. \tag{4.52}$$

The generalized inverse operating on such a data vector gives

$$\mathbf{m}_\dagger = \mathbf{V}_p \mathbf{S}_p^{-1} \mathbf{U}_p^T \mathbf{d}_0 \tag{4.53}$$

$$= \mathbf{V}_p \mathbf{S}_p^{-1} \sum_{i=p+1}^{n} \beta_i \mathbf{U}_p^T \mathbf{U}_{\cdot,i} \tag{4.54}$$

$$= \mathbf{0} \tag{4.55}$$

because the $\mathbf{U}_{.,i}$ are orthogonal. $N(\mathbf{G}^T)$ is a subspace of R^m consisting of all vectors $\mathbf{d}_0$ that have no influence on the generalized inverse model, $\mathbf{m}_\dagger$. If $p < n$ there are an infinite number of potential data sets that will produce the same model when (4.22) is applied.

4.2 COVARIANCE AND RESOLUTION OF THE GENERALIZED INVERSE SOLUTION

The generalized inverse always gives us a solution, $\mathbf{m}_\dagger$, with well-determined properties, but it is essential to investigate how faithful a representation any model is likely to be of the true situation.

In Section 2.2, we found that, under the assumption of independent and normally distributed measurement errors, the least squares solution was an unbiased estimator of the true model, and that the estimated model parameters had a multivariate normal distribution with covariance

$$\text{Cov}(\mathbf{m}_{L_2}) = \sigma^2 (\mathbf{G}^T \mathbf{G})^{-1}. \tag{4.56}$$

We can attempt the same analysis for the generalized inverse solution $\mathbf{m}_\dagger$. The covariance matrix would be given by

$$\text{Cov}(\mathbf{m}_\dagger) = \mathbf{G}^\dagger \text{Cov}(\mathbf{d})(\mathbf{G}^\dagger)^T \tag{4.57}$$

$$= \sigma^2 \mathbf{G}^\dagger (\mathbf{G}^\dagger)^T \tag{4.58}$$

$$= \sigma^2 \mathbf{V}_p \mathbf{S}_p^{-2} \mathbf{V}_p^T \tag{4.59}$$

$$= \sigma^2 \sum_{i=1}^{p} \frac{V_{.,i} V_{.,i}^T}{s_i^2}. \tag{4.60}$$

Unfortunately, unless $p = n$, the generalized inverse solution is *not* an unbiased estimator of the true solution. This occurs because the true model may have nonzero projections onto those basis vectors in $\mathbf{V}$ that are unused in the generalized inverse solution. In practice, the bias introduced by restricting the solution to the subspace spanned by the columns of $\mathbf{V}_p$ is frequently far larger than the uncertainty due to measurement error.

The concept of **model resolution** is an important way to characterize the bias of the generalized inverse solution. In this approach we see how closely the generalized inverse solution matches a given model, assuming that there are no errors in the data. We begin with any model $\mathbf{m}$. By multiplying $\mathbf{G}$ times $\mathbf{m}$, we can find a corresponding data vector $\mathbf{d}$. If we then multiply $\mathbf{G}^\dagger$ times $\mathbf{d}$, we get back a generalized inverse solution $\mathbf{m}_\dagger$:

$$\mathbf{m}_\dagger = \mathbf{G}^\dagger \mathbf{G} \mathbf{m}. \tag{4.61}$$

We would obviously like to get back our original model so that $\mathbf{m}_\dagger = \mathbf{m}$. Since the original model may have had a nonzero projection onto the model null space $N(\mathbf{G})$, $\mathbf{m}_\dagger$ will not in general be equal to **m**. The **model resolution matrix** is

$$\mathbf{R}_{\mathrm{m}} = \mathbf{G}^{\dagger}\mathbf{G} \tag{4.62}$$

$$= \mathbf{V}_p \mathbf{S}_p^{-1} \mathbf{U}_p^T \mathbf{U}_p \mathbf{S}_p \mathbf{V}_p^T \tag{4.63}$$

$$= \mathbf{V}_p \mathbf{V}_p^T. \tag{4.64}$$

If $N(\mathbf{G})$ is trivial, then $\mathrm{rank}(\mathbf{G}) = p = n$, and $\mathbf{R}_{\mathrm{m}}$ is the n by n identity matrix. In this case the original model is recovered exactly, and we say that the resolution is perfect. If $N(\mathbf{G})$ is a nontrivial subspace of R^n, then $p = \mathrm{rank}(\mathbf{G}) < n$, so that $\mathbf{R}_{\mathrm{m}}$ is not the identity matrix. The model resolution matrix is instead a symmetric matrix describing how the generalized inverse solution smears out the original model, **m**, into a recovered model, $\mathbf{m}_\dagger$. The trace of $\mathbf{R}_{\mathrm{m}}$ is often used as a simple quantitative measure of the resolution. If $\mathrm{Tr}\,(\mathbf{R}_{\mathrm{m}})$ is close to n, then $\mathbf{R}_{\mathrm{m}}$ is relatively close to the identity matrix.

The model resolution matrix can be used to quantify the bias introduced by the pseudoinverse when **G** does not have full column rank. We begin by showing that the expected value of $\mathbf{m}_\dagger$ is $\mathbf{R}_{\mathrm{m}}\mathbf{m}_{\mathrm{true}}$.

$$E[\mathbf{m}_\dagger] = E[\mathbf{G}^{\dagger}\mathbf{d}] \tag{4.65}$$

$$= \mathbf{G}^{\dagger} E[\mathbf{d}] \tag{4.66}$$

$$= \mathbf{G}^{\dagger}\mathbf{G}\mathbf{m}_{\mathrm{true}} \tag{4.67}$$

$$= \mathbf{R}_{\mathrm{m}}\mathbf{m}_{\mathrm{true}}. \tag{4.68}$$

Thus the bias in the generalized inverse solution is

$$E[\mathbf{m}_\dagger] - \mathbf{m}_{\mathrm{true}} = \mathbf{R}_{\mathrm{m}}\mathbf{m}_{\mathrm{true}} - \mathbf{m}_{\mathrm{true}} \tag{4.69}$$

$$= (\mathbf{R}_{\mathrm{m}} - \mathbf{I})\mathbf{m}_{\mathrm{true}} \tag{4.70}$$

where

$$\mathbf{R}_{\mathrm{m}} - \mathbf{I} = \mathbf{V}_p \mathbf{V}_p^T - \mathbf{V}\mathbf{V}^T \tag{4.71}$$

$$= -\mathbf{V}_0 \mathbf{V}_0^T. \tag{4.72}$$

Notice that as p increases, $\mathbf{R}_{\mathrm{m}}$ approaches **I**. Equations (4.60) and (4.72) reveal an important trade-off associated with the value of p. As p increases, the variance in the generalized inverse solution increases (4.60), but bias decreases.

We can formulate a bound on the norm of the bias (4.70):

$$\|E[\mathbf{m}_\dagger] - \mathbf{m}_{\mathrm{true}}\| \le \|\mathbf{R}_{\mathrm{m}} - \mathbf{I}\|\|\mathbf{m}_{\mathrm{true}}\|. \tag{4.73}$$

Computing $\|\mathbf{R}_{\mathrm{m}} - \mathbf{I}\|$ can give us some idea of how much bias has been introduced by the generalized inverse solution. The bound is not very useful, since we typically have no *a priori* knowledge of $\|\mathbf{m}_{\mathrm{true}}\|$.

In practice, the model resolution matrix is commonly used in two different ways. First, we can examine the diagonal elements of $\mathbf{R}_m$. Diagonal elements that are close to one correspond to parameters for which we can claim good resolution. Conversely, if any of the diagonal elements are small, then the corresponding model parameters will be poorly resolved. Second, we can multiply $\mathbf{R}_m$ times a particular test model $\mathbf{m}$ to see how that model would be resolved in the inverse solution. This strategy is called a **resolution test**. One commonly used test model is a **spike model**, which is a vector with all zero elements, except for a single entry which is one. Multiplying $\mathbf{R}_m$ times a spike model effectively picks out the corresponding column of the resolution matrix. These columns of the resolution matrix are called **resolution kernels**. These are similar to the averaging kernels in the method of Backus and Gilbert (see Section 3.5).

We can multiply $\mathbf{G}^{\dagger}$ and $\mathbf{G}$ in the opposite order from (4.64) to obtain the **data space resolution matrix**, $\mathbf{R}_d$:

$$\mathbf{d}_{\dagger} = \mathbf{Gm}_{\dagger} \tag{4.74}$$

$$= \mathbf{GG}^{\dagger}\mathbf{d} \tag{4.75}$$

$$= \mathbf{R}_d\mathbf{d} \tag{4.76}$$

where

$$\mathbf{R}_d = \mathbf{U}_p\mathbf{S}_p\mathbf{V}_p^T\mathbf{V}_p\mathbf{S}_p^{-1}\mathbf{U}_p^T \tag{4.77}$$

$$= \mathbf{U}_p\mathbf{U}_p^T. \tag{4.78}$$

If $N(\mathbf{G}^T)$ contains only the zero vector, then $p = m$, and $\mathbf{R}_d = \mathbf{I}$. In this case, $\mathbf{d}_{\dagger} = \mathbf{d}$, and the generalized inverse solution $\mathbf{m}_{\dagger}$ fits the data exactly. However, if $N(\mathbf{G}^T)$ is nontrivial, then $p < m$, and $\mathbf{R}_d$ is not the identity matrix. In this case $\mathbf{m}_{\dagger}$ does not exactly fit the data.

Note that model and data space resolution matrices (4.64) and (4.78) do *not* depend on specific data or models, but are exclusively properties of $\mathbf{G}$. They reflect the physics and geometry of a problem, and can thus be assessed during the design phase of an experiment.

4.3 INSTABILITY OF THE GENERALIZED INVERSE SOLUTION

The generalized inverse solution $\mathbf{m}_{\dagger}$ has zero projection onto $N(\mathbf{G})$. However, it may include terms involving column vectors in $\mathbf{V}_p$ with very small nonzero singular values. In analyzing the generalized inverse solution it is useful to examine the **singular value spectrum**, which is simply the range of singular values. Small singular values cause the generalized inverse solution to be extremely sensitive to small amounts of noise in the data. As a practical matter, it can also be difficult to distinguish between zero singular values and extremely small singular values. We can quantify the instabilities created by small singular values by recasting the generalized inverse solution to make the effect of small singular values explicit. We start with the formula for the generalized inverse solution

$$\mathbf{m}_{\dagger} = \mathbf{V}_p\mathbf{S}_p^{-1}\mathbf{U}_p^T\mathbf{d}. \tag{4.79}$$

The elements of the vector $\mathbf{U}_p^T\mathbf{d}$ are the dot products of the first p columns of $\mathbf{U}$ with $\mathbf{d}$:

$$\mathbf{U}_p^T\mathbf{d} = \begin{bmatrix} (\mathbf{U}_{\cdot,1})^T\mathbf{d} \\ (\mathbf{U}_{\cdot,2})^T\mathbf{d} \\ \vdots \\ (\mathbf{U}_{\cdot,p})^T\mathbf{d} \end{bmatrix}. \tag{4.80}$$

When we left-multiply $\mathbf{S}_p^{-1}$ times (4.80), we obtain

$$\mathbf{S}_p^{-1}\mathbf{U}_p^T\mathbf{d} = \begin{bmatrix} \dfrac{(\mathbf{U}_{\cdot,1})^T\mathbf{d}}{s_1} \\ \dfrac{(\mathbf{U}_{\cdot,2})^T\mathbf{d}}{s_2} \\ \vdots \\ \dfrac{(\mathbf{U}_{\cdot,p})^T\mathbf{d}}{s_p} \end{bmatrix}. \tag{4.81}$$

Finally, when we left-multiply $\mathbf{V}_p$ times (4.81), we obtain a linear combination of the columns of $\mathbf{V}_p$ that can be written as

$$\mathbf{m}_\dagger = \mathbf{V}_p\mathbf{S}_p^{-1}\mathbf{U}_p^T\mathbf{d} = \sum_{i=1}^p \frac{\mathbf{U}_{\cdot,i}^T\mathbf{d}}{s_i}\mathbf{V}_{\cdot,i}. \tag{4.82}$$

In the presence of random noise, $\mathbf{d}$ will generally have a nonzero projection onto each of the directions specified by the columns of $\mathbf{U}$. The presence of a very small s_i in the denominator of (4.82) can thus give us a very large coefficient for the corresponding model space basis vector $\mathbf{V}_{\cdot,i}$, and these basis vectors can dominate the solution. In the worst case, the generalized inverse solution is just a noise amplifier, and the answer is practically useless. A measure of the instability of the solution is the **condition number**. Note that the condition number considered here for an m by n matrix is a generalization of the condition number for an n by n matrix in (A.109), and that the two formulations are equivalent when $m = n$.

Suppose that we have a data vector $\mathbf{d}$ and an associated generalized inverse solution $\mathbf{m}_\dagger = \mathbf{G}^\dagger\mathbf{d}$. If we consider a slightly perturbed data vector $\mathbf{d}'$ and its associated generalized inverse solution $\mathbf{m}'_\dagger = \mathbf{G}^\dagger\mathbf{d}'$, then

$$\mathbf{m}_\dagger - \mathbf{m}'_\dagger = \mathbf{G}^\dagger(\mathbf{d} - \mathbf{d}') \tag{4.83}$$

and

$$\|\mathbf{m}_\dagger - \mathbf{m}'_\dagger\|_2 \le \|\mathbf{G}^\dagger\|_2\|\mathbf{d} - \mathbf{d}'\|_2. \tag{4.84}$$

From (4.82), it is clear that the largest difference in the inverse models will occur when $\mathbf{d} - \mathbf{d}'$ is in the direction $\mathbf{U}_{\cdot,p}$. If

$$\mathbf{d} - \mathbf{d}' = \alpha \mathbf{U}_{\cdot,p} \tag{4.85}$$

then

$$\|\mathbf{d} - \mathbf{d}'\|_2 = \alpha. \tag{4.86}$$

We can then compute the effect on the generalized inverse solution as

$$\mathbf{m}_\dagger - \mathbf{m}'_\dagger = \frac{\alpha}{s_p} \mathbf{V}_{\cdot,p} \tag{4.87}$$

with

$$\|\mathbf{m}_\dagger - \mathbf{m}'_\dagger\|_2 = \frac{\alpha}{s_p}. \tag{4.88}$$

Thus, we have a bound on the instability of the generalized inverse solution

$$\|\mathbf{m}_\dagger - \mathbf{m}'_\dagger\|_2 \leq \frac{1}{s_p} \|\mathbf{d} - \mathbf{d}'\|_2. \tag{4.89}$$

Similarly, we can see that the generalized inverse model is smallest in norm when $\mathbf{d}$ points in a direction parallel to $\mathbf{V}_{\cdot,1}$. Thus

$$\|\mathbf{m}_\dagger\|_2 \geq \frac{1}{s_1} \|\mathbf{d}\|_2. \tag{4.90}$$

Combining these inequalities, we obtain

$$\frac{\|\mathbf{m}_\dagger - \mathbf{m}'_\dagger\|_2}{\|\mathbf{m}_\dagger\|_2} \leq \frac{s_1}{s_p} \frac{\|\mathbf{d} - \mathbf{d}'\|_2}{\|\mathbf{d}\|_2}. \tag{4.91}$$

The bound (4.91) is applicable to pseudoinverse solutions, regardless of what value of p we use. If we decrease p and thus eliminate model space vectors associated with small singular values, the solution becomes more stable. However, this stability comes at the expense of reducing the dimension of the subspace of R^n where the solution lies. As a result, the model resolution matrix for the stabilized solution obtained by decreasing p becomes less like the identity matrix, and the fit to the data worsens.

The condition number of $\mathbf{G}$ is the coefficient in (4.91)

$$\text{cond}(\mathbf{G}) = \frac{s_1}{s_k} \tag{4.92}$$

where $k = \min(m, n)$. The MATLAB command **cond** can be used to compute (4.92). If $\mathbf{G}$ is of full rank, and we use all of the singular values in the pseudoinverse solution ($p = k$), then the condition number is exactly (4.92). If **G** is of less than full rank, then the condition number is effectively infinite. As with the model and data resolution matrices [(4.64) and (4.78)], the condition number is a property of **G** that can be computed in the design phase of an experiment before any data are collected.

A condition that insures solution stability and arises naturally from consideration of (4.82) is the **discrete Picard condition** [59]. The discrete Picard condition is satisfied when the dot products of the columns of **U** and the data vector decay to zero more quickly than the singular values, s_i. Under this condition, we should not see instability due to small singular values. The discrete Picard condition can be assessed by plotting the ratios of $\mathbf{U}_i^T \mathbf{d}$ to s_i across the singular value spectrum.

If the discrete Picard condition is not satisfied, we may still be able to recover a useful model by truncating (4.82) at some highest term $p' < p$, to produce a **truncated SVD** or **TSVD** solution. One way to decide when to truncate (4.82) in this case is to apply the **discrepancy principle**. In the discrepancy principle, we pick the smallest value of p' so that the model fits the data to some tolerance based on the length of the residual vector

$$\|\mathbf{G}_w \mathbf{m} - \mathbf{d}_w\|_2 \leq \delta \tag{4.93}$$

where $\mathbf{G}_w$ and $\mathbf{d}_w$ are the weighted system matrix and data vector, respectively.

How should we select δ? We discussed in Chapter 2 that when we estimate the solution to a full column rank least squares problem, $\|\mathbf{G}_w \mathbf{m}_{L_2} - \mathbf{d}_w\|_2^2$ has a χ^2 distribution with $m - n$ degrees of freedom. Unfortunately, when the number of model parameters n is greater than or equal to the number of data m, this formulation fails because there is no χ^2 distribution with fewer than one degree of freedom. In practice, a common heuristic is to require $\|\mathbf{G}_w \mathbf{m} - \mathbf{d}_w\|_2$ to be smaller than $\sqrt{m}$, because the approximate median of a χ^2 distribution with m degrees of freedom is m.

A TSVD solution will not fit the data as well as solutions that do include the model space basis vectors with small singular values. Perhaps surprisingly, this is an example of the general approach for solving ill-posed problems with noise. If we fit the data vector exactly or nearly exactly, we are in fact **overfitting** the data and may be letting the noise control major features of the model.

The TSVD solution is but one example of **regularization**, where solutions are selected to sacrifice fit to the data in exchange for solution stability. Understanding the trade-off between fitting the data and solution stability involved in regularization is of fundamental importance.

4.4 AN EXAMPLE OF A RANK-DEFICIENT PROBLEM

A linear least squares problem is said to be **rank-deficient** if there is a clear distinction between the nonzero and zero singular values and rank(**G**) is less than n. Numerically computed singular values will often include some that are extremely small but not quite zero, because of round-off errors. If there is a substantial gap between the largest of these tiny singular values and the first truly nonzero singular value, then it can be easy to distinguish between the two populations. Rank deficient problems can often be solved in a straightforward manner by applying the generalized inverse solution. After truncating the effectively zero singular values, a least squares model of limited resolution will be produced, and stability will seldom be an issue.

■ **Example 4.1** Using the SVD, let us revisit the straight ray path tomography example that we considered earlier in Example 1.6 (see Figure 4.2). We introduced a rank-deficient system in which we were constraining a nine-parameter slowness model with eight travel time observations:

$$\mathbf{Gm} = \begin{bmatrix} 1 & 0 & 0 & 1 & 0 & 0 & 1 & 0 & 0 \\ 0 & 1 & 0 & 0 & 1 & 0 & 0 & 1 & 0 \\ 0 & 0 & 1 & 0 & 0 & 1 & 0 & 0 & 1 \\ 1 & 1 & 1 & 0 & 0 & 0 & 0 & 0 & 0 \\ 0 & 0 & 0 & 1 & 1 & 1 & 0 & 0 & 0 \\ 0 & 0 & 0 & 0 & 0 & 0 & 1 & 1 & 1 \\ \sqrt{2} & 0 & 0 & 0 & \sqrt{2} & 0 & 0 & 0 & \sqrt{2} \\ 0 & 0 & 0 & 0 & 0 & 0 & 0 & 0 & \sqrt{2} \end{bmatrix} \begin{bmatrix} s_{11} \\ s_{12} \\ s_{13} \\ s_{21} \\ s_{22} \\ s_{23} \\ s_{31} \\ s_{32} \\ s_{33} \end{bmatrix} = \begin{bmatrix} t_1 \\ t_2 \\ t_3 \\ t_4 \\ t_5 \\ t_6 \\ t_7 \\ t_8 \end{bmatrix}. \qquad (4.94)$$

The eight singular values of **G** are, numerically evaluated,

$$\mathrm{diag}(\mathbf{S}) = \begin{bmatrix} 3.180 \\ 2.000 \\ 1.732 \\ 1.732 \\ 1.732 \\ 1.607 \\ 0.553 \\ 4.230 \times 10^{-16} \end{bmatrix}. \qquad (4.95)$$

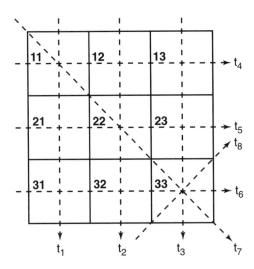

Figure 4.2 A simple tomography example (revisited).

The smallest singular value, s_8, is nonzero in numerical evaluation only because of round-off error in the SVD algorithm. It is zero in an analytical solution. s_8 is clearly effectively zero relative to the other singular values. The ratio of the largest to smallest nonzero singular values is about 6, and the generalized inverse solution (4.82) will thus be stable in the presence of noise. Because rank$(\mathbf{G}) = p = 7$, the problem is rank-deficient. The model null space, $N(\mathbf{G})$, is spanned by the two orthonormal vectors that form the eighth and ninth columns of $\mathbf{V}$:

$$\mathbf{V}_0 = \begin{bmatrix} \mathbf{V}_{.,8}, \mathbf{V}_{.,9} \end{bmatrix} = \begin{bmatrix} -0.136 & -0.385 \\ 0.385 & -0.136 \\ -0.249 & 0.521 \\ -0.385 & 0.136 \\ 0.136 & 0.385 \\ 0.249 & -0.521 \\ 0.521 & 0.249 \\ -0.521 & -0.249 \\ 0.000 & 0.000 \end{bmatrix}. \tag{4.96}$$

To obtain a geometric appreciation for the two model null space vectors, we can reshape them into 3 by 3 matrices corresponding to the geometry of the blocks (e.g., by using the MATLAB **reshape** command) to plot their elements in proper physical positions:

$$\text{reshape}(\mathbf{V}_{.,8}, 3, 3) = \begin{bmatrix} -0.136 & -0.385 & 0.521 \\ 0.385 & 0.136 & -0.521 \\ -0.249 & 0.249 & 0.000 \end{bmatrix} \tag{4.97}$$

$$\text{reshape}(\mathbf{V}_{.,9}, 3, 3) = \begin{bmatrix} -0.385 & 0.136 & 0.249 \\ -0.136 & 0.385 & -0.249 \\ 0.521 & -0.521 & 0.000 \end{bmatrix}. \tag{4.98}$$

See Figures 4.3 and 4.4.

Recall that if $\mathbf{m}_0$ is in the model null space, then (because $\mathbf{Gm}_0 = 0$) we can add such a model to any solution and not change the fit to the data. When mapped to their physical locations, three common features of the model null space basis vector elements in this example stand out:

1. The sums along all rows and columns are zero.
2. The upper left to lower right diagonal sum is zero.
3. There is no projection in the $m_9 = s_{33}$ model space direction.

The zero sum conditions (1) and (2) arise because paths passing through any three horizontal or vertical sets of blocks can only constrain the sum of those block values. The condition of zero value for m_9 (3) occurs because that model element is uniquely constrained by the eighth ray, which passes exclusively through the $s_{3,3}$ block. Thus, any variation in m_9 will

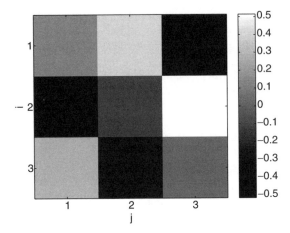

Figure 4.3 Image of the null space model $\mathbf{V}_{.,8}$.

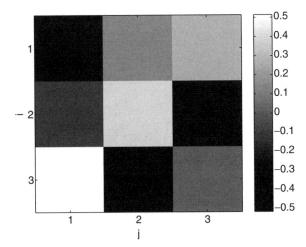

Figure 4.4 Image of the null space model $\mathbf{V}_{.,9}$.

clearly affect the predicted data, and any vector in the model null space must have a value of 0 in m_9.

The single basis vector spanning the data null space in this example is

$$
\mathbf{U}_0 = \mathbf{U}_{.,8} = \begin{bmatrix} -0.408 \\ -0.408 \\ -0.408 \\ 0.408 \\ 0.408 \\ 0.408 \\ 0.000 \\ 0.000 \end{bmatrix}.
\tag{4.99}
$$

Recall that, even for noise-free data, we will not recover a general $\mathbf{m}_{\text{true}}$ in a rank-deficient problem using (4.22), but will instead recover a "smeared" model $\mathbf{R}_m\mathbf{m}_{\text{true}}$. Because $\mathbf{R}_m$ for a rank-deficient problem is itself rank-deficient, this smearing is irreversible. The full $\mathbf{R}_m$ matrix dictates precisely how this smearing occurs. The elements of $\mathbf{R}_m$ for this example are shown in Figure 4.5.

Examining the entire n by n model resolution matrix becomes cumbersome in large problems. The n diagonal elements of $\mathbf{R}_m$ can be examined more easily to provide basic information on how well recovered each model parameter will be. The reshaped diagonal of $\mathbf{R}_m$ from Figure 4.5 is

$$\text{reshape}(\text{diag}(\mathbf{R}_m), 3, 3) = \begin{bmatrix} 0.833 & 0.833 & 0.667 \\ 0.833 & 0.833 & 0.667 \\ 0.667 & 0.667 & 1.000 \end{bmatrix}. \tag{4.100}$$

See Figure 4.6.

Figure 4.6 and (4.100) tell us that m_9 is perfectly resolved, but that we can expect loss of resolution (and hence smearing of the true model into other blocks) for all of the other solution parameters.

We next assess the smoothing effects of limited model resolution by performing a resolution test using synthetic data for a test model of interest. The resolution test assesses the recovery of the test model by examining the corresponding inverse solution. One synthetic model that is commonly used in resolution tests is uniformly zero except for a single perturbed model element. Examining the inverse recovery using data generated by such a model is commonly referred to as a **spike** or **impulse** resolution test. For this example, consider the spike model consisting of the vector with its fifth element equal to one and zeros elsewhere. This model is shown in Figure 4.7. Forward modeling gives the predicted data set for $\mathbf{m}_{\text{test}}$:

$$\mathbf{d}_{\text{test}} = \mathbf{G}\mathbf{m}_{\text{test}} = \begin{bmatrix} 0 & 1 & 0 & 0 & 1 & 0 & 0 & \sqrt{2} & 0 \end{bmatrix}^T, \tag{4.101}$$

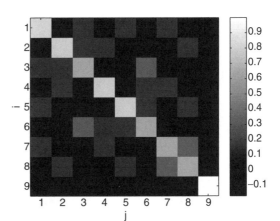

Figure 4.5 The model resolution matrix elements, $\mathrm{R}_{m i,j}$ for the generalized inverse solution.

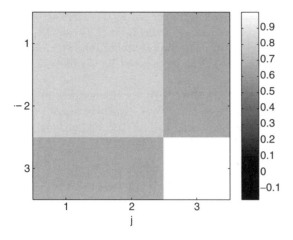

Figure 4.6   Diagonal elements of the resolution matrix plotted in their respective geometric model locations.

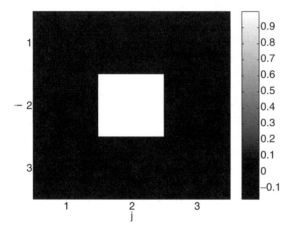

Figure 4.7   A spike test model.

and the corresponding (reshaped) generalized inverse model is the fifth column of $\mathbf{R_m}$, which is

$$\mathrm{reshape}(\mathbf{m}_\dagger, 3, 3) = \begin{bmatrix} 0.167 & 0 & -0.167 \\ 0 & 0.833 & 0.167 \\ -0.167 & 0.167 & 0.000 \end{bmatrix}. \tag{4.102}$$

See Figure 4.8. The recovered model in this spike test shows that limited resolution causes information about the central block slowness to smear into some, but not all, of the adjacent blocks *even for noise-free data*, with the exact form of the smearing dictated by the model resolution matrix.

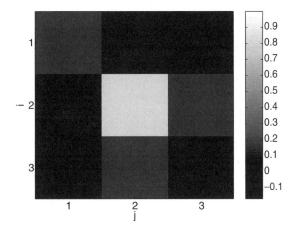

Figure 4.8   The generalized inverse solution for the noise-free spike test.

It is important to reemphasize that the ability to recover the true model in practice is affected both by the bias caused by limited resolution, which is a characteristic of the matrix **G** and hence applies even to noise-free data, and by the mapping of any data noise into the model parameters. In practice, the error due to noise in the data can also be very significant. ∎

4.5   DISCRETE ILL-POSED PROBLEMS

In many problems the singular values decay gradually toward zero and do not show an obvious jump between nonzero and zero singular values. This happens frequently when we discretize Fredholm integral equations of the first kind as in Chapter 3. In particular, as we increase the number of points in the discretization, we typically find that **G** becomes more and more poorly conditioned. Discrete inverse problems such as these cannot formally be called ill-posed, because the condition number remains finite although very large. We will refer to these as **discrete ill-posed problems**.

The rate of singular value spectrum decay can be used to characterize a discrete ill-posed problem as mildly, moderately, or severely ill-posed. If $s_j = O(j^{-\alpha})$ for $\alpha \leq 1$, then we call the problem mildly ill-posed. If $s_j = O(j^{-\alpha})$ for $\alpha > 1$, then the problem is moderately ill-posed. If $s_j = O(e^{-\alpha j})$ then the problem is severely ill-posed.

In addition to the general pattern of singular values which decay to 0, discrete ill-posed problems are typically characterized by differences in the character of the singular vectors $\mathbf{V}_{\cdot, j}$ [59]. For large singular values, the corresponding singular vectors are smooth, while for smaller singular values, the corresponding singular vectors may be highly oscillatory. These oscillations become apparent in the generalized inverse solution as more singular values and vectors are included.

When we attempt to solve such a problem with the TSVD, it is difficult to decide where to truncate (4.82). If we truncate the sum too early, then our solution will lack details that correspond to model vectors associated with the smaller singular values. If we include too many of the terms, then the solution becomes unstable in the presence of noise. In particular we can expect that more oscillatory components of the generalized inverse solution may be most strongly affected by noise [59]. Regularization is required to address this fundamental issue.

■ **Example 4.2** Consider an inverse problem where we have a physical process (e.g., seismic ground motion) recorded by a linear instrument of limited bandwidth (e.g., a vertical seismometer). The response of such a device is commonly characterized by an **instrument impulse response**, which is the response of the system to a delta function input. Consider the instrument impulse response

$$g(t) = \begin{cases} g_0 t e^{-t/T_0} & (t \geq 0) \\ 0 & (t < 0) \end{cases}.$$ (4.103)

Figure 4.9 shows the displacement response of a critically damped seismometer with a characteristic time constant T_0 to a unit area (1 m/s^2 · s) impulsive ground acceleration input, where g_0 is a gain constant. Assuming that the displacement of the seismometer is electronically converted to output volts, we conveniently choose g_0 to be $T_0 e^{-1}$ V/m · s to produce a 1 V maximum output value for the impulse response, and $T_0 = 10$ s.

The seismometer output (or seismogram), $v(t)$, is a voltage record given by the convolution of the true ground acceleration, $m_{\text{true}}(t)$, with (4.103):

$$v(t) = \int_{-\infty}^{\infty} g(t - \tau)\, m_{\text{true}}(\tau)\, d\tau.$$ (4.104)

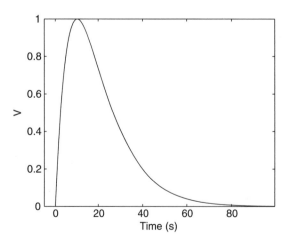

Figure 4.9 Example instrument response; seismometer output voltage in response to a unit area ground acceleration impulse.

We are interested in the inverse **deconvolution** operation that will remove the smoothing effect of $g(t)$ in (4.104) and allow us to recover the true ground acceleration m_{true}.

Discretizing (4.104) using the midpoint rule with a time interval Δt, we obtain

$$\mathbf{d} = \mathbf{Gm} \tag{4.105}$$

where

$$G_{i,j} = \begin{cases} (t_i - t_j)e^{-(t_i - t_j)/T_0}\Delta t & (t_j \geq t_i) \\ 0 & (t_j < t_i) \end{cases}. \tag{4.106}$$

The rows of $\mathbf{G}$ in (4.106) are time reversed, and the columns of $\mathbf{G}$ are non-time-reversed, sampled versions of the impulse response $g(t)$, lagged by i and j, respectively. Using a time interval of $[-5, 100]$ s, outside of which (4.103) and any model, $\mathbf{m}$, of interest are assumed to be very small or zero, and a discretization interval of $\Delta t = 0.5$ s, we obtain a discretized m by n system matrix $\mathbf{G}$ with $m = n = 210$.

The singular values of $\mathbf{G}$ are all nonzero and range from about 25.3 to 0.017, giving a condition number of ≈ 1480, and showing that this discretization has produced a discrete system that is mildly ill-posed. See Figure 4.10. However, adding noise at the level of 1 part in 1000 will be sufficient to make the generalized inverse solution unstable. The reason for the large condition number can be seen by examining successive rows of $\mathbf{G}$, which are nearly but not quite identical, with

$$\frac{\mathbf{G}_{i,\cdot}\mathbf{G}_{i+1,\cdot}^T}{\|\mathbf{G}_{i,\cdot}\|_2\|\mathbf{G}_{i+1,\cdot}\|_2} \approx 0.999. \tag{4.107}$$

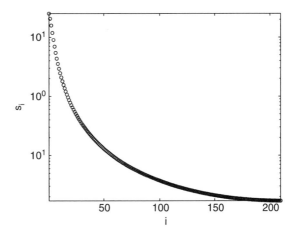

Figure 4.10 Singular values for the discretized convolution matrix.

Now, consider a true ground acceleration signal that consists of two acceleration pulses with widths of $\sigma = 2$ s, centered at $t = 8$ s and $t = 25$ s (Figure 4.11):

$$m_{\text{true}}(t) = e^{-(t-8)^2/(2\sigma^2)} + 0.5e^{-(t-25)^2/(2\sigma^2)}. \tag{4.108}$$

We sample $\mathbf{m}_{\text{true}}(t)$ on the time interval $[-5, 100]$ s to obtain a 210-element vector $\mathbf{m}_{\text{true}}$, and generate the noise-free data set

$$\mathbf{d}_{\text{true}} = \mathbf{G}\mathbf{m}_{\text{true}} \tag{4.109}$$

and a second data set with independent $N(0, (0.05 \text{ V})^2)$ noise added. The data set with noise is shown in Figure 4.12.

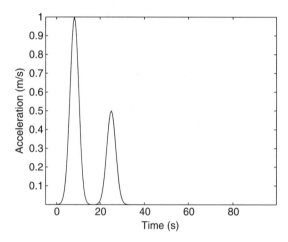

Figure 4.11   The true model.

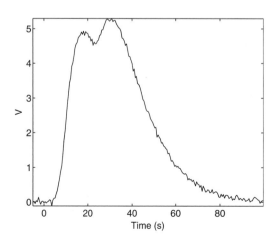

Figure 4.12   Predicted data from the true model plus independent $N(0, (0.05 \text{ V})^2)$ noise.

The recovered least squares model from the full ($p = 210$) generalized inverse solution

$$\mathbf{m} = \mathbf{V}\mathbf{S}^{-1}\mathbf{U}^T\mathbf{d}_{\text{true}} \tag{4.110}$$

is shown in Figure 4.13. The model fits its noiseless data vector, $\mathbf{d}_{\text{true}}$, perfectly, and is essentially identical to the true model.

The least squares solution for the noisy data vector, $\mathbf{d}_{\text{true}} + \boldsymbol{\eta}$,

$$\mathbf{m} = \mathbf{V}\mathbf{S}^{-1}\mathbf{U}^T(\mathbf{d}_{\text{true}} + \boldsymbol{\eta}) \tag{4.111}$$

is shown in Figure 4.14.

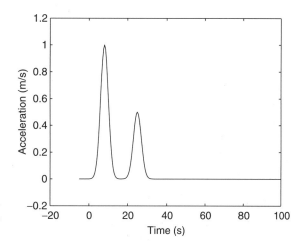

Figure 4.13 Generalized inverse solution using all 210 singular values for the noise-free data.

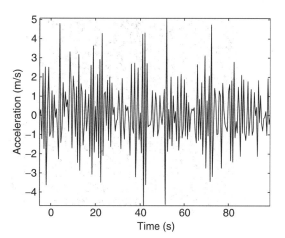

Figure 4.14 Generalized inverse solution using all 210 singular values for the noisy data of Figure 4.12.

Although this solution fits its particular data vector, $\mathbf{d}_{\text{true}} + \boldsymbol{\eta}$, exactly, it is worthless in divining information about the true ground motion. Information about m_{true} is overwhelmed by the small amount of added noise, amplified enormously by the inversion process.

Can a useful model be recovered by the truncated SVD? Using the discrepancy principle as our guide and selecting a range of solutions with varying p', we can in fact obtain an appropriate solution when we keep $p' = 26$ columns in $\mathbf{V}$. See Figure 4.15.

Essential features of the true model are resolved in the solution of Figure 4.15, but the solution technique introduces oscillations and loss of resolution. Specifically, we see that the widths of the inferred pulses are somewhat wider, and the inferred amplitudes somewhat less, than those of the true ground acceleration. These effects are both hallmarks of limited resolution, as characterized by a nonidentity model resolution matrix. An image of the model resolution matrix in Figure 4.16 shows a finite-width central band and oscillatory side lobes.

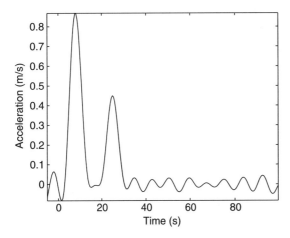

Figure 4.15   Solution using the 26 largest singular values for noisy data shown in Figure 4.12.

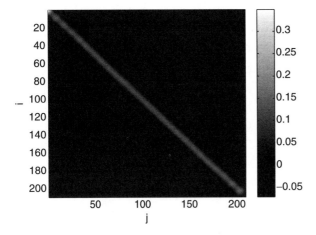

Figure 4.16   The model resolution matrix elements $R_{m i, j}$ for the truncated SVD solution including the 26 largest singular values.

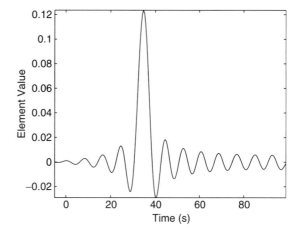

Figure 4.17   A column from the model resolution matrix $\mathbf{R}_m$ for the truncated SVD solution including the 26 largest singular values.

A typical (80th) column of the model resolution matrix quantifies the smearing of the true model into the recovered model for the choice of the $p = 26$ inverse operator. See Figure 4.17. The smoothing is over a characteristic width of about 5 seconds, which is why our recovered model, although it does a decent job of rejecting noise, underestimates the amplitude and narrowness of the true model. The oscillatory behavior of the resolution matrix is attributable to our abrupt truncation of the model space. Each of the n columns of $\mathbf{V}$ is an oscillatory model basis function, with $j - 1$ zero crossings, where j is the column number.

When we truncate (4.82) after 26 terms to stabilize the inverse solution, we place a limit on the most oscillatory model space basis vectors that we will allow in our solution. This truncation gives us a model, and model resolution, that contain oscillatory structure with around $p - 1 = 25$ zero crossings. We will examine this perspective further in Chapter 8, where issues associated with highly oscillatory model basis functions will be revisited in the context of Fourier theory.   ■

■ **Example 4.3**   Recall the Shaw problem from Example 3.2. The MATLAB Regularization Tools contains a routine **shaw** that computes the $\mathbf{G}$ matrix and an example model and data for this problem [58]. We computed the $\mathbf{G}$ matrix for $n = 20$ and examined the singular values. Figure 4.18 shows the singular value spectrum, which is characterized by very rapid singular value decay to zero in an exponential fashion.

This is a severely ill-posed problem, and there is no obvious break point above which the singular values can reasonably be considered to be nonzero and below which the singular values can be considered to be 0. The MATLAB rank command gives $p = 18$, suggesting that the last two singular values are effectively 0. The condition number of this problem is enormous (larger than 10^{14}).

The 18th column of $\mathbf{V}$, which corresponds to the smallest nonzero singular value, is shown in Figure 4.19. In contrast, the first column of $\mathbf{V}$, which corresponds to the largest singular value, represents a much smoother model. See Figure 4.20. This behavior is typical of discrete ill-posed problems.

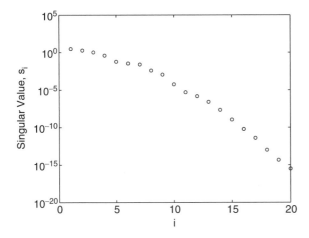

Figure 4.18 Singular values of **G** for the Shaw example ($n = 20$).

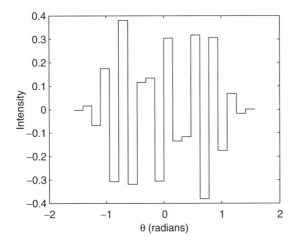

Figure 4.19 $\mathbf{V}_{\cdot,18}$.

Next, we will perform a simple resolution test. Suppose that the input to the system is given by

$$m_i = \begin{cases} 1 & i = 10 \\ 0 & \text{otherwise.} \end{cases} \tag{4.112}$$

See Figure 4.21. We use the model to obtain noise-free data and then apply the generalized inverse (4.22) with various values of p to obtain TSVD inverse solutions. The corresponding data are shown in Figure 4.22. If we compute the generalized inverse from these data using MATLAB's double-precision algorithms, we get fairly good recovery of (4.112). See Figure 4.23.

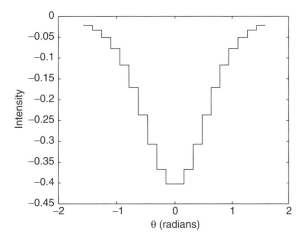

Figure 4.20 $V_{\cdot,1}$.

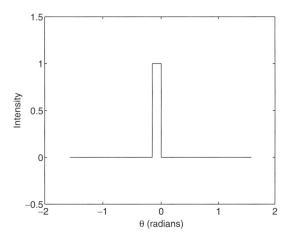

Figure 4.21   The spike model.

However, if we add a very small amount of noise to the data in Figure 4.22, things change dramatically. Adding $N(0, (10^{-6})^2)$ noise to the data of Figure 4.22 and computing a generalized inverse solution using $p = 18$ produces the wild solution of Figure 4.24, which bears no resemblance to the true model. Note that the vertical scale in Figure 4.24 is multiplied by 10^6! Furthermore, the solution involves negative intensities, which are not physically possible. This inverse solution is even more sensitive to noise than that of the previous deconvolution example, to the extent that even noise on the order of 1 part in 10^6 will destabilize the solution.

Next, we consider what happens when we use only the 10 largest singular values and their corresponding model space vectors to construct a TSVD solution. Figure 4.25 shows the solution using 10 singular values with the same noise as Figure 4.24. Because we have

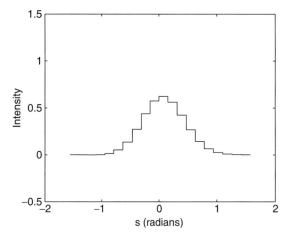

Figure 4.22 Noise-free data predicted for the spike model.

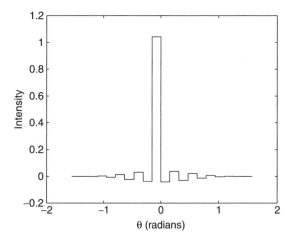

Figure 4.23 The generalized inverse solution for the spike model, no noise.

cut off a number of singular values, we have reduced the model resolution. The inverse solution is smeared out, but it is still possible to conclude that there is some significant spike-like feature near $\theta = 0$. In contrast to the situation that we observed in Figure 4.24, the model recovery is now not visibly affected by the noise. The trade-off is that we must now accept the imperfect resolution of this solution and its attendant bias towards smoother models.

What happens if we discretize the problem with a larger number of intervals? Figure 4.26 shows the singular values for the **G** matrix with $n = 100$ intervals. The first 20 or so singular values are apparently nonzero, whereas the last 80 or so singular values are effectively zero.

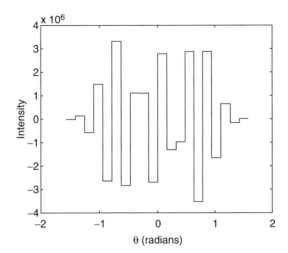

Figure 4.24 Recovery of the spike model with noise ($p = 18$).

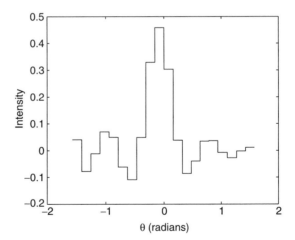

Figure 4.25 Recovery of the spike model with noise ($p = 10$).

Figure 4.27 shows the inverse solution for the spike model with $n = 100$ and $p = 10$. This solution is very similar to the solution shown in Figure 4.25. In general, discretizing over more intervals does not hurt as long as the solution is appropriately regularized and the additional computation time is acceptable.

What about a smaller number of intervals? Figure 4.28 shows the singular values of the **G** matrix with $n = 6$. In this case there are no terribly small singular values. However, with only six elements in this coarse model vector, we cannot hope to resolve the details of a source intensity distribution with a complex intensity structure. This is an example of regularization by discretization.

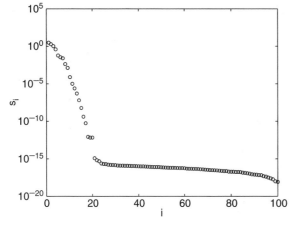

Figure 4.26   Singular values of **G** for the Shaw example ($n = 100$).

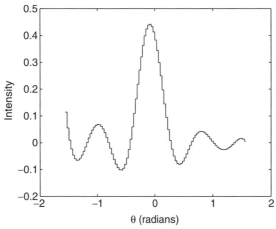

Figure 4.27   Recovery of the spike model with noise ($n = 100$, $p = 10$).

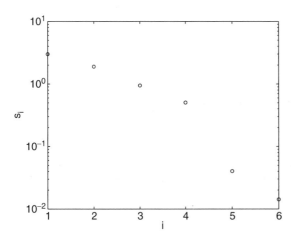

Figure 4.28   Singular values of **G** for the Shaw example ($n = 6$).

This example demonstrates the dilemma posed by small singular values. If we include the small singular values, then our inverse solution becomes unstable in the presence of data noise. If we do not include the smaller singular values, then our solution is not as sensitive to noise in the data, but we lose resolution and introduce bias. ■

4.6 EXERCISES

4.1 The pseudoinverse of a matrix $\mathbf{G}$ was originally defined by Moore and Penrose as the unique matrix $\mathbf{G}^\dagger$ with the properties

   (a) $\mathbf{G}\mathbf{G}^\dagger\mathbf{G} = \mathbf{G}$.
   (b) $\mathbf{G}^\dagger\mathbf{G}\mathbf{G}^\dagger = \mathbf{G}^\dagger$.
   (c) $(\mathbf{G}\mathbf{G}^\dagger)^T = \mathbf{G}\mathbf{G}^\dagger$.
   (d) $(\mathbf{G}^\dagger\mathbf{G})^T = \mathbf{G}^\dagger\mathbf{G}$.

Show that $\mathbf{G}^\dagger$ as given by (4.20) satisfies these four properties.

4.2 Another resolution test commonly performed in tomography studies is a **checkerboard test**, which consists of using a test model composed of alternating positive and negative perturbations. Perform a checkerboard test on the tomography problem in Example 4.1. Evaluate the difference between the true model and the recovered model, and interpret the pattern of differences.

4.3 A large north–south by east–west oriented, nearly square plan view, sandstone quarry block (16 m by 16 m) with a bulk P-wave seismic velocity of approximately 3000 m/s is suspected of harboring higher-velocity dinosaur remains. An ultrasonic P-wave travel-time tomography scan is performed in a horizontal plane bisecting the boulder, producing a data set consisting of 16 E→W, 16 N→S, 31 SW→NE, and 31 NW→SE travel times. See Figure 4.29. Each travel-time measurement has statistically independent errors with estimated standard deviations of 15 μs.

The data files that you will need to load from your working directory into your MAT-LAB program are **rowscan**, **colscan**, **diag1scan**, **diag2scan** containing the travel-time data, and **std** containing the standard deviations of the data measurements. The travel time contribution from a uniform background model (velocity of 3000 m/s) has been subtracted from each travel-time measurement for you, so you will be solving for perturbations from a uniform slowness model of 3000 m/s. The row format of each data file is (x_1, y_1, x_2, y_2, t) where the starting-point coordinate of each shot is (x_1, y_1), the end-point coordinate is (x_2, y_2), and the travel time along a ray path between the start and end points is a path integral (in seconds)

$$t = \int_l s(\mathbf{x})dl, \tag{4.113}$$

where s is the slowness along the path, l, between source and receiving points, and Δl_{block} is the length of the ray in each block.

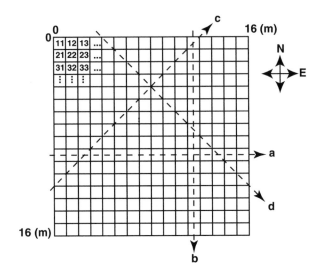

Figure 4.29 Tomography exercise, showing block discretization, block numbering convention, and representative ray paths going east–west (a), north–south (b), southwest–northeast (c), and northwest–southeast (d).

Parameterize the slowness structure in the plane of the survey by dividing the boulder into a 16 by 16 grid of 256 1-m-square, N by E blocks to construct a linear system for the problem. See Figure 4.29. Assume that the ray paths through each homogeneous block can be well approximated by straight lines, so that the travel time expression is

$$t = \int_l s(\mathbf{x})\, dl \tag{4.114}$$

$$= \sum_{\text{blocks}} s_{\text{block}} \cdot \Delta l_{\text{block}} \tag{4.115}$$

where Δl_{block} is 1 m for the row and column scans and $\sqrt{2}$ m for the diagonal scans.

Use the SVD to find a minimum-length/least-squares solution, $\mathbf{m}_\dagger$, for the 256 block slowness perturbations which fit the data as exactly as possible. Perform two inversions:

(A)   Using the row and column scans only, and
(B)   Using the complete data set.

For each inversion:

(a)   State and discuss the significance of the elements and dimensions of the data and model null spaces.
(b)   Note if there any model parameters that have perfect resolution.
(c)   Note the condition number of your **G** matrix relating the data and model.
(d)   Note the condition number of your generalized inverse matrix.

(e)   Produce a 16 by 16 element contour or other plot of your slowness perturbation model, displaying the maximum and minimum slowness perturbations in the title of each plot. Anything in there? If so, how fast or slow is it (in m/s)?

(f)   Show the model resolution by contouring or otherwise displaying the 256 diagonal elements of the model resolution matrix, reshaped into an appropriate 16 by 16 grid.

(g)   Construct and contour or otherwise display a nonzero model which fits the trivial data set $\mathbf{d} = \mathbf{0}$ exactly.

(h)   Describe how one could use solutions of the type discussed in (g) to demonstrate that very rough models exist which will fit any data set just as well as a generalized inverse model. Show one such wild model.

4.4   Find the singular value decomposition of the $\mathbf{G}$ matrix from Exercise 3.1. Taking into account the fact that the measured data are only accurate to about four digits, use the truncated SVD to compute a solution to this problem.

4.5   Revisiting Example 3.4, apply the generalized inverse to estimate the density of the Earth as a function of radius, using the given values of mass and moment of inertia. Obtain a density model composed of 20 spherical shells of equal thickness, and compare your results to a standard model.

4.7   NOTES AND FURTHER READING

The Moore–Penrose generalized inverse was independently discovered by Moore in 1920 and Penrose in 1955 [102, 125]. Penrose is generally credited with first showing that the SVD can be used to compute the generalized inverse [125]. Books that discuss the linear algebra of the generalized inverse in more detail include [10, 20].

There was significant early work on the SVD in the 19th century by Beltrami, Jordan, Sylvester, Schmidt, and Weyl [154]. However, the singular value decomposition in matrix form is typically credited to Eckart and Young [34]. Some books that discuss the properties of the SVD and prove its existence include [49, 101, 155]. Lanczos presents an alternative derivation of the SVD [89]. Algorithms for the computation of the SVD are discussed in [31, 49, 167]. Books that discuss the use of the SVD and truncated SVD in solving discrete linear inverse problems include [59, 100, 156].

Resolution tests with spike and checkerboard models as in Example 4.1 are very commonly used in practice. However, Leveque, Rivera, and Wittlinger discuss some serious problems with such resolution tests [93].

Matrices like those in Example 4.2 in which the elements along diagonals are constant are called **Toeplitz matrices** [69]. Specialized methods for regularization of problems involving Toeplitz matrices are available [60].

As we have seen, it is possible to effectively regularize the solution to a discretized version of a continuous inverse problem by selecting a coarse discretization. This approach is analyzed in [38]. However, in doing so we lose the ability to analyze the bias introduced by the regularization. In general, we prefer to user a fine discretization and then explicitly regularize the discretized problem.

5

TIKHONOV REGULARIZATION

Synopsis: The method of Tikhonov regularization for stabilizing the solution of inverse problems is introduced and illustrated with examples. Zeroth-order Tikhonov regularization is explored, including its resolution, bias, and uncertainty properties. The concepts of filter factors (which control the contribution of singular values and their corresponding singular vectors to the solution) and the L-curve criterion (a strategy for selecting the regularization parameter) are presented. Higher-order Tikhonov regularization techniques and their computation by the generalized SVD (GSVD) and truncated GSVD are discussed. Generalized cross validation is introduced as an alternative method for selecting the regularization parameter. Schemes for bounding the error in the regularized solution are discussed.

We saw in Chapter 4 that, given the SVD of $\mathbf{G}$ (4.1), we can express a generalized inverse solution by (4.82):

$$\mathbf{m}_\dagger = \mathbf{V}_p \mathbf{S}_p^{-1} \mathbf{U}_p^T \mathbf{d} = \sum_{i=1}^{p} \frac{\mathbf{U}_{\cdot,i}^T \mathbf{d}}{s_i} \mathbf{V}_{\cdot,i}. \tag{5.1}$$

We also saw that the generalized inverse solution can become extremely unstable when one or more of the singular values, s_i, is small. One fix for this difficulty was to drop terms in the sum that were associated with smaller singular values. This stabilized, or regularized, the solution in the sense that it made the result less sensitive to data noise. We paid a price for this stability in that the regularized solution had reduced resolution and was no longer unbiased.

In this chapter we will discuss Tikhonov regularization, which is perhaps the most widely used technique for regularizing discrete ill-posed problems. The Tikhonov solution can be expressed quite easily in terms of the SVD of $\mathbf{G}$. We will derive a formula for the Tikhonov solution and see how it is a variant on the generalized inverse solution that effectively gives greater weight to large singular values in the SVD solution and gives less weight to small singular values.

5.1 SELECTING A GOOD SOLUTION

For a general linear least squares problem there may be infinitely many least squares solutions. If we consider that the data contain noise, and that there is no point in fitting such noise exactly, it becomes evident that there can be many solutions that adequately fit the data in the sense that $\|\mathbf{Gm} - \mathbf{d}\|_2$ is small enough.

In Tikhonov regularization, we consider all solutions with $\|\mathbf{Gm} - \mathbf{d}\|_2 \leq \delta$, and select the one that minimizes the norm of $\mathbf{m}$:

$$\min \quad \|\mathbf{m}\|_2$$
$$\|\mathbf{Gm} - \mathbf{d}\|_2 \quad \leq \quad \delta. \tag{5.2}$$

Why select the minimum norm solution from among those solutions that adequately fit the data? One intuitive explanation is that any nonzero feature that appears in the regularized solution increases the norm of $\mathbf{m}$. Such features appear in the solution because they are necessary to fit the data. Conversely, the minimization of $\|\mathbf{m}\|_2$ should ensure that unneeded features will not appear in the regularized solution.

Note that as δ increases, the set of feasible models expands, and the minimum value of $\|\mathbf{m}\|_2$ decreases. We can thus trace out a curve of minimum values of $\|\mathbf{m}\|_2$ versus δ. See Figure 5.1. It is also possible to trace out this curve by considering problems of the form

$$\min \quad \|\mathbf{Gm} - \mathbf{d}\|_2$$
$$\|\mathbf{m}\|_2 \quad \leq \quad \epsilon. \tag{5.3}$$

As ϵ decreases, the set of feasible solutions becomes smaller, and the minimum value of $\|\mathbf{Gm} - \mathbf{d}\|_2$ increases. Again, as we adjust ϵ we trace out the curve of optimal values of $\|\mathbf{m}\|_2$ and $\|\mathbf{Gm} - \mathbf{d}\|_2$. See Figure 5.2.

A third option is to consider the damped least squares problem

$$\min \quad \|\mathbf{Gm} - \mathbf{d}\|_2^2 + \alpha^2 \|\mathbf{m}\|_2^2 \tag{5.4}$$

which arises when we apply the method of Lagrange multipliers to (5.2), where α is a **regularization parameter**. It can be shown that for appropriate choices of δ, ϵ, and α, the three problems (5.2), (5.3), and (5.4) yield the same solution [59]. We will concentrate on solving the damped least squares form of the problem (5.4). Solutions to (5.2) and (5.3) can be

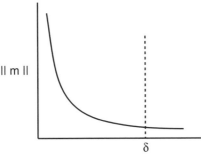

Figure 5.1   A particular misfit norm, $\delta = \|\mathbf{Gm} - \mathbf{d}\|_2$, and its associated model norm, $\|\mathbf{m}\|_2$.

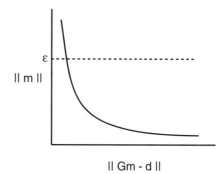

Figure 5.2   A particular model norm, $\epsilon = \|\mathbf{m}\|_2$, and its associated misfit norm, $\|\mathbf{Gm} - \mathbf{d}\|_2$.

obtained using (5.4) by adjusting the regularization parameter α until the constraints are just satisfied.

When plotted on a log–log scale, the curve of optimal values of $\|\mathbf{m}\|_2$ versus $\|\mathbf{Gm} - \mathbf{d}\|_2$ often takes on a characteristic L shape. This happens because $\|\mathbf{m}\|_2$ is a strictly decreasing function of α and $\|\mathbf{Gm} - \mathbf{d}\|_2$ is a strictly increasing function of α. The sharpness of the "corner" varies from problem to problem, but it is frequently well-defined. For this reason, the curve is called an **L-curve** [57]. In addition to the discrepancy principle, another popular criterion for picking the value of α is the **L-curve criterion** in which the value of α that gives the solution closest to the corner of the L-curve is selected.

5.2   SVD IMPLEMENTATION OF TIKHONOV REGULARIZATION

The damped least squares problem (5.4) is equivalent to the ordinary least squares problem obtained by augmenting the least squares problem for $\mathbf{Gm} = \mathbf{d}$ in the following manner:

$$\min \; \left\| \begin{bmatrix} \mathbf{G} \\ \alpha\mathbf{I} \end{bmatrix} \mathbf{m} - \begin{bmatrix} \mathbf{d} \\ \mathbf{0} \end{bmatrix} \right\|_2^2 . \tag{5.5}$$

As long as α is nonzero, the last n rows of the augmented matrix in (5.5) are obviously linearly independent. Equation (5.5) is thus a full-rank least squares problem that can be solved by the method of normal equations, i.e.,

$$\begin{bmatrix} \mathbf{G}^T & \alpha\mathbf{I} \end{bmatrix} \begin{bmatrix} \mathbf{G} \\ \alpha\mathbf{I} \end{bmatrix} \mathbf{m} = \begin{bmatrix} \mathbf{G}^T & \alpha\mathbf{I} \end{bmatrix} \begin{bmatrix} \mathbf{d} \\ \mathbf{0} \end{bmatrix} . \tag{5.6}$$

Equation (5.6) simplifies to

$$(\mathbf{G}^T\mathbf{G} + \alpha^2\mathbf{I})\mathbf{m} = \mathbf{G}^T\mathbf{d}, \tag{5.7}$$

which is the set of constraint equations for a **zeroth-order Tikhonov regularization** solution of $\mathbf{Gm} = \mathbf{d}$. Employing the SVD of $\mathbf{G}$, (5.7) can be written as

$$(\mathbf{VS}^T\mathbf{U}^T\mathbf{USV}^T + \alpha^2\mathbf{I})\mathbf{m} = \mathbf{VS}^T\mathbf{U}^T\mathbf{d} \tag{5.8}$$

$$(\mathbf{VS}^T\mathbf{SV}^T + \alpha^2\mathbf{I})\mathbf{m} = \mathbf{VS}^T\mathbf{U}^T\mathbf{d}. \tag{5.9}$$

Since (5.9) is nonsingular, it has a unique solution. We will show that this solution is

$$\mathbf{m}_\alpha = \sum_{i=1}^{k} \frac{s_i^2}{s_i^2 + \alpha^2} \frac{(\mathbf{U}_{\cdot,i})^T\mathbf{d}}{s_i} \mathbf{V}_{\cdot,i} \tag{5.10}$$

where $k = \min(m, n)$, so that all singular values are included. To show that (5.10) is the solution to (5.9), we substitute (5.10) into the left-hand side of (5.9) to obtain

$$(\mathbf{VS}^T\mathbf{SV}^T + \alpha^2\mathbf{I})\sum_{i=1}^{k} \frac{s_i^2}{s_i^2 + \alpha^2} \frac{(\mathbf{U}_{\cdot,i})^T\mathbf{d}}{s_i} \mathbf{V}_{\cdot,i} = \sum_{i=1}^{k} \frac{s_i^2}{s_i^2 + \alpha^2} \frac{(\mathbf{U}_{\cdot,i})^T\mathbf{d}}{s_i} (\mathbf{VS}^T\mathbf{SV}^T + \alpha^2\mathbf{I})\mathbf{V}_{\cdot,i}. \tag{5.11}$$

$(\mathbf{VS}^T\mathbf{SV}^T + \alpha^2\mathbf{I})\mathbf{V}_{\cdot,i}$ can be simplified by noting that $\mathbf{V}^T\mathbf{V}_{\cdot,i}$ is a standard basis vector, $\mathbf{e}_i$. When we multiply $\mathbf{S}^T\mathbf{S}$ times a standard basis vector, we get a vector with s_i^2 in position i and zeros elsewhere. When we multiply $\mathbf{V}$ times this vector, we get $s_i^2\mathbf{V}_{\cdot,i}$. Thus

$$(\mathbf{VS}^T\mathbf{SV}^T + \alpha^2\mathbf{I})\sum_{i=1}^{k} \frac{s_i^2}{s_i^2 + \alpha^2} \frac{(\mathbf{U}_{\cdot,i})^T\mathbf{d}}{s_i} \mathbf{V}_{\cdot,i} = \tag{5.12}$$

$$= \sum_{i=1}^{k} \frac{s_i^2}{s_i^2 + \alpha^2} \frac{\mathbf{U}_{\cdot,i}^T\mathbf{d}}{s_i} (s_i^2 + \alpha^2)\mathbf{V}_{\cdot,i} \tag{5.13}$$

$$= \sum_{i=1}^{k} s_i(\mathbf{U}_{\cdot,i}^T\mathbf{d})\mathbf{V}_{\cdot,i} \tag{5.14}$$

$$= \mathbf{VS}^T\mathbf{U}^T\mathbf{d} \tag{5.15}$$

$$= \mathbf{G}^T\mathbf{d}. \tag{5.16}$$

The quantities

$$f_i = \frac{s_i^2}{s_i^2 + \alpha^2} \tag{5.17}$$

are called **filter factors**. For $s_i \gg \alpha$, $f_i \approx 1$, and for $s_i \ll \alpha$, $f_i \approx 0$. For singular values between these two extremes, as the s_i decrease, the f_i decrease monotonically. A similar method (called the **damped SVD method**) uses the filter factors

$$\hat{f_i} = \frac{s_i}{s_i + \alpha}. \tag{5.18}$$

This has a similar effect to (5.17), but transitions more slowly between including large and rejecting small singular values and their associated model space vectors.

The MATLAB Regularization Tools [58] contains a number of useful commands for performing Tikhonov regularization. These commands include **l_curve** for plotting the L-curve and estimating its corner using a smoothed spline interpolation method, **tikhonov** for computing the solution for a particular value of α, **lsqi** for solving (5.3), **discrep** for solving (5.2), **picard** for plotting and interpreting singular values, and **fil_fac** for computing filter factors. The package also includes a routine **regudemo** that leads the user   through a tour of the features of the package. Note that the Regularization Tools package uses notation that is somewhat different from that used in this book.

■ **Example 5.1** We will now revisit the Shaw problem, which was previously introduced in Example 3.2 and analyzed using the SVD in Example 4.3. We begin by computing the L-curve and finding its corner. Figure 5.3 shows the L-curve. The corner of the curve is at $\alpha = 5.35 \times 10^{-6}$.

Next, we compute the Tikhonov regularization solution corresponding to this value of α. This solution is shown in Figure 5.4. Note that this solution is much better than the wild solution obtained by the TSVD with $p = 18$ shown in Figure 4.24.

We can also use the **discrep** command to find the appropriate α to obtain a Tikhonov regularized solution. Because independent $N(0, (1 \times 10^{-6})^2)$ noise was added to these $m = 20$ data points, we search for a solution where the square of the norm of the residuals is roughly 20×10^{-12}, which   corresponds to a misfit norm $\|\mathbf{Gm} - \mathbf{d}\|_2$ of roughly $\sqrt{20} \times 10^{-6} \approx 4.47 \times 10^{-6}$.

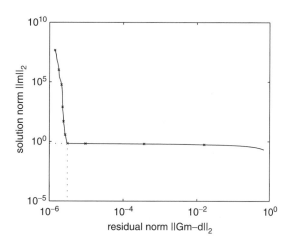

Figure 5.3   L-curve for the Shaw problem.

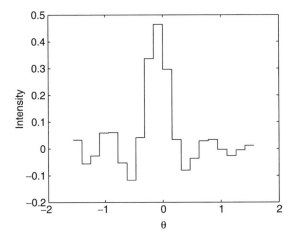

Figure 5.4   Recovery of the spike model with noise, Tikhonov solution ($\alpha = 5.35 \times 10^{-6}$).

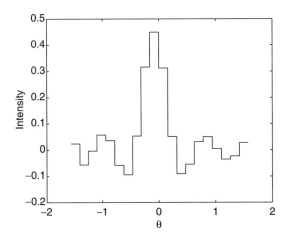

Figure 5.5   Recovery of the spike model with noise, Tikhonov solution ($\alpha = 4.29 \times 10^{-5}$).

The discrepancy principle results in a somewhat larger value of the regularization parameter, $\alpha = 4.29 \times 10^{-5}$, than that obtained using the L-curve technique. The corresponding solution, shown in Figure 5.5, thus has a smaller model norm. However, the two models are virtually indistinguishable.

It is interesting to note that the misfit of the original spike model, approximately 4.42×10^{-7}, is actually smaller than the tolerance that we specified in finding a solution by the discrepancy principle. Why did **discrep** not recover the original spike model? This is because the spike model has a norm of 1, while the solution obtained by **discrep** has a norm of only 0.67. Since Tikhonov regularization prefers solutions with smaller norms, we ended up with the solution in Figure 5.5.

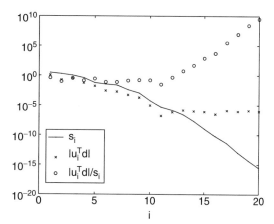

Figure 5.6 Picard plot for the Shaw problem.

The Regularization Tools command **picard** can be used to produce a plot of the singular values s_i, the values of $|(\mathbf{U}_{\cdot,i})^T\mathbf{d}|$, and the ratios $|(\mathbf{U}_{\cdot,i})^T\mathbf{d}|/s_i$. Figure 5.6 shows the values for our problem. $|(\mathbf{U}_{\cdot,i})^T\mathbf{d}|$ reaches a noise floor of about 1×10^{-6} after $i = 11$. The singular values continue to decay. As a consequence, the ratios increase rapidly. It is clear from this plot that we cannot expect to obtain useful information from the singular values beyond $p = 11$. The 11th singular value is $\approx 5.1 \times 10^{-6}$, which is comparable to the values of α in Figures 5.4 and 5.5. ∎

5.3 RESOLUTION, BIAS, AND UNCERTAINTY IN THE TIKHONOV SOLUTION

As in our earlier TSVD approach, we can compute a model resolution matrix for the Tikhonov regularization method. Using equation (5.7) and the SVD, the solution can be written as

$$\mathbf{m}_\alpha = (\mathbf{G}^T\mathbf{G} + \alpha^2\mathbf{I})^{-1}\mathbf{G}^T\mathbf{d} \tag{5.19}$$

$$= \mathbf{G}^\sharp\mathbf{d} \tag{5.20}$$

$$= \mathbf{VFS}^\dagger\mathbf{U}^T\mathbf{d}. \tag{5.21}$$

$\mathbf{F}$ is an n by n diagonal matrix with diagonal elements given by the filter factors f_i of (5.17), and $\mathbf{S}^\dagger$ is the generalized inverse of $\mathbf{S}$. $\mathbf{G}^\sharp$ is a generalized inverse matrix that can be used to construct model and data resolution matrices as was done for the SVD solution in (4.64) and (4.78). The resolution matrices are

$$\mathbf{R}_{\mathrm{m},\alpha} = \mathbf{G}^\sharp\mathbf{G} = \mathbf{VFV}^T \tag{5.22}$$

and

$$\mathbf{R}_{d,\alpha} = \mathbf{G}\mathbf{G}^{\sharp} = \mathbf{U}\mathbf{F}\mathbf{U}^{T}. \tag{5.23}$$

Note that $\mathbf{R}_m$ and $\mathbf{R}_d$ are dependent on the particular value of α used in (5.21).

■ **Example 5.2** In our Shaw example, with $\alpha = 4.29 \times 10^{-5}$, the model resolution matrix has the following diagonal elements:

$$\text{diag}(\mathbf{R}) \approx \begin{bmatrix} 0.9076 \\ 0.4879 \\ 0.4484 \\ 0.3892 \\ 0.4209 \\ 0.4093 \\ 0.4279 \\ 0.4384 \\ 0.4445 \\ 0.4506 \\ 0.4506 \\ 0.4445 \\ 0.4384 \\ 0.4279 \\ 0.4093 \\ 0.4209 \\ 0.3892 \\ 0.4484 \\ 0.4879 \\ 0.9076 \end{bmatrix}, \tag{5.24}$$

indicating that most model parameters are rather poorly resolved. Figure 5.7 displays this poor resolution by applying $\mathbf{R}$ to the (true) spike model (4.64). Recall that this is the model recovered when the true model is a spike and there is no noise added to the data vector; in practice, noise will result in a solution which is worse than this ideal.   ■

As in Chapter 2, we can compute a covariance matrix for the estimated model parameters. Since

$$\mathbf{m}_{\alpha} = \mathbf{G}^{\sharp}\mathbf{d} \tag{5.25}$$

the covariance is

$$\text{Cov}(\mathbf{m}_{\alpha}) = \mathbf{G}^{\sharp}\text{Cov}(\mathbf{d})(\mathbf{G}^{\sharp})^{T}. \tag{5.26}$$

Note that, as with the SVD solution of Chapter 4, the Tikhonov regularized solution will generally be biased, and differences between the regularized solution values and the true

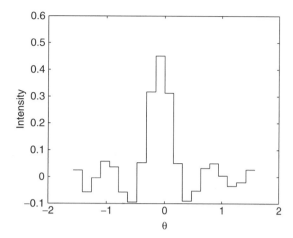

Figure 5.7 Resolution of the spike model, $\alpha = 5.36 \times 10^{-5}$.

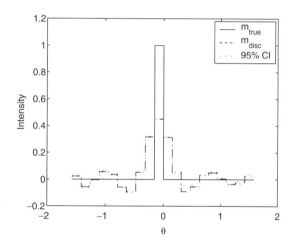

Figure 5.8 Tikhonov solution and confidence intervals for the Shaw problem, estimated using (5.26), where the true model is a spike. The regularization parameter $\alpha = 4.29 \times 10^{-5}$ was chosen using the discrepancy principle.

model may actually be much larger than the confidence intervals obtained from the covariance matrix of the model parameters.

■ **Example 5.3** Recall our earlier example of the Shaw problem with the spike model input. Figure 5.8 shows the true model, the solution obtained using $\alpha = 4.29 \times 10^{-5}$ chosen using the discrepancy principle, and 95% confidence intervals for the estimated parameters. Note that the confidence intervals are extremely tight, and that very few of the true model parameters are included within the confidence intervals. In this case, the regularization bias, which is *not* estimated by the covariance matrix, is far larger than the propagated data uncertainty.

In fact what we are seeing here is caused almost entirely by regularization. The solution shown in Figure 5.8 is essentially identical to the product of $\mathbf{R}_m$ and $\mathbf{m}_{true}$ shown in Figure 5.7. ∎

5.4 HIGHER-ORDER TIKHONOV REGULARIZATION

So far in our discussions of Tikhonov regularization we have minimized an objective function involving $\|\mathbf{m}\|_2$. In many situations, we would prefer to obtain a solution that minimizes some other measure of $\mathbf{m}$, such as the norm of the first or second derivative.

For example, if we have discretized our problem using simple collocation and our model is one-dimensional, then we can approximate, to a multiplicative constant, the first derivative of the model by $\mathbf{Lm}$, where

$$\mathbf{L} = \begin{bmatrix} -1 & 1 & & & \\ & -1 & 1 & & \\ & & \cdots & & \\ & & & -1 & 1 \\ & & & & -1 & 1 \end{bmatrix}. \tag{5.27}$$

Matrices that are used to differentiate $\mathbf{m}$ for the purposes of regularization are referred to as **roughening matrices**. In (5.27), $\mathbf{Lm}$ is a finite-difference approximation that is proportional to the first derivative of $\mathbf{m}$. By minimizing $\|\mathbf{Lm}\|_2$, we will favor solutions that are relatively flat. Note that $\|\mathbf{Lm}\|_2$ is a seminorm because it is zero for any constant model, not just for $\mathbf{m} = \mathbf{0}$. In **first-order Tikhonov regularization**, we solve the damped least squares problem

$$\min \|\mathbf{Gm} - \mathbf{d}\|_2^2 + \alpha^2 \|\mathbf{Lm}\|_2^2 \tag{5.28}$$

using an $\mathbf{L}$ matrix like (5.27). In **second-order Tikhonov regularization**, we use

$$\mathbf{L} = \begin{bmatrix} 1 & -2 & 1 & & & \\ & 1 & -2 & 1 & & \\ & & \cdots & & & \\ & & & 1 & -2 & 1 \\ & & & & 1 & -2 & 1 \end{bmatrix}. \tag{5.29}$$

Here $\mathbf{Lm}$ is a finite-difference approximation proportional to the second derivative of $\mathbf{m}$, and minimizing the seminorm $\|\mathbf{Lm}\|_2$ penalizes solutions that are rough in a second derivative sense.

If our model is two- or three-dimensional then the roughening matrices described here would not be appropriate. In such cases a finite-difference approximation to the Laplacian operator is often used. This is discussed in Exercise 5.3.

We have already seen how to apply zeroth-order Tikhonov regularization to solve (5.28), where $\mathbf{L} = \mathbf{I}$, using the singular value decomposition (5.10). To solve higher-order Tikhonov regularization problems, we employ the **generalized singular value decomposition**, or

GSVD [59, 56]. The GSVD enables the solution to (5.28) to be expressed as a sum of filter factors times generalized singular vectors.

Unfortunately, the definition of the GSVD and associated notation are not presently standardized. In the following, we will follow the conventions used by the MATLAB Regularization Tools and its **cgsvd** command [58]. One notational difference is that we will use γ_i for the generalized singular value, while the Regularization Tools use σ_i for the generalized singular value. Note that MATLAB also has a built-in command, **gsvd**, which employs a different definition of the GSVD.

Here, we will assume that **G** is an m by n matrix, and that **L** is a p by n matrix, with $m \geq n \geq p$, that rank(**L**) $= p$, and that the null spaces of **G** and **L** intersect only at the zero vector.

Under the foregoing assumptions there exist matrices **U**, **V**, **Λ**, **M**, and **X** with the following properties and relationships:

- **U** is m by n with orthonormal columns.
- **V** is p by p and orthogonal.
- **X** is n by n and nonsingular.
- **Λ** is a p by p diagonal matrix, with

$$0 \leq \lambda_1 \leq \lambda_2 \leq \cdots \leq \lambda_p \leq 1. \tag{5.30}$$

- **M** is a p by p diagonal matrix with

$$1 \geq \mu_1 \geq \mu_2 \geq \cdots \geq \mu_p > 0. \tag{5.31}$$

- The λ_i and μ_i are normalized so that

$$\lambda_i^2 + \mu_i^2 = 1, \qquad i = 1, 2, \ldots, p. \tag{5.32}$$

- The **generalized singular values** are

$$\gamma_i = \frac{\lambda_i}{\mu_i} \tag{5.33}$$

  where

$$0 \leq \gamma_1 \leq \gamma_2 \leq \cdots \leq \gamma_p. \tag{5.34}$$

- $$\mathbf{G} = \mathbf{U} \begin{bmatrix} \mathbf{\Lambda} & \mathbf{0} \\ \mathbf{0} & \mathbf{I} \end{bmatrix} \mathbf{X}^{-1} \tag{5.35}$$

- $$\mathbf{L} = \mathbf{V} [\mathbf{M} \quad \mathbf{0}] \mathbf{X}^{-1} \tag{5.36}$$

-
$$\mathbf{X}^T\mathbf{G}^T\mathbf{G}\mathbf{X} = \begin{bmatrix} \Lambda^2 & 0 \\ 0 & \mathbf{I} \end{bmatrix} \tag{5.37}$$

-
$$\mathbf{X}^T\mathbf{L}^T\mathbf{L}\mathbf{X} = \begin{bmatrix} \mathbf{M}^2 & 0 \\ 0 & 0 \end{bmatrix} \tag{5.38}$$

- When $p < n$, the matrix $\mathbf{L}$ will have a nontrivial null space for which the vectors $\mathbf{X}_{\cdot,p+1}$, $\mathbf{X}_{\cdot,p+2}, \ldots, \mathbf{X}_{\cdot,n}$ form a basis.

When $\mathbf{G}$ comes from an IFK, the GSVD typically has two properties that were also characteristic of the SVD. First, the generalized singular values γ_i (5.33) tend to zero without any obvious break in the sequence. Second, the vectors $\mathbf{U}_{\cdot,i}$, $\mathbf{V}_{\cdot,i}$, and $\mathbf{X}_{\cdot,i}$ tend to become rougher as i increases and γ_i decreases.

Using the GSVD, the solution to (5.28) is

$$\mathbf{m}_{\alpha,L} = \sum_{i=1}^{p} \frac{\gamma_i^2}{\gamma_i^2 + \alpha^2} \frac{\mathbf{U}_{\cdot,i}^T\mathbf{d}}{\lambda_i}\mathbf{X}_{\cdot,i} + \sum_{i=p+1}^{n} (\mathbf{U}_{\cdot,i}^T\mathbf{d})\mathbf{X}_{\cdot,i} \tag{5.39}$$

where

$$f_i = \frac{\gamma_i^2}{\gamma_i^2 + \alpha^2} \tag{5.40}$$

are filter factors for the GSVD, analogous to those obtained earlier in the expression for the zeroth-order Tikhonov regularized solution (5.17).

■ **Example 5.4** We return now to the vertical seismic profiling example previously discussed in Example 1.3 and Example 3.1. In this case, for a 1-km-deep borehole experiment, the problem is discretized using $m = n = 50$ data and model points, corresponding to sensors every 20 m, and 20-m-thick, constant-slowness model layers. Figure 5.9 shows the test model that we will try to recover. A synthetic data set was generated with $N(0, (2 \times 10^{-4}\,\mathrm{s})^2)$ noise added.

The discretized system of equations $\mathbf{Gm} = \mathbf{d}$ has a small condition number of 64. This happens in part because we have chosen a very coarse discretization, which effectively regularizes the problem by discretization. Another reason is that the vertical seismic profiling problem is only mildly ill posed [38]. Figure 5.10 shows the least squares solution, together with 95% confidence intervals.

From the statistical point of view, this solution is completely acceptable. However, suppose that from other information, we believe that the slowness should vary smoothly with depth. We will use higher-order Tikhonov regularization to obtain smooth solutions to this problem.

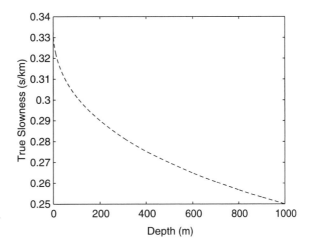

Figure 5.9   A smooth test model for the VSP problem.

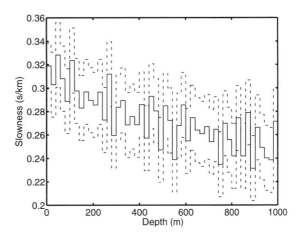

Figure 5.10   Least squares solution for the VSP problem, with 95% confidence intervals.

Figure 5.11 shows the first-order Tikhonov regularization L-curve for this problem obtained using the Regularization Tools commands **get_l**, **cgsvd**, **l_curve**, and **tikhonov**. The L-curve has a distinct corner near $\alpha \approx 137$. Figure 5.12 shows the corresponding solution. The first-order regularized solution is much smoother than the least squares solution and is much closer to the true solution.

Figure 5.13 shows the L-curve for second–order Tikhonov regularization, which has a corner near $\alpha \approx 2325$. Figure 5.14 shows the corresponding solution. This solution is smoother still compared to the first-order regularized solution. Both the first- and second-order solutions depart most from the true solution at shallow depths where the true slowness has the greatest slope and curvature. This happens because the regularized solutions are biased toward smoothness.

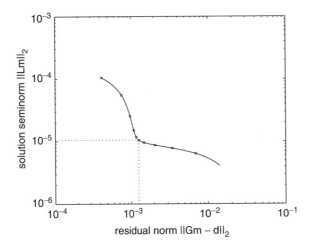

Figure 5.11 L-curve for the VSP problem, first-order regularization.

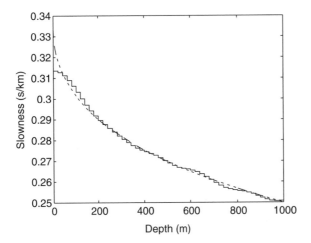

Figure 5.12 Tikhonov solution for the VSP problem, first-order regularization, $\alpha = 137$, shown in comparison with the true model.

Figure 5.15 shows filter factors corresponding to these first- and second-order solutions. Higher-order terms in (5.39) are severely downweighted in both cases, particularly in the second-order case. The rapid attenuation of the basis functions for these solutions arises because the true solution is in fact quite smooth. Because of the smoothness of the true model, the model seminorms can be reduced considerably through the selection of relatively large regularization parameters α without attendant large data misfit increases. In this example the 2-norms of the difference between the first- and second-order solutions and the true model (discretized into 50 values) are approximately 1.21×10^{-2} s/km and 1.18×10^{-2} s/km, respectively.   ■

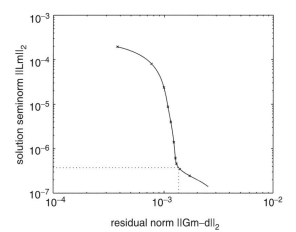

Figure 5.13   L-curve for the VSP problem, second-order regularization.

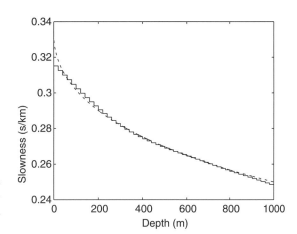

Figure 5.14   Tikhonov solution for the VSP problem, second-order regularization, $\alpha = 2325$, shown in comparison with the true model.

5.5   RESOLUTION IN HIGHER-ORDER TIKHONOV REGULARIZATION

As with zeroth-order Tikhonov regularization, we can compute a resolution matrix for higher-order Tikhonov regularization. For particular values of $\mathbf{L}$ and α, the Tikhonov regularization solution can be written as

$$\mathbf{m}_{\alpha, L} = \mathbf{G}^{\sharp}\mathbf{d} \tag{5.41}$$

where

$$\mathbf{G}^{\sharp} = (\mathbf{G}^{T}\mathbf{G} + \alpha^{2}\mathbf{L}^{T}\mathbf{L})^{-1}\mathbf{G}^{T}. \tag{5.42}$$

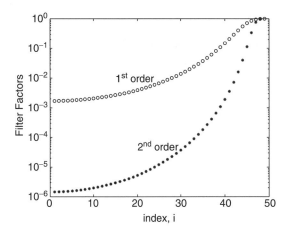

Figure 5.15 Filter factors (5.17) for optimal first- and second-order Tikhonov solutions to the VSP problem shown in Figures 5.12 and 5.14.

Using properties of the GSVD we can simplify this expression to

$$\mathbf{G}^{\sharp} = \mathbf{X}\begin{bmatrix} \mathbf{\Lambda}^2 + \alpha^2\mathbf{M}^2 & \mathbf{0} \\ \mathbf{0} & \mathbf{I} \end{bmatrix}^{-1} \mathbf{X}^T\mathbf{X}^{-T}\begin{bmatrix} \mathbf{\Lambda}^T & \mathbf{0} \\ \mathbf{0} & \mathbf{I} \end{bmatrix}\mathbf{U}^T \tag{5.43}$$

$$= \mathbf{X}\begin{bmatrix} \mathbf{\Lambda}^2 + \alpha^2\mathbf{M}^2 & \mathbf{0} \\ \mathbf{0} & \mathbf{I} \end{bmatrix}^{-1}\begin{bmatrix} \mathbf{\Lambda}^T & \mathbf{0} \\ \mathbf{0} & \mathbf{I} \end{bmatrix}\mathbf{U}^T \tag{5.44}$$

$$= \mathbf{X}\begin{bmatrix} \mathbf{F}\mathbf{\Lambda}^{-1} & \mathbf{0} \\ \mathbf{0} & \mathbf{I} \end{bmatrix}\mathbf{U}^T \tag{5.45}$$

where $\mathbf{F}$ is a diagonal matrix of GSVD filter factors

$$F_{i,i} = \frac{\gamma_i^2}{\gamma_i^2 + \alpha^2}. \tag{5.46}$$

The resolution matrix is then

$$\mathbf{R}_{\mathrm{m},\alpha,L} = \mathbf{G}^{\sharp}\mathbf{G} = \mathbf{X}\begin{bmatrix} \mathbf{F}\mathbf{\Lambda}^{-1} & \mathbf{0} \\ \mathbf{0} & \mathbf{I} \end{bmatrix}\mathbf{U}^T\mathbf{U}\begin{bmatrix} \mathbf{\Lambda} & \mathbf{0} \\ \mathbf{0} & \mathbf{I} \end{bmatrix}\mathbf{X}^{-1} \tag{5.47}$$

$$= \mathbf{X}\begin{bmatrix} \mathbf{F} & \mathbf{0} \\ \mathbf{0} & \mathbf{I} \end{bmatrix}\mathbf{X}^{-1}. \tag{5.48}$$

■ **Example 5.5** To examine the resolution of the Tikhonov-regularized inversions of Example 5.4, we performed a spike test using (5.48). Figure 5.16 shows the effect of multiplying $\mathbf{R}_{\alpha,L}$ times a unit amplitude spike model (spike depth 500 m) under first- and second-order Tikhonov regularization using α values of 137 and 2325, respectively. The shapes of these curves shown in Figure 5.16 show how the regularized solutions cannot

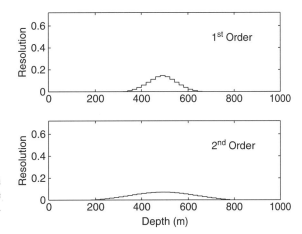

Figure 5.16 The model resolution matrix $\mathbf{R}_{\alpha,L}$ multiplied times the spike model for each of the regularized solutions of Example 5.4.

correctly recover abrupt changes in slowness halfway through the model. Under first- or second-order regularization, the resolution of various model features will depend critically on how smooth or rough these features are in the true model. In this example, the higher-order solutions recover the true model better because the true model is smooth. ■

5.6 THE TGSVD METHOD

In the Chapter 4 discussion of the SVD, we examined the TSVD method of regularization. In the construction of a solution, the TSVD method rejects model space basis vectors associated with smaller singular values. Equivalently, this can be thought of as a damped SVD solution in which filter factors of one are used for basis vectors associated with larger singular values and filter factors of zero are used for basis vectors associated with smaller singular values. This approach can be extended to the GSVD solution (5.39) to produce a **truncated generalized singular value decomposition** or **TGSVD** solution. In the TGSVD solution we simply assign filter factors (5.40) of one to the k largest generalized singular values in (5.39) to obtain

$$\mathbf{m}_{k,L} = \sum_{i=p-k+1}^{p} \frac{(\mathbf{U}_{\cdot,i})^T \mathbf{d}}{\lambda_i} \mathbf{X}_{\cdot,i} + \sum_{i=p+1}^{n} ((\mathbf{U}_{\cdot,i})^T \mathbf{d}) \mathbf{X}_{\cdot,i}. \tag{5.49}$$

■ **Example 5.6** Applying the TGSVD method to the VSP problem from Example 1.3, we find L-curve corners near $k = 8$ in the first-order case shown in Figure 5.17 and $k = 3$ in the second-order case shown in Figure 5.18. Examining the filter factors obtained for the corresponding Tikhonov solutions shown in Figure 5.15 we find that the filter factors decline precipitously with decreasing index near these locations. Figures 5.19 and 5.20 show the corresponding TGSVD solutions. The model recovery is comparable to that obtained with the Tikhonov method. The 2-norms of the difference between the first- and second-order

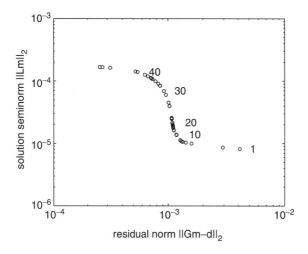

Figure 5.17 TGSVD L-curve for the VSP problem as a function of k, first-order regularization.

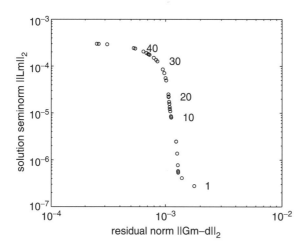

Figure 5.18 TGSVD L-curve for the VSP problem as a function of k, second-order regularization.

solutions and the true model are approximately 1.21×10^{-2} s/km and 1.18×10^{-2} s/km, respectively, which are similar to the Tikhonov solution. ∎

5.7 GENERALIZED CROSS VALIDATION

Generalized cross validation (GCV) is an alternative method for selecting a regularization parameter, α, that has a number of desirable statistical properties.

In ordinary or "leave-one-out" cross validation, we consider the models that are obtained by leaving one of the m data points out of the fitting process. Consider the modified Tikhonov

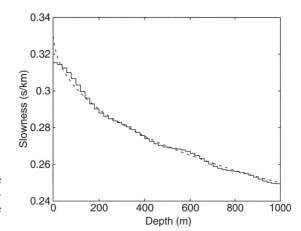

Figure 5.19 TGSVD solution of the VSP problem, $k = 8$, first-order regularization, shown in comparison with the true model.

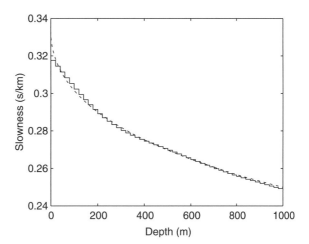

Figure 5.20 TGSVD solution of the VSP problem, $k = 3$, second-order regularization.

regularization problem in which we ignore a data point d_k,

$$\min \sum_{i \neq k}((\mathbf{Gm})_i - d_i)^2 + \alpha^2\|\mathbf{Lm}\|_2^2. \tag{5.50}$$

Call the solution to this problem $\mathbf{m}_{\alpha,L}^{[k]}$, where the superscript indicates that d_k was left out of the computation. Ideally, the model $\mathbf{m}_{\alpha,L}^{[k]}$ would accurately predict the missing data value d_k. In the leave-one-out approach, we select the regularization parameter α so as to minimize the predictive errors for all k:

$$\min \quad V_0(\alpha) = \frac{1}{m} \sum_{k=1}^{m} ((\mathbf{Gm}_{\alpha,L}^{[k]})_k - d_k)^2. \tag{5.51}$$

Unfortunately, computing $V_0(\alpha)$ involves solving m problems of the form (5.50). Generalized cross validation is a way to simplify this computation.

First, let

$$\tilde{d}_i = \begin{cases} (\mathbf{Gm}_{\alpha,L}^{[k]})_k & i = k \\ d_i & i \neq k. \end{cases} \tag{5.52}$$

Note that because $(\mathbf{Gm}_{\alpha,L}^{[k]})_k = \tilde{d}_k$, $\mathbf{m}_{\alpha,L}^{[k]}$ also solves

$$\min \quad ((\mathbf{Gm})_k - \tilde{d}_k)^2 + \sum_{i \neq k} ((\mathbf{Gm})_i - \tilde{d}_i)^2 + \alpha^2 \|\mathbf{Lm}\|_2^2, \tag{5.53}$$

which is equivalent to

$$\min \quad \|\mathbf{Gm} - \tilde{\mathbf{d}}\|_2^2 + \alpha^2 \|\mathbf{Lm}\|_2^2. \tag{5.54}$$

This result is known as the **leave-one-out lemma**. By the leave-one-out lemma,

$$\mathbf{m}_{\alpha,L}^{[k]} = \mathbf{G}^\sharp \tilde{\mathbf{d}}. \tag{5.55}$$

We will use (5.55) to eliminate $\mathbf{m}_{\alpha,L}^{[k]}$ from (5.51). It is easy to see that

$$\frac{(\mathbf{GG}^\sharp \tilde{\mathbf{d}})_k - (\mathbf{GG}^\sharp \mathbf{d})_k}{\tilde{d}_k - d_k} = (\mathbf{GG}^\sharp)_{k,k}. \tag{5.56}$$

Subtracting both sides of the equation from 1 gives

$$\frac{\tilde{d}_k - d_k - (\mathbf{GG}^\sharp \tilde{\mathbf{d}})_k + (\mathbf{GG}^\sharp \mathbf{d})_k}{\tilde{d}_k - d_k} = 1 - (\mathbf{GG}^\sharp)_{k,k}. \tag{5.57}$$

Since $(\mathbf{GG}^\sharp \mathbf{d})_k = (\mathbf{Gm}_{\alpha,L})_k$, $(\mathbf{GG}^\sharp \tilde{\mathbf{d}})_k = \tilde{d}_k$, and $\mathbf{Gm}_{\alpha,L}^{[k]} = \tilde{d}_k$, (5.57) simplifies to

$$\frac{(\mathbf{Gm}_{\alpha,L})_k - d_k}{(\mathbf{Gm}_{\alpha,L}^{[k]})_k - d_k} = 1 - (\mathbf{GG}^\sharp)_{k,k}. \tag{5.58}$$

Substituting this formula into (5.51), we obtain

$$V_0(\alpha) = \frac{1}{m} \sum_{k=1}^{m} \left(\frac{(\mathbf{Gm}_{\alpha,L})_k - d_k}{1 - (\mathbf{GG}^{\sharp})_{k,k}} \right)^2. \tag{5.59}$$

We can simplify the formula further by replacing the $(\mathbf{GG}^{\sharp})_{k,k}$ with the average value

$$(\mathbf{GG}^{\sharp})_{k,k} \approx \frac{1}{m} \mathrm{Tr}(\mathbf{GG}^{\sharp}). \tag{5.60}$$

Using (5.60), we have

$$V_0(\alpha) \approx \frac{1}{m} \sum_{i=1}^{m} \frac{((\mathbf{Gm}_{\alpha,L})_k - d_k)^2}{(\frac{1}{m}(m - \mathrm{Tr}(\mathbf{GG}^{\sharp})))^2} \tag{5.61}$$

$$= \frac{m \|\mathbf{Gm}_{\alpha,L} - \mathbf{d}\|_2^2}{\mathrm{Tr}(\mathbf{I} - \mathbf{GG}^{\sharp})^2}. \tag{5.62}$$

In GCV we minimize (5.62).

It can be shown that under reasonable assumptions regarding the noise and smoothness of $\mathbf{m}_{\mathrm{true}}$, the value of α that minimizes (5.62) approaches the value that minimizes $E[\|\mathbf{Gm}_{\alpha,L} - \mathbf{d}\|]$ as the number of data points m goes to infinity, and that under the same assumptions, $E[\|\mathbf{m}_{\mathrm{true}} - \mathbf{m}_{\alpha,L}\|_2]$ goes to 0 as m goes to infinity [30, 176]. In practice, the size of the data set is fixed in advance, so the limit is not directly applicable. However, these results provide a theoretical justification for using the GCV method to select the Tikhonov regularization parameter.

■ **Example 5.7** Figures 5.21 and 5.22 show $V_0(\alpha)$ for the VSP test problem, using first- and second-order Tikhonov regularization, respectively. The respective minima occur near $\alpha = 43$ and $\alpha = 816$, which are significantly less than the α values estimated previously using the L-curve. Figures 5.23 and 5.24 show the corresponding solutions. The 2-norms of the difference between the first- and second-order solutions and the true model are approximately 1.53×10^{-2} s/km and 9.72×10^{-3} s/km, respectively, making the second-order solution the closest to the true model of any of the models obtained here.   ■

5.8 ERROR BOUNDS

We next present two theoretical results that help to address the accuracy of Tikhonov regularization solutions. We will present these results in a simplified form, covering only zeroth-order Tikhonov regularization.

The first question is whether for a particular value of the regularization parameter α, we can establish a bound on the sensitivity of the regularized solution to the noise in the observed data $\mathbf{d}$ and/or errors in the system matrix $\mathbf{G}$. This would provide a sort of condition number for the

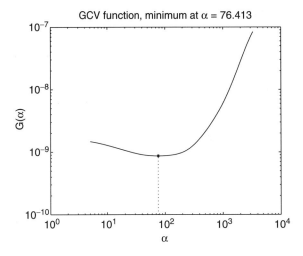

Figure 5.21 GCV curve for the VSP problem, first-order regularization.

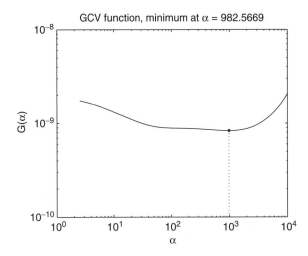

Figure 5.22 GCV curve for the VSP problem, second-order regularization.

inverse problem. Note that this does not tell us how far the regularized solution is from the true model, since Tikhonov regularization has introduced a bias in the solution. Under Tikhonov regularization with a nonzero α, we would not obtain the true model even if the noise was 0.

The following theorem gives a bound for zeroth-order Tikhonov regularization [59]. A slightly more complicated formula is available for higher-order Tikhonov regularization [59].

■ **Theorem 5.1** Suppose that the problems

$$\min \|\mathbf{Gm} - \mathbf{d}\|_2^2 + \alpha^2 \|\mathbf{m}\|_2^2 \tag{5.63}$$

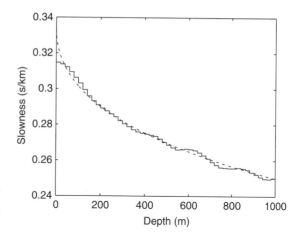

Figure 5.23 GCV solution for the VSP problem, first-order, $\alpha = 43.4$, shown in comparison with the true model.

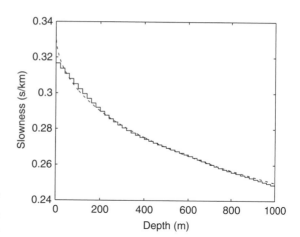

Figure 5.24 GCV solution for the VSP problem, second-order, $\alpha = 815.9$, shown in comparison with the true model.

and

$$\min \|\bar{\mathbf{G}}\mathbf{m} - \bar{\mathbf{d}}\|_2^2 + \alpha^2 \|\mathbf{m}\|_2^2 \tag{5.64}$$

are solved to obtain $\mathbf{m}_\alpha$ and $\bar{\mathbf{m}}_\alpha$. Then

$$\frac{\|\mathbf{m}_\alpha - \bar{\mathbf{m}}_\alpha\|_2}{\|\mathbf{m}_\alpha\|_2} \le \frac{\bar{\kappa}_\alpha}{1 - \epsilon\bar{\kappa}_\alpha} \left(2\epsilon + \frac{\|\mathbf{e}\|_2}{\|\mathbf{d}_\alpha\|_2} + \epsilon\bar{\kappa}_\alpha \frac{\|\mathbf{r}_\alpha\|_2}{\|\mathbf{d}_\alpha\|_2} \right) \tag{5.65}$$

where

$$\bar{\kappa}_\alpha = \frac{\|\mathbf{G}\|_2}{\alpha} \qquad (5.66)$$

$$\mathbf{E} = \mathbf{G} - \bar{\mathbf{G}} \qquad (5.67)$$

$$\mathbf{e} = \mathbf{d} - \bar{\mathbf{d}} \qquad (5.68)$$

$$\epsilon = \frac{\|\mathbf{E}\|_2}{\|\mathbf{G}\|_2} \qquad (5.69)$$

$$\mathbf{d}_\alpha = \mathbf{G}\mathbf{m}_\alpha \qquad (5.70)$$

and

$$\mathbf{r}_\alpha = \mathbf{d} - \mathbf{d}_\alpha. \qquad (5.71)$$

∎

In the particular case when $\mathbf{G} = \bar{\mathbf{G}}$, and the only difference between the two problems is $\mathbf{e} = \mathbf{d} - \bar{\mathbf{d}}$, the inequality becomes even simpler:

$$\frac{\|\mathbf{m}_\alpha - \bar{\mathbf{m}}_\alpha\|_2}{\|\mathbf{m}_\alpha\|_2} \le \bar{\kappa}_\alpha \frac{\|\mathbf{e}\|_2}{\|\mathbf{d}_\alpha\|_2}. \qquad (5.72)$$

The condition number $\bar{\kappa}_\alpha$ is inversely proportional to α. Thus increasing α will decrease the sensitivity of the solution to perturbations in the data. Of course, increasing α also increases the error in the solution due to regularization bias and decreases resolution.

The second question is whether we can establish any sort of bound on the norm of the difference between the regularized solution and the true model. This bound would incorporate both sensitivity to noise and the bias introduced by Tikhonov regularization. Such a bound must of course depend on the magnitude of the noise in the data. It must also depend on the particular regularization parameter chosen. Tikhonov developed a beautiful theorem that addresses this question in the context of inverse problems involving IFKs [164]. More recently, Neumaier has developed a version of Tikhonov's theorem that can be applied directly to discretized problems [111].

Recall that in a discrete ill-posed linear inverse problem, the matrix $\mathbf{G}$ has a smoothing effect, in that when we multiply $\mathbf{G}\mathbf{m}$, the result is smoother than $\mathbf{m}$. Similarly, if we multiply $\mathbf{G}^T$ times $\mathbf{G}\mathbf{m}$, the result will be smoother than $\mathbf{G}\mathbf{m}$. This smoothing is a consequence of the fact that the singular vectors corresponding to the larger singular values of $\mathbf{G}$ are smooth. This is not necessarily a property of matrices that do not arise from discrete ill-posed linear inverse problems. For example, if $\mathbf{G}$ is a matrix that approximates the differentiation operator, then $\mathbf{G}\mathbf{m}$ will be rougher than $\mathbf{m}$.

It should be clear that models in the range of $\mathbf{G}^T$ form a relatively smooth subspace of all possible models. Models in this subspace of R^n can be written as $\mathbf{m} = \mathbf{G}^T\mathbf{w}$, for some weights $\mathbf{w}$. Furthermore, models in the range of $\mathbf{G}^T\mathbf{G}$ form a subspace of $R(\mathbf{G}^T)$, since any model in $R(\mathbf{G}^T\mathbf{G})$ can be written as $\mathbf{m} = \mathbf{G}^T(\mathbf{G}\mathbf{w})$ which is a linear combination of columns of $\mathbf{G}^T$. Because of the smoothing effect of $\mathbf{G}$ and $\mathbf{G}^T$, we would expect these models to be even smoother than the models in $R(\mathbf{G}^T)$. We could construct smaller subspaces of R^n that contain even smoother models, but it turns out that with zeroth-order Tikhonov regularization these are the only subspaces of interest.

There is another way to see that models in $R(\mathbf{G}^T)$ will be relatively smooth. Recall that the vectors $\mathbf{V}_{.,1}, \mathbf{V}_{.,2}, \ldots, \mathbf{V}_{.,p}$ from the SVD of $\mathbf{G}$ form an orthonormal basis for $R(\mathbf{G}^T)$. For discrete ill-posed problems, we know from Chapter 4 that these basis vectors will be relatively smooth, so linear combinations of these vectors in $R(\mathbf{G}^T)$ should be smooth.

The following theorem gives a bound on the total error including bias due to regularization and error due to noise in the data for zeroth-order Tikhonov regularization [111]. In the following $p = 1$ indicates that $\mathbf{m}_{\text{true}}$ is in $R(\mathbf{G}^T)$ while $p = 2$ indicates that $\mathbf{m}_{\text{true}}$ is in $R(\mathbf{G}^T\mathbf{G})$.

■ **Theorem 5.2** Suppose that we use zeroth-order Tikhonov regularization to solve $\mathbf{Gm} = \mathbf{d}$ and that $\mathbf{m}_{\text{true}}$ can be expressed as

$$\mathbf{m}_{\text{true}} = \begin{cases} \mathbf{G}^T\mathbf{w} & p = 1 \\ \mathbf{G}^T\mathbf{Gw} & p = 2 \end{cases} \tag{5.73}$$

and that

$$\|\mathbf{Gm}_{\text{true}} - \mathbf{d}\|_2 \le \Delta \|\mathbf{w}\|_2 \tag{5.74}$$

for some $\Delta > 0$. Then

$$\|\mathbf{m}_{\text{true}} - \mathbf{G}^\sharp\mathbf{d}\|_2 \le \left(\frac{\Delta}{2\alpha} + \gamma\alpha^p \right) \|\mathbf{w}\|_2 \tag{5.75}$$

where

$$\gamma = \begin{cases} 1/2 & p = 1 \\ 1 & p = 2 \end{cases}. \tag{5.76}$$

Furthermore, if we begin with the bound

$$\|\mathbf{Gm}_{\text{true}} - \mathbf{d}\|_2 \le \delta \tag{5.77}$$

we can let

$$\Delta = \frac{\delta}{\|\mathbf{w}\|_2}. \tag{5.78}$$

Under this condition the optimal value of α is

$$\hat{\alpha} = \left(\frac{\Delta}{2\gamma p} \right)^{\frac{1}{p+1}} = O(\Delta^{\frac{1}{p+1}}). \tag{5.79}$$

With this choice of α,

$$\Delta = 2\gamma p \hat{\alpha}^{p+1} \tag{5.80}$$

and the error bound simplifies to

$$\|\mathbf{m}_{\text{true}} - \mathbf{G}^{\sharp}_{\hat{\alpha}}\mathbf{d}\|_2 \le \gamma(p+1)\hat{\alpha}^p = O(\Delta^{\frac{p}{p+1}}). \tag{5.81}$$

∎

This theorem tells us that the error in the Tikhonov regularization solution depends on both the noise level Δ and on the regularization parameter α. For very large values of α, the error due to regularization will be dominant. For very small values of α, the error due to noise in the data will be dominant. There is an optimal value of α which balances these effects. Using the optimal α, we can obtain an error bound of $O(\Delta^{2/3})$ if $p = 2$, and an error bound of $O(\Delta^{1/2})$ if $p = 1$.

Of course, the foregoing result can only be applied when our true model lives in the restricted subspace of models in $R(\mathbf{G}^T)$. In practice, even if the model does lie in $R(\mathbf{G}^T)$, the vector $\mathbf{w}$ may have a very large norm, making the bound useless.

Applying this theorem in a quantitative fashion is typically impractical. However, the theorem does provide some useful rules of thumb. The first point is that the accuracy of the regularized solution depends very much on the smoothness of the true model. If $\mathbf{m}_{\text{true}}$ is not smooth, then Tikhonov regularization simply will not give an accurate solution. Furthermore, if the model $\mathbf{m}_{\text{true}}$ is smooth, then we can hope for an error in the Tikhonov regularized solution that is $O(\delta^{1/2})$ or $O(\delta^{2/3})$. Another way of saying this is that we can hope *at best* for an answer with about two-thirds as many correct significant digits as the data.

∎ **Example 5.8** Recall the Shaw problem previously considered in Examples 3.2, 4.3, and 5.1. Since $\mathbf{G}^T$ is a nonsingular matrix, the spike model should lie in $R(\mathbf{G}^T)$. However, $\mathbf{G}^T$ is numerically singular, and since the spike model lies outside of the effective range of $\mathbf{G}^T$, any attempt to find $\mathbf{w}$ simply produces a meaningless answer, and Theorem 5.2 cannot be applied.

Figure 5.25 shows a smooth model that does lie in the range of $\mathbf{G}^T$. For this model we constructed a synthetic data set with noise as before at $\delta = 4.47 \times 10^{-6}$. The theorem suggests using $\alpha = 8.0 \times 10^{-4}$. The resulting error bound is 8.0×10^{-4}, while the actual norm of the error is 6.6×10^{-4}. Here the data were accurate to roughly six digits, while the solution was accurate to roughly four digits. Figure 5.26 shows the reconstruction of the model from the noisy data. This example demonstrates the importance of smoothness of the true model in determining how well it can be reconstructed through Tikhonov regularization. ∎

5.9 EXERCISES

5.1 Use the method of Lagrange multipliers to derive the damped least squares problem (5.4) from the discrepancy principle problem (5.2), and demonstrate that (5.4) can be written as (5.5).

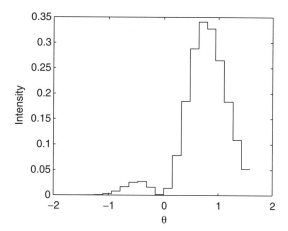

Figure 5.25 A smooth model in $R(\mathbf{G}^T)$.

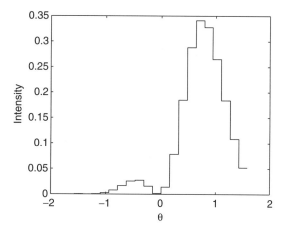

Figure 5.26 Reconstruction of the smooth model.

5.2 Consider the integral equation and data set from Problem 3.1. You can find a copy of this data set in **ifk.mat**.

(a) Discretize the problem using simple collocation.

(b) Using the data supplied, and assuming that the numbers are accurate to four significant figures, determine a reasonable bound δ for the misfit.

(c) Use zeroth-order Tikhonov regularization to solve the problem. Use GCV, the discrepancy principle, and the L-curve criterion to   pick the regularization parameter. Estimate the norm of the difference between your solutions and $\mathbf{m}_{\text{true}}$.

(d) Use first-order Tikhonov regularization to solve the problem. Use GCV, the discrepancy principle, and the L-curve criterion to pick the regularization parameter.

(e) Use second-order Tikhonov regularization to solve the problem. Use GCV, the discrepancy principle, and the L-curve criterion to pick the regularization parameter.

(f) Analyze the resolution of your solutions. Are the features you see in your inverse
solutions unambiguously real? Interpret your results. Describe the size and location
of any significant features in the solution.

5.3 Consider the following problem in **cross-well tomography**. Two vertical wells
are located 1600 meters apart. A seismic source is inserted in one well at depths
of 50, 150, ..., 1550 m. A string of receivers is inserted in the other well at depths
of 50, 150, ..., 1550 m. See Figure 5.27. For each source–receiver pair, a travel time is
recorded, with a measurement standard deviation of 0.5 msec. There are 256 ray paths
and 256 corresponding data points. We wish to determine the velocity structure in the
two-dimensional plane between the two wells.

Discretizing the problem into a 16 by 16 grid of 100 meter by 100 meter blocks gives
256 model parameters. The **G** matrix and noisy data, **d**, for this problem (assuming
straight ray paths) are in the file **crosswell.mat**.

(a) Use the truncated SVD to solve this inverse problem. Plot the result.

(b) Use zeroth-order Tikhonov regularization to solve this problem and plot your
solution. Explain why it is hard to use the discrepancy principle to select the regularization parameter. Use the L-curve criterion to select your regularization parameter.
Plot the L-curve as well as your solution.

(c) Use second-order Tikhonov regularization to solve this problem and plot your
solution. Because this is a two-dimensional problem, you will need to implement
a finite-difference approximation to the Laplacian (second derivative in the horizontal direction plus the second derivative in the vertical direction) in the roughening
matrix. One such **L** matrix can be generated using the following MATLAB code:

```
L=zeros(14*14,256);
k=1;
for i=2:15,
  for j=2:15,
    M=zeros(16,16);
    M(i,j)=-4;
    M(i,j+1)=1;
    M(i,j-1)=1;
    M(i+1,j)=1;
    M(i-1,j)=1;
    L(k,:)=(reshape(M,256,1))';
    k=k+1;
  end
end
```

What, if any, problems did you have in using the L-curve criterion on this problem?
Plot the L-curve as well as your solution.

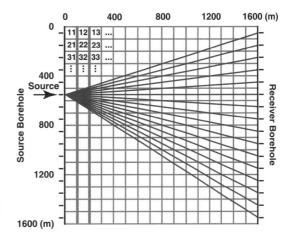

Figure 5.27 Cross-well tomography problem, showing block discretization, block numbering convention, and one set of straight source–receiver ray paths.

(d)   Discuss your results. Can you explain the characteristic vertical bands that appeared in some of your solutions?

5.4   Returning to the problem in Exercise 4.5, solve for the density distribution within the Earth using Tikhonov regularization. Use the L-curve to select an optimal solution. How does it compare to your solution to Exercise 4.5?

5.5   In some situations it is appropriate to bias the regularized solution toward a particular model $\mathbf{m}_0$. In this case, we would solve the damped least squares problem

$$\min \|\mathbf{Gm} - \mathbf{d}\|_2^2 + \alpha^2 \|\mathbf{L}(\mathbf{m} - \mathbf{m}_0)\|_2^2. \tag{5.82}$$

Write this as an ordinary linear least squares problem. What are the normal equations? Can you find a solution for this problem using the GSVD?

5.10   NOTES AND FURTHER READING

Hansen's book [59] is a very complete reference on the linear algebra of Tikhonov regularization. Arnold Neumaier's tutorial [111] is also a very useful reference. Two other surveys of Tikhonov regularization are [37, 38]. Vogel [175] includes an extensive discussion of methods for selecting the regularization parameter. Hansen's MATLAB Regularization Tools [58] is a collection of software for performing regularization within MATLAB.

   The GSVD was first defined by Van Loan [173]. References on the GSVD and algorithms for computing the GSVD include [4, 49, 56, 59].

   Selecting the regularization parameter is an important problem in both theory and practice. Much of the literature on functional analytic approaches assumes that the noise level is known. When the noise level is known, the discrepancy principle provides a scheme for selecting the regularization parameter for ill-posed problems, which is convergent in the sense that in the limit as the noise level goes to zero, the regularized solution goes to $\mathbf{m}_{\text{true}}$ [38].

In practice, the noise level is often unknown, so there has been a great deal of interest in schemes for selecting the regularization parameter without prior knowledge of the noise level. The two most popular approaches are the L-curve method and GCV. The L-curve method was introduced by Hansen [57, 59]. GCV was introduced by Craven and Wahba [30, 176]. The formula for GCV given here is very expensive to compute for large problems. A GCV algorithm for large scale problems is given by Golub and von Matt [50]. Vogel has shown that the L-curve method can fail to converge [174]. It can be shown that no scheme that depends only on the noisy data without knowledge of the noise level can be convergent in the limit as the noise level goes to 0 [38].

Within statistics, badly conditioned linear regression problems are said to suffer from "multi-collinearity." A method called "ridge regression," which is identical to Tikhonov regularization, is often used to deal with such problems [33]. Statisticians also use a method called "principal components regression" (PCR) which is identical to the TSVD method [106].

6

ITERATIVE METHODS

Synopsis: Several techniques for solving inverse problems that are far too large for the methods previously discussed to be practical are presented. These methods are iterative in that a sequence of solutions is generated that converges to a solution to the inverse problem. Kaczmarz's algorithm and the related ART and SIRT methods form one class, while methods based on conjugate gradients form a second class. When the method of conjugate gradients is applied to the normal equations, the resulting CGLS method regularizes the solution of the inverse problem. Illustrative examples involving tomography and image deblurring are given.

6.1 INTRODUCTION

SVD-based pseudoinverse and Tikhonov regularization solutions become impractical when we consider larger problems in which $\mathbf{G}$ has tens of thousands of rows and columns. Storing all of the elements in a large $\mathbf{G}$ matrix can require a great deal of memory. If the majority of the elements in the $\mathbf{G}$ matrix are 0, then $\mathbf{G}$ is a **sparse matrix**, and we can save storage by only storing the nonzero elements of $\mathbf{G}$ and their locations. The **density** of $\mathbf{G}$ is the percentage of nonzero elements in the matrix. **Dense matrices** contain enough nonzero elements that sparse storage schemes are not efficient.

Methods for the solution of linear systems of equations that are based on matrix factorizations such as the Cholesky factorization, QR factorization, or SVD do not tend to work well with sparse matrices. The problem is that the matrices that occur in the factorization of $\mathbf{G}$ are often more dense than $\mathbf{G}$ itself. In particular, the $\mathbf{U}$ and $\mathbf{V}$ matrices in the SVD and the $\mathbf{Q}$ matrix in the QR factorization are required to be orthogonal matrices. This typically makes these matrices fully dense.

The iterative methods discussed in this chapter do not require the storage of additional dense matrices. Instead, they work by generating a sequence of models $\mathbf{m}$ that converge to an optimal solution. These steps typically involve multiplying $\mathbf{G}$ and $\mathbf{G}^T$ times vectors, which can be done without additional storage. Because iterative methods can take advantage of the sparsity commonly found in the $\mathbf{G}$ matrix, they are often used for very large problems.

For example, consider a tomography problem in which the model is of size 256 by 256 (65,536 model elements), and there are 100,000 ray paths. Most of the ray paths miss most of the model cells, so the majority of the elements in $\mathbf{G}$ are zero. The $\mathbf{G}$ matrix might have a density of less than 1%. If we stored $\mathbf{G}$ as a dense matrix, it would require about 50 gigabytes of storage. Furthermore, the **U** matrix would require 80 gigabytes of storage, and the **V** matrix would require about 35 gigabytes. Using a sparse storage technique, **G** can be stored in less than 1 gigabyte.

At the time this book was written, computers with 1 gigabyte of main memory were quite common, whereas only computers classified as supercomputers would have hundreds of gigabytes of main storage. The point at which it becomes necessary to use sparse matrix storage depends on the computer that we are using. The memory capacity of computers has been increasing steadily for many years. However, we can safely say that there will always be problems for which sparse matrix storage is required.

6.2 ITERATIVE METHODS FOR TOMOGRAPHY PROBLEMS

We will concentrate in this section on Kaczmarz's algorithm and its algebraic reconstruction technique (ART) and simultaneous iterative reconstruction technique (SIRT) variants. These algorithms were originally developed for tomographic applications and are particularly effective for such problems.

Kaczmarz's algorithm is an easy to implement algorithm for solving a linear system of equations $\mathbf{Gm} = \mathbf{d}$. To understand the algorithm, note that each of the m rows of the system $\mathbf{G}_{i,.}\mathbf{m} = d_i$ defines an n-dimensional hyperplane in R^m. Kaczmarz's algorithm starts with an initial solution $\mathbf{m}^{(0)}$, and then moves to a solution $\mathbf{m}^{(1)}$ by projecting the initial solution onto the hyperplane defined by the first row in $\mathbf{G}$. Next $\mathbf{m}^{(1)}$ is similarly projected onto the hyperplane defined by the second row in $\mathbf{G}$, and so forth. The process is repeated until the solution has been projected onto all m hyperplanes defined by the system of equations. At that point, a new cycle of projections begins. These cycles are repeated until the solution has converged sufficiently. Figure 6.1 shows an example in which Kaczmarz's algorithm is used to solve the system of equations

$$y = 1$$
$$-x + y = -1. \tag{6.1}$$

To implement the algorithm, we need a formula to compute the projection of a vector onto the hyperplane defined by equation i. Let $\mathbf{G}_{i,.}$ be the ith row of $\mathbf{G}$. Consider the hyperplane defined by $\mathbf{G}_{i+1,.}\mathbf{m} = d_{i+1}$. Because the vector $\mathbf{G}_{i+1,.}^T$ is perpendicular to this hyperplane, the update to $\mathbf{m}^{(i)}$ from the constraint due to row $i + 1$ of $\mathbf{G}$ will be proportional to $\mathbf{G}_{i+1,.}^T$:

$$\mathbf{m}^{(i+1)} = \mathbf{m}^{(i)} + \beta\mathbf{G}_{i+1,.}^T. \tag{6.2}$$

Using the fact that $\mathbf{G}_{i+1,.}\mathbf{m}^{(i+1)} = d_{i+1}$ to solve for β, we obtain

$$\mathbf{G}_{i+1,.}\left(\mathbf{m}^{(i)} + \beta\mathbf{G}_{i+1,.}^T\right) = d_{i+1} \tag{6.3}$$

$$\mathbf{G}_{i+1,.}\mathbf{m}^{(i)} - d_{i+1} = -\beta\mathbf{G}_{i+1,.}\mathbf{G}_{i+1,.}^T. \tag{6.4}$$

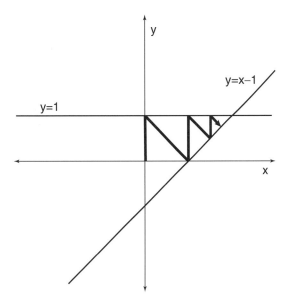

Figure 6.1 Kaczmarz's algorithm on a system of two equations.

$$\beta = -\frac{\mathbf{G}_{i+1,\cdot}\mathbf{m}^{(i)} - d_{i+1}}{\mathbf{G}_{i+1,\cdot}\mathbf{G}_{i+1,\cdot}^T}. \tag{6.5}$$

Thus, the update formula is

$$\mathbf{m}^{(i+1)} = \mathbf{m}^{(i)} - \frac{\mathbf{G}_{i+1,\cdot}\mathbf{m}^{(i)} - d_{i+1}}{\mathbf{G}_{i+1,\cdot}\mathbf{G}_{i+1,\cdot}^T}\mathbf{G}_{i+1,\cdot}^T. \tag{6.6}$$

■ **Algorithm 6.1 Kaczmarz's Algorithm** Given a system of equations $\mathbf{Gm} = \mathbf{d}$.

1. Let $\mathbf{m}^{(0)} = \mathbf{0}$.
2. For $i = 0, 1, \ldots, m$, let

$$\mathbf{m}^{(i+1)} = \mathbf{m}^{(i)} - \frac{\mathbf{G}_{i+1,\cdot}\mathbf{m}^{(i)} - d_{i+1}}{\mathbf{G}_{i+1,\cdot}\mathbf{G}_{i+1,\cdot}^T}\mathbf{G}_{i+1,\cdot}^T. \tag{6.7}$$

3. If the solution has not yet converged, go back to step 2. ■

It can be shown that if the system of equations $\mathbf{Gm} = \mathbf{d}$ has a unique solution, then Kaczmarz's algorithm will converge to this solution. If the system of equations has many solutions, then the algorithm will converge to the solution that is closest to the point $\mathbf{m}^{(0)}$. In particular, if we start with $\mathbf{m}^{(0)} = \mathbf{0}$, we will obtain a minimum-length solution. If there is no exact solution to the system of equations, then the algorithm will fail to converge, but will typically bounce around near an approximate solution.

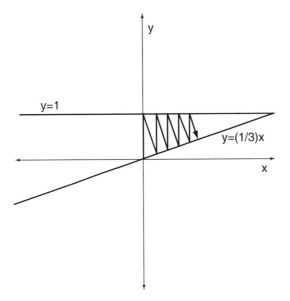

Figure 6.2 Slow convergence occurs when hyperplanes are nearly parallel.

A second important question is how quickly Kaczmarz's algorithm will converge to a solution. If the hyperplanes described by the system of equations are nearly orthogonal, then the algorithm will converge very quickly. However, if two or more hyperplanes are nearly parallel to each other, convergence can be extremely slow. Figure 6.2 shows a typical situation in which the algorithm zigzags back and forth without making much progress toward a solution. As the two lines become more nearly parallel, the problem becomes worse. This problem can be alleviated by picking an ordering of the equations such that adjacent equations describe hyperplanes that are nearly orthogonal to each other. In the context of tomography, this can be done by ordering the equations so that successive equations do not share common model cells.

■ **Example 6.1** Consider a tomographic reconstruction problem with the same geometry used in Exercise 4.3, in which the slowness structure is parameterized in homogeneous blocks of size l by l. The true model is shown in Figure 6.3. Synthetic data were generated, with normally distributed random noise added. The random noise had standard deviation 0.01. Figure 6.4 shows the truncated SVD solution. The two anomalies are apparent, but it is not possible to distinguish the small hole within the larger of the two.

Figure 6.5 shows the solution obtained after 200 iterations of Kaczmarz's algorithm. This solution is extremely similar to the truncated SVD solution, and both solutions are about the same distance from the true model.   ■

ART is a version of Kaczmarz's algorithm that has been modified especially for the tomographic reconstruction problem. In (6.6), the updates to the solution always consist of adding a multiple of a row of **G** to the current solution. The numerator in the fraction

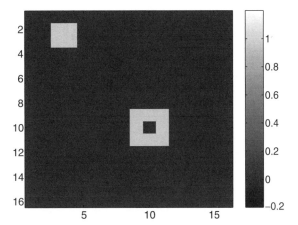

Figure 6.3   True model.

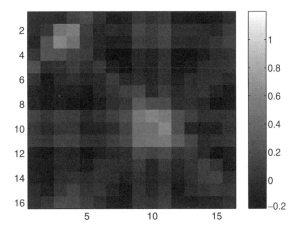

Figure 6.4   Truncated SVD solution.

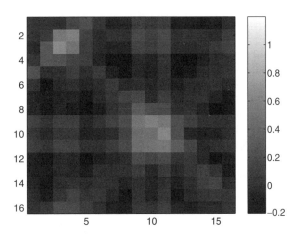

Figure 6.5   Kaczmarz's algorithm solution.

is the difference between the right-hand side of equation $i + 1$ and the value of the $i + 1$ component of $\mathbf{Gm}$. The denominator in (6.6) is simply the square of the norm of $\mathbf{G}_{i+1,\cdot}$. Effectively, Kaczmarz's algorithm is determining the error in equation $i + 1$, and then adjusting the solution by spreading the required correction over the elements of $\mathbf{m}$ which appear in equation $i + 1$.

A crude approximation to the Kaczmarz update, used in ART, is to replace all of the nonzero elements in row $i + 1$ of $\mathbf{G}$ with ones. We define

$$q_{i+1} = \sum_{\text{cell } j \text{ in ray path } i + 1} \mathbf{m}_j \, l \tag{6.8}$$

as an approximation to the travel time along ray path $i + 1$. The difference between q_{i+1} and d_{i+1} is roughly the residual for the predicted travel time of ray $i + 1$.

Examining (6.6) for the ART-modified $\mathbf{G}$, we see that ART simply takes the total error in the travel time for ray $i + 1$ and divides it by the number of cells in ray path $i + 1$, N_{i+1}, and by the cell dimension, l. This correction factor is then multiplied by a vector that has ones in cells along the ray path $i + 1$. This procedure has the effect of smearing the needed correction in travel time equally over all of the cells in ray path $i + 1$.

The ART approximate update formula can thus be written as

$$m_j^{(i+1)} = \begin{cases} m_j^{(i)} - \dfrac{q_{i+1} - d_{i+1}}{lN_{i+1}} & \text{cell } j \text{ in ray path } i + 1 \\ m_j^{(i)} & \text{cell } j \text{ not in ray path } i + 1 \end{cases}. \tag{6.9}$$

The approximation can be improved by taking into account that the ray path lengths actually will vary from cell to cell. If L_{i+1} is the length of ray path $i + 1$, the corresponding improved update formula from (6.6) for the tomography problem is

$$m_j^{(i+1)} = \begin{cases} m_j^{(i)} + \dfrac{d_{i+1}}{L_{i+1}} - \dfrac{q_{i+1}}{lN_{i+1}} & \text{cell } j \text{ in ray path } i + 1 \\ m_j^{(i)} & \text{cell } j \text{ not in ray path } i + 1 \end{cases}. \tag{6.10}$$

■ **Algorithm 6.2 ART** Given a system of equations $\mathbf{Gm} = \mathbf{d}$ arising from a tomography problem.

1. Let $\mathbf{m}^{(0)} = \mathbf{0}$.
2. For $i = 0, 1, \ldots, m$, let N_i be the number of cells touched by ray path i.
3. For $i = 0, 1, \ldots, m$, let L_i be the length of ray path i.
4. For $i = 0, 1, \ldots, m - 1$, $j = 1, 2, \ldots, n$, let

$$m_j^{(i+1)} = \begin{cases} m_j^{(i)} + \dfrac{d_{i+1}}{L_{i+1}} - \dfrac{q_{i+1}}{lN_{i+1}} & \text{cell } j \text{ in ray path } i + 1 \\ m_j^{(i)} & \text{cell } j \text{ not in ray path } i + 1 \end{cases}. \tag{6.11}$$

5. If the solution has not yet converged, let $\mathbf{m}^{(0)} = \mathbf{m}^{(m)}$ and go back to step 4. Otherwise, return the solution $\mathbf{m} = \mathbf{m}^{(m)}$. ∎

The main advantage of ART is that it saves storage. We need only store information about which rays pass through which cells, and we do not need to record the length of each ray in each cell. A second advantage of the method is that it reduces the number of floating-point multiplications required by Kaczmarz's algorithm. Although in current computers floating-point multiplications and additions require roughly the same amount of time, during the 1970s when ART was first developed, multiplication was slower than addition.

One problem with ART is that the resulting tomographic images tend to be noisier than images produced by Kaczmarz's algorithm (6.7). SIRT is a variation on ART which gives slightly better images in practice, at the expense of a slightly slower algorithm. In the SIRT algorithm, all (up to m nonzero) updates using (6.10) are computed for each cell j of the model, for each ray that passes through cell j. The set of updates for cell j are then averaged before updating the model element m_j.

∎ **Algorithm 6.3 SIRT** Given a system of equations $\mathbf{Gm} = \mathbf{d}$ arising from a tomography problem.

1. Let $\mathbf{m} = \mathbf{0}$.
2. For $j = 0, 1, \ldots, n$, let K_j be the number of ray paths that pass through cell j.
3. For $i = 0, 1, \ldots, m$, let L_i be the length of ray path i.
4. For $i = 0, 1, \ldots, m$, let N_i be the number of cells touched by ray path i.
5. Let $\Delta\mathbf{m} = \mathbf{0}$.
6. For $i = 0, 1, \ldots, m - 1$, $j = 1, 2, \ldots, n$, let

$$\Delta m_j = \Delta m_j + \begin{cases} \dfrac{d_{i+1}}{L_{i+1}} - \dfrac{q_{i+1}}{lN_{i+1}} & \text{cell } j \text{ in ray path } i + 1 \\ 0 & \text{cell } j \text{ not in ray path } i + 1 \end{cases}. \quad (6.12)$$

7. For $j = 1, 2, \ldots, n$, let

$$m_j = m_j + \frac{\Delta m_j}{K_j}. \quad (6.13)$$

8. If the solution has not yet converged, go back to step 5. Otherwise, return the current solution. ∎

∎ **Example 6.2** Returning to our earlier tomography example, Figure 6.6 shows the ART solution obtained after 200 iterations. Again, the solution is very similar to the truncated SVD solution.

Figure 6.7 shows the SIRT solution for our example tomography problem after 200 iterations. This solution is similar to the Kaczmarz's and ART solutions. ∎

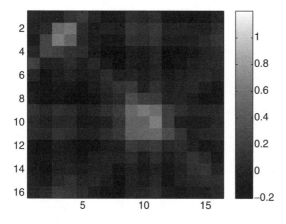

Figure 6.6   ART solution.

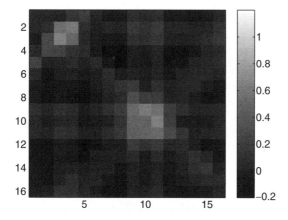

Figure 6.7   SIRT solution.

6.3 THE CONJUGATE GRADIENT METHOD

We next consider the **conjugate gradient** (CG) method for solving a symmetric and positive definite system of equations $\mathbf{Ax} = \mathbf{b}$. We will later apply the CG method to solving the normal equations for $\mathbf{Gm} = \mathbf{d}$. Consider the quadratic optimization problem

$$\min \quad \phi(\mathbf{x}) = \frac{1}{2}\mathbf{x}^T \mathbf{Ax} - \mathbf{b}^T \mathbf{x} \tag{6.14}$$

where $\mathbf{A}$ is an n by n symmetric and positive definite matrix. We require $\mathbf{A}$ be positive definite so that the function $\phi(\mathbf{x})$ will be convex and have a unique minimum. We can calculate $\nabla\phi(\mathbf{x}) = \mathbf{Ax} - \mathbf{b}$ and set it equal to zero to find the minimum. The minimum occurs at a point $\mathbf{x}$ that satisfies the equation

$$\mathbf{Ax} - \mathbf{b} = \mathbf{0} \tag{6.15}$$

or

$$\mathbf{Ax} = \mathbf{b}. \tag{6.16}$$

Thus solving the system of equations $\mathbf{Ax} = \mathbf{b}$ is equivalent to minimizing $\phi(\mathbf{x})$.

The conjugate gradient method approaches the problem of minimizing $\phi(\mathbf{x})$ by constructing a basis for R^n in which the minimization problem is extremely simple. The basis vectors $\mathbf{p}_0, \mathbf{p}_1, \ldots, \mathbf{p}_{n-1}$ are selected so that

$$\mathbf{p}_i^T \mathbf{A} \mathbf{p}_j = 0 \quad \text{when} \quad i \neq j. \tag{6.17}$$

A collection of vectors with this property is said to be **mutually conjugate** with respect to $\mathbf{A}$. We express $\mathbf{x}$ in terms of these basis vectors as

$$\mathbf{x} = \sum_{i=0}^{n-1} \alpha_i \mathbf{p}_i \tag{6.18}$$

so that

$$\phi(\boldsymbol{\alpha}) = \frac{1}{2} \left(\sum_{i=0}^{n-1} \alpha_i \mathbf{p}_i \right)^T \mathbf{A} \left(\sum_{i=0}^{n-1} \alpha_i \mathbf{p}_i \right) - \mathbf{b}^T \left(\sum_{i=0}^{n-1} \alpha_i \mathbf{p}_i \right). \tag{6.19}$$

The product $\mathbf{x}^T \mathbf{Ax}$ can be written as a double sum:

$$\phi(\boldsymbol{\alpha}) = \frac{1}{2} \sum_{i=0}^{n-1} \sum_{j=0}^{n-1} \alpha_i \alpha_j \mathbf{p}_i^T \mathbf{A} \mathbf{p}_j - \mathbf{b}^T \left(\sum_{i=0}^{n-1} \alpha_i \mathbf{p}_i \right). \tag{6.20}$$

Since the vectors are mutually conjugate with respect to $\mathbf{A}$, this simplifies to

$$\phi(\boldsymbol{\alpha}) = \frac{1}{2} \sum_{i=0}^{n-1} \alpha_i^2 \mathbf{p}_i^T \mathbf{A} \mathbf{p}_i - \mathbf{b}^T \left(\sum_{i=0}^{n-1} \alpha_i \mathbf{p}_i \right) \tag{6.21}$$

or

$$\phi(\boldsymbol{\alpha}) = \frac{1}{2} \sum_{i=0}^{n-1} \left(\alpha_i^2 \mathbf{p}_i^T \mathbf{A} \mathbf{p}_i - 2\alpha_i \mathbf{b}^T \mathbf{p}_i \right). \tag{6.22}$$

Equation (6.22) shows that $\phi(\boldsymbol{\alpha})$ consists of n terms, each of which is independent of the other terms. Thus we can minimize $\phi(\boldsymbol{\alpha})$ by selecting each α_i to minimize the ith term,

$$\alpha_i^2 \mathbf{p}_i^T \mathbf{A} \mathbf{p}_i - 2\alpha_i \mathbf{b}^T \mathbf{p}_i. \tag{6.23}$$

Differentiating with respect to α_i and setting the derivative equal to zero, we find that the optimal value for α_i is

$$\alpha_i = \frac{\mathbf{b}^T \mathbf{p}_i}{\mathbf{p}_i^T \mathbf{A} \mathbf{p}_i}. \tag{6.24}$$

This shows that if we have a basis of vectors that are mutually conjugate with respect to $\mathbf{A}$, then minimizing $\phi(\mathbf{x})$ is very easy. We have not yet shown how to construct the mutually conjugate basis vectors.

Our algorithm will actually construct a sequence of solution vectors $\mathbf{x}_i$, residual vectors $\mathbf{r}_i = \mathbf{b} - \mathbf{A}\mathbf{x}_i$, and basis vectors $\mathbf{p}_i$. The algorithm begins with $\mathbf{x}_0 = \mathbf{0}$, $\mathbf{r}_0 = \mathbf{b}$, $\mathbf{p}_0 = \mathbf{r}_0$, and $\alpha_0 = (\mathbf{r}_0^T \mathbf{r}_0)/(\mathbf{p}_0^T \mathbf{A} \mathbf{p}_0)$.

Suppose that at the start of iteration k of the algorithm we have constructed $\mathbf{x}_0, \mathbf{x}_1, \ldots, \mathbf{x}_k$, $\mathbf{r}_0, \mathbf{r}_1, \ldots, \mathbf{r}_k, \mathbf{p}_0, \mathbf{p}_1, \ldots, \mathbf{p}_k$, and $\alpha_0, \alpha_1, \ldots, \alpha_k$. We assume that the first $k+1$ basis vectors $\mathbf{p}_i$ are mutually conjugate with respect to $\mathbf{A}$, that the first $k+1$ residual vectors $\mathbf{r}_i$ are mutually orthogonal, and that $\mathbf{r}_i^T \mathbf{p}_j = 0$ when $i \neq j$.

We let

$$\mathbf{x}_{k+1} = \mathbf{x}_k + \alpha_k \mathbf{p}_k. \tag{6.25}$$

This effectively adds one more term of (6.18) into the solution. Next, we let

$$\mathbf{r}_{k+1} = \mathbf{r}_k - \alpha_k \mathbf{A} \mathbf{p}_k. \tag{6.26}$$

This correctly updates the residual, because

$$\mathbf{r}_{k+1} = \mathbf{b} - \mathbf{A}\mathbf{x}_{k+1} \tag{6.27}$$

$$= \mathbf{b} - \mathbf{A}(\mathbf{x}_k + \alpha_k \mathbf{p}_k) \tag{6.28}$$

$$= (\mathbf{b} - \mathbf{A}\mathbf{x}_k) - \alpha_k \mathbf{A} \mathbf{p}_k \tag{6.29}$$

$$= \mathbf{r}_k - \alpha_k \mathbf{A} \mathbf{p}_k. \tag{6.30}$$

We let

$$\beta_{k+1} = \frac{\mathbf{r}_{k+1}^T \mathbf{r}_{k+1}}{\mathbf{r}_k^T \mathbf{r}_k} \tag{6.31}$$

and

$$\mathbf{p}_{k+1} = \mathbf{r}_{k+1} + \beta_{k+1} \mathbf{p}_k. \tag{6.32}$$

In the following calculations, it will be useful to know that $\mathbf{b}^T \mathbf{p}_k = \mathbf{r}_k^T \mathbf{r}_k$. This is shown by

$$\mathbf{b}^T \mathbf{p}_k = (\mathbf{r}_k + \mathbf{A}\mathbf{x}_k)^T \mathbf{p}_k \tag{6.33}$$

$$= \mathbf{r}_k^T \mathbf{p}_k + \mathbf{p}_k^T \mathbf{A}\mathbf{x}_k \tag{6.34}$$

$$= \mathbf{r}_k^T (\mathbf{r}_k + \beta_k \mathbf{p}_{k-1}) + \mathbf{p}_k^T \mathbf{A}\mathbf{x}_k \tag{6.35}$$

$$= \mathbf{r}_k^T \mathbf{r}_k + \beta_k \mathbf{r}_k^T \mathbf{p}_{k-1} + \mathbf{p}_k^T \mathbf{A} (\alpha_0 \mathbf{p}_0 + \ldots \alpha_{k-1}\mathbf{p}_{k-1}) \tag{6.36}$$

$$= \mathbf{r}_k^T \mathbf{r}_k + 0 + 0 \tag{6.37}$$

$$= \mathbf{r}_k^T \mathbf{r}_k. \tag{6.38}$$

We will now show that $\mathbf{r}_{k+1}$ is orthogonal to $\mathbf{r}_i$ for $i \leq k$. For every $i < k$,

$$\mathbf{r}_{k+1}^T \mathbf{r}_i = (\mathbf{r}_k - \alpha_k \mathbf{A}\mathbf{p}_k)^T \mathbf{r}_i \tag{6.39}$$

$$= \mathbf{r}_k^T \mathbf{r}_i - \alpha_k \mathbf{r}_i^T \mathbf{A}\mathbf{p}_k. \tag{6.40}$$

Since $\mathbf{r}_k$ is orthogonal to all of the earlier $\mathbf{r}_i$ vectors,

$$\mathbf{r}_{k+1}^T \mathbf{r}_i = 0 - \alpha_k \mathbf{p}_k^T \mathbf{A}\mathbf{r}_k. \tag{6.41}$$

Because $\mathbf{A}$ is symmetric, $\mathbf{p}_k^T \mathbf{A}\mathbf{r}_k = \mathbf{r}_k^T \mathbf{A}\mathbf{p}_k$. Also, since $\mathbf{p}_i = \mathbf{r}_i + \beta_i \mathbf{p}_{i-1}$,

$$\mathbf{r}_{k+1}^T \mathbf{r}_i = 0 - \alpha_k (\mathbf{p}_i - \beta_i \mathbf{p}_{i-1})^T \mathbf{A}\mathbf{p}_k. \tag{6.42}$$

Both $\mathbf{p}_i$ and $\mathbf{p}_{i-1}$ are conjugate with $\mathbf{p}_k$. Thus

$$\mathbf{r}_{k+1}^T \mathbf{r}_i = 0. \tag{6.43}$$

We also have to show that $\mathbf{r}_{k+1}^T \mathbf{r}_k = 0$:

$$\mathbf{r}_{k+1}^T \mathbf{r}_k = (\mathbf{r}_k - \alpha_k \mathbf{A}\mathbf{p}_k)^T \mathbf{r}_k \tag{6.44}$$

$$= \mathbf{r}_k^T \mathbf{r}_k - \alpha_k (\mathbf{p}_k - \beta_k \mathbf{p}_{k-1})^T \mathbf{A}\mathbf{p}_k \tag{6.45}$$

$$= \mathbf{r}_k^T \mathbf{r}_k - \alpha_k \mathbf{p}_k^T \mathbf{A}\mathbf{p}_k + \alpha_k \beta_k \mathbf{p}_{k-1}^T \mathbf{A}\mathbf{p}_k \tag{6.46}$$

$$= \mathbf{r}_k^T \mathbf{r}_k - \mathbf{r}_k^T \mathbf{r}_k + \alpha_k \beta_k 0 \tag{6.47}$$

$$= 0. \tag{6.48}$$

Next, we will show that $\mathbf{r}_{k+1}$ is orthogonal to $\mathbf{p}_i$ for $i \leq k$:

$$\mathbf{r}_{k+1}^T \mathbf{p}_i = \mathbf{r}_{k+1}^T (\mathbf{r}_i + \beta_i \mathbf{p}_{i-1}) \tag{6.49}$$

$$= \mathbf{r}_{k+1}^T \mathbf{r}_i + \beta_i \mathbf{r}_{k+1}^T \mathbf{p}_{i-1} \tag{6.50}$$

$$= 0 + \beta_i \mathbf{r}_{k+1}^T \mathbf{p}_{i-1} \tag{6.51}$$

$$= \beta_i (\mathbf{r}_k - \alpha_k \mathbf{A} \mathbf{p}_k)^T \mathbf{p}_{i-1} \tag{6.52}$$

$$= \beta_i \mathbf{r}_k^T \mathbf{p}_{i-1} - \alpha_k \mathbf{p}_{i-1}^T \mathbf{A} \mathbf{p}_k \tag{6.53}$$

$$= 0 - 0 = 0. \tag{6.54}$$

Finally, we need to show that $\mathbf{p}_{k+1}^T \mathbf{A} \mathbf{p}_i = 0$ for $i \leq k$. For $i < k$,

$$\mathbf{p}_{k+1}^T \mathbf{A} \mathbf{p}_i = (\mathbf{r}_{k+1} + \beta_{k+1} \mathbf{p}_k)^T \mathbf{A} \mathbf{p}_i \tag{6.55}$$

$$= \mathbf{r}_{k+1}^T \mathbf{A} \mathbf{p}_i + \beta_{k+1} \mathbf{p}_k^T \mathbf{A} \mathbf{p}_i \tag{6.56}$$

$$= \mathbf{r}_{k+1}^T \mathbf{A} \mathbf{p}_i + 0 \tag{6.57}$$

$$= \mathbf{r}_{k+1}^T \left(\frac{1}{\alpha_i} (\mathbf{r}_i - \mathbf{r}_{i+1}) \right) \tag{6.58}$$

$$= \frac{1}{\alpha_i} \left(\mathbf{r}_{k+1}^T \mathbf{r}_i - \mathbf{r}_{k+1}^T \mathbf{r}_{i+1} \right) \tag{6.59}$$

$$= 0. \tag{6.60}$$

For $i = k$,

$$\mathbf{p}_{k+1}^T \mathbf{A} \mathbf{p}_k = (\mathbf{r}_{k+1} + \beta_{k+1} \mathbf{p}_k)^T \left(\frac{1}{\alpha_k} (\mathbf{r}_k - \mathbf{r}_{k+1}) \right) \tag{6.61}$$

$$= \frac{1}{\alpha_k} \left(\beta_{k+1} (\mathbf{r}_k + \beta_k \mathbf{p}_{k-1})^T \mathbf{r}_k - \mathbf{r}_{k+1}^T \mathbf{r}_{k+1} \right) \tag{6.62}$$

$$= \frac{1}{\alpha_k} \left(\beta_{k+1} \mathbf{r}_k^T \mathbf{r}_k + \beta_{k+1} \beta_k \mathbf{p}_{k-1}^T \mathbf{r}_k - \mathbf{r}_{k+1}^T \mathbf{r}_{k+1} \right) \tag{6.63}$$

$$= \frac{1}{\alpha_k} \left(\mathbf{r}_{k+1}^T \mathbf{r}_{k+1} + \beta_{k+1} \beta_k 0 - \mathbf{r}_{k+1}^T \mathbf{r}_{k+1} \right) \tag{6.64}$$

$$= 0. \tag{6.65}$$

We have now shown that the algorithm generates a sequence of mutually conjugate basis vectors. In theory, the algorithm will find an exact solution to the system of equations in n iterations. In practice, because of round-off errors in the computation, the exact solution may not be obtained in n iterations. In practical implementations of the algorithm, we iterate until the residual is smaller than some tolerance that we specify. The algorithm can be summarized as follows.

■ **Algorithm 6.4 Conjugate Gradient Method** Given a positive definite and symmetric system of equations $\mathbf{A}\mathbf{x} = \mathbf{b}$, and an initial solution $\mathbf{x}_0$, let $\beta_0 = 0$, $\mathbf{p}_{-1} = \mathbf{0}$, $\mathbf{r}_0 = \mathbf{b} - \mathbf{A}\mathbf{x}_0$, and $k = 0$. Repeat the following steps until convergence.

1. If $k > 0$, let $\beta_k = \dfrac{\mathbf{r}_k^T \mathbf{r}_k}{\mathbf{r}_{k-1}^T \mathbf{r}_{k-1}}$.

2. Let $\mathbf{p}_k = \mathbf{r}_k + \beta_k \mathbf{p}_{k-1}$.

3. Let $\alpha_k = \dfrac{\mathbf{r}_k^T \mathbf{r}_k}{\mathbf{p}_k^T \mathbf{A} \mathbf{p}_k}$.

4. Let $\mathbf{x}_{k+1} = \mathbf{x}_k + \alpha_k \mathbf{p}_k$.

5. Let $\mathbf{r}_{k+1} = \mathbf{r}_k - \alpha_k \mathbf{A} \mathbf{p}_k$.

6. Let $k = k + 1$.   ■

A major advantage of the CG method is that it requires storage only for the vectors $\mathbf{x}_k$, $\mathbf{p}_k$, $\mathbf{r}_k$ and the matrix $\mathbf{A}$. If $\mathbf{A}$ is large and sparse, then sparse matrix techniques can be used to store $\mathbf{A}$. Unlike factorization methods such as QR, SVD, or Cholesky factorization, there will be no fill-in of the zero elements in $\mathbf{A}$ at any stage in the solution process. Thus it is possible to solve extremely large systems using CG in cases where direct factorization would require far too much storage. In fact, the only way in which the algorithm uses $\mathbf{A}$ is in multiplications of $\mathbf{A}\mathbf{p}_k$. One such matrix vector multiplication must be performed in each iteration. In some applications of the CG method, it is possible to perform these matrix vector multiplications without explicitly constructing $\mathbf{A}$.

6.4 THE CGLS METHOD

The CG method by itself can only be applied to positive definite systems of equations, and is thus not directly applicable to general least squares problems. In the **conjugate gradient least squares** (CGLS) method, we solve a least squares problem

$$\min \|\mathbf{G}\mathbf{m} - \mathbf{d}\|_2 \tag{6.66}$$

by applying CG to the normal equations

$$\mathbf{G}^T \mathbf{G} \mathbf{m} = \mathbf{G}^T \mathbf{d}. \tag{6.67}$$

In implementing this algorithm it is important to avoid round-off errors. One important source of error is the evaluation of the residual, $\mathbf{G}^T \mathbf{d} - \mathbf{G}^T \mathbf{G} \mathbf{m}$. It turns out that this calculation is more accurate when we factor out $\mathbf{G}^T$ and compute $\mathbf{G}^T(\mathbf{d} - \mathbf{Gm})$. We will use the notation $\mathbf{s}_k = \mathbf{d} - \mathbf{Gm}_k$, and $\mathbf{r}_k = \mathbf{G}^T \mathbf{s}_k$. Note that we can compute $\mathbf{s}_{k+1}$ recursively from $\mathbf{s}_k$ as follows

$$\mathbf{s}_{k+1} = \mathbf{d} - \mathbf{Gm}_{k+1} \tag{6.68}$$

$$= \mathbf{d} - \mathbf{G}(\mathbf{m}_k + \alpha_k \mathbf{p}_k) \tag{6.69}$$

$$= (\mathbf{d} - \mathbf{Gm}_k) - \alpha_k \mathbf{Gp}_k \tag{6.70}$$

$$= \mathbf{s}_k - \alpha_k \mathbf{Gp}_k. \tag{6.71}$$

With this trick, we can now state the CGLS algorithm.

■ **Algorithm 6.5 CGLS** Given a system of equations $\mathbf{Gm} = \mathbf{d}$, let $k = 0$, $\mathbf{m}_0 = \mathbf{0}$, $\mathbf{p}_{-1} = \mathbf{0}$, $\beta_0 = 0$, $\mathbf{s}_0 = \mathbf{d}$, and $\mathbf{r}_0 = \mathbf{G}^T \mathbf{s}_0$. Repeat the following iterations until convergence

1. If $k > 0$, let $\beta_k = \dfrac{\mathbf{r}_k^T \mathbf{r}_k}{\mathbf{r}_{k-1}^T \mathbf{r}_{k-1}}$.

2. Let $\mathbf{p}_k = \mathbf{r}_k + \beta_k \mathbf{p}_{k-1}$.

3. Let $\alpha_k = \dfrac{\mathbf{r}_k^T \mathbf{r}_k}{(\mathbf{Gp}_k)^T (\mathbf{Gp}_k)}$.

4. Let $\mathbf{x}_{k+1} = \mathbf{x}_k + \alpha_k \mathbf{p}_k$.

5. Let $\mathbf{s}_{k+1} = \mathbf{s}_k - \alpha_k \mathbf{Gp}_k$.

6. Let $\mathbf{r}_{k+1} = \mathbf{G}^T \mathbf{s}_{k+1}$.

7. Let $k = k + 1$.   ■

Notice that this algorithm requires only one multiplication of $\mathbf{Gp}_k$ and one multiplication of $\mathbf{G}^T \mathbf{s}_{k+1}$ per iteration. We never explicitly compute $\mathbf{G}^T \mathbf{G}$, which might require considerable time, and which might have far more nonzero elements than $\mathbf{G}$ itself.

The CGLS algorithm has an important property that makes it particularly useful for ill-posed problems. It can be shown that, at least for exact arithmetic, $\|\mathbf{m}_k\|_2$ increases monotonically, and that $\|\mathbf{Gm}_k - \mathbf{d}\|_2$ decreases monotonically [59]. We can use the discrepancy principle together with this property to obtain a regularized solution. Simply stop the CGLS algorithm as soon as $\|\mathbf{Gm}_k - \mathbf{d}\|_2 < \delta$. In practice, this algorithm typically gives good solutions after a very small number of iterations.

An alternative way to use CGLS is to solve the Tikhonov regularization problem (5.4) by applying CGLS to

$$\min \left\| \begin{bmatrix} \mathbf{G} \\ \alpha \mathbf{L} \end{bmatrix} \mathbf{m} - \begin{bmatrix} \mathbf{d} \\ \mathbf{0} \end{bmatrix} \right\|_2^2. \tag{6.72}$$

For nonzero values of the regularization parameter α, this least squares problem should be reasonably well-conditioned. By solving this problem for several values of α, we can compute an L-curve. The disadvantage of this approach is that the number of CGLS iterations for each value of α may be large, and we need to solve the problem for several values of α. Thus the computational effort is far greater.

■ **Example 6.3** A commonly used mathematical model of image blurring involves the two-dimensional convolution of the true image $I_{true}(x, y)$ with a **point spread function**, $\Psi(u, v)$ [14]:

$$I_{blurred}(x, y) = \int_{-\infty}^{\infty} \int_{-\infty}^{\infty} I_{true}(x - u, y - v)\Psi(u, v) \, du \, dv. \tag{6.73}$$

Here the point spread function shows how a point in the true image is altered in the blurred image. A point spread function that is commonly used to represent the blurring that occurs because an image is out of focus is the **Gaussian point spread function**

$$\Psi(u, v) = e^{-\frac{u^2 + v^2}{2\sigma^2}}. \tag{6.74}$$

Here the parameter σ controls the relative width of the point spread function. In practice, the blurred image and point spread function are discretized into pixels. In theory, Ψ is nonzero for all values of u and v. However, it becomes small quickly as u and v increase. If we set small values of Ψ to 0, then the **G** matrix in the discretized problem will be sparse. The Regularization Tools includes a command **blur** that can be used to construct the discretized blurring operator.

Figure 6.8 shows an image that has been blurred and also has a small amount of added noise. This image is of size 200 pixels by 200 pixels, so the **G** matrix for the blurring operator is of size 40,000 by 40,000. Fortunately, the blurring matrix **G** is quite sparse, with less than 0.1% nonzero elements. The sparse matrix requires about 12 megabytes of storage. A dense matrix of this size would require about 13 gigabytes of storage. Using the SVD approach to Tikhonov regularization would require far more storage than most current computers have. However, CGLS works quite well on this problem.

Figure 6.8   Blurred image.

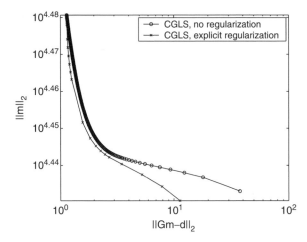

Figure 6.9 L-curves for CGLS deblurring.

Figure 6.10 CGLS solution after 30 iterations, no explicit regularization.

Figure 6.9 shows the L-curve for the solution of this problem by CGLS without explicit regularization and by CGLS with explicit regularization. The line with circles shows the solutions obtained by CGLS without explicit regularization. For the first 30 or so iterations of CGLS without explicit regularization, $\|\mathbf{Gm} - \mathbf{d}\|_2$ decreases quickly. After that point, the improvement in misfit slows down, while $\|\mathbf{m}\|_2$ increases rapidly.

Figure 6.10 shows the CGLS solution without explicit regularization after 30 iterations. The blurring has been greatly improved. Note that 30 iterations is far less than the size of the matrix ($n = 40{,}000$). Unfortunately, further CGLS iterations do not significantly improve the image. In fact, noise builds up rapidly, both because of the accumulation of round-off errors and because the algorithm is converging slowly toward an unregularized least squares solution. Figure 6.11 shows the CGLS solution after 100 iterations. In this image the noise has been greatly amplified, with little or no improvement in the clarity of the image.

Figure 6.11 CGLS solution after 100 iterations, no explicit regularization.

Figure 6.12 CGLS solution, explicit regularization, $\alpha = 7.0 \times 10^{-4}$.

We also computed CGLS solutions with explicit Tikhonov regularization for 22 values of α. For each value of α, 200 iterations of CGLS were performed. The resulting L-curve is shown in Figure 6.9 with "x" markers for each regularized solution that was obtained. This L-curve is slightly better than the L-curve from the CGLS solution without explicit regularization in that the values of $\|\mathbf{m}\|_2$ and $\|\mathbf{Gm} - \mathbf{d}\|_2$ are smaller. However, it required 40 times as much computational effort. The corner solution for $\alpha = 7.0 \times 10^{-4}$ is shown in Figure 6.12. This solution is similar to the solution shown in Figure 6.10. ■

6.5 EXERCISES

6.1 Consider the cross-well tomography problem of Exercise 5.3.

    (a)   Apply Kaczmarz's algorithm to this problem.
    (b)   Apply ART to this problem.
    (c)   Apply SIRT to this problem.
    (d)   Comment on the solutions that you obtained.

6.2 A very simple iterative regularization method is the **Landweber iteration** [59]. The algorithm begins with $\mathbf{m}^0 = \mathbf{0}$, and then follows the iteration

$$\mathbf{m}^{k+1} = \mathbf{m}^k - \omega \mathbf{G}^T (\mathbf{G}\mathbf{m}^k - \mathbf{d}). \tag{6.75}$$

To ensure convergence, the parameter ω must be selected so that $0 < \omega < 2/s_1^2$, where s_1 is the largest singular value of $\mathbf{G}$.

   In practice, the CGLS method generally works better than the Landweber iteration. However, it is easier to analyze the performance of the Landweber iteration. It can be shown that the kth iterate of the Landweber iteration is exactly the same as the SVD solution with filter factors of

$$f_i^k = 1 - (1 - \omega s_i^2)^k. \tag{6.76}$$

   (a) Implement the Landweber iteration and apply it to the Shaw problem from Example 4.3.
   (b) Verify that $\mathbf{m}^{10}$ from the Landweber iteration matches the SVD solution with filter factors given by (6.76).
   (c) Derive (6.76).

6.3 The Regularization Tools command **blur** computes the system matrix for the problem of deblurring an image that has been blurred by a Gaussian point spread function. The file **blur.mat** contains a particular $\mathbf{G}$ matrix and a data vector $\mathbf{d}$.

   (a) How large is the $\mathbf{G}$ matrix? How many nonzero elements does it have? How much storage would be required for the $\mathbf{G}$ matrix if all of its elements were nonzero? How much storage would the SVD of $\mathbf{G}$ require?
   (b) Plot the raw image.
   (c) Using CGLS, deblur the image. Plot your solution.

6.4 Show that if $\mathbf{p}_0, \mathbf{p}_1, \ldots, \mathbf{p}_{n-1}$ are nonzero and mutually conjugate with respect to an n by n symmetric and positive definite matrix $\mathbf{A}$, then the vectors are also linearly independent. Hint: Use the definition of linear independence.

6.6 NOTES AND FURTHER READING

Iterative methods for the solution of linear systems of equations are an important topic in numerical analysis. Some basic references include [6, 82, 136, 145].

   Iterative methods for tomography problems including Kaczmarz's algorithm, ART, and SIRT are discussed in [78, 110, 166]. Parallel algorithms based on ART and SIRT are discussed in [23]. These methods are often referred to as **row action methods** because they access only one row of the matrix at a time. This makes these methods relatively easy to implement in parallel. In practice, the conjugate gradient method generally provides better performance than the row action methods. There are some interesting connections between SIRT and the conjugate gradient method [116, 150, 171].

Hestenes and Stiefel are generally credited with the invention of the conjugate gradient method [66]. However, credit is also due to Lanczos [88]. The history of the conjugate gradient method is discussed in [48, 65].

Shewchuk's technical report [145] provides an introduction to the conjugate gradient method with illustrations that help to make the geometry of the method very clear. Filter factors for the CGLS method similar to those in Exercise 6.2 can be determined. These are derived in [59]. The LSQR method of Paige and Saunders [59, 120, 119] is an alternative way to apply the CG method to the normal equations. The resolution of LSQR solutions is discussed in [12, 13].

Schemes have been developed for using CGLS with explicit regularization and dynamic adjustment of the regularization parameter α [80, 81, 99]. This can potentially remove the computational burden of solving the problem for many values of α. An alternative approach can be used to compute regularized solutions for several values of α at once [43].

The performance of the CG algorithm degrades dramatically on poorly conditioned systems of equations. In such situations a technique called **preconditioning** can be used to improve the performance of CG. Essentially, preconditioning involves a change of variables $\bar{\mathbf{x}} = \mathbf{C}\mathbf{x}$. The matrix $\mathbf{C}$ is selected so that the resulting system of equations will be better conditioned than the original system of equations [31, 49, 167].

The conjugate gradient method can be generalized to nonlinear minimization problems [108, 145]. This approach has been used to find 1-norm solutions to very large linear systems of equations [134].

Inverse problems in image processing are a very active area of research. Some books on inverse problems in imaging include [14, 110].

Several authors have developed specialized algorithms [36, 128, 132, 133, 180] for the solution of quadratic optimization problems of the form

$$\min \quad \frac{1}{2}\mathbf{x}^T\mathbf{A}\mathbf{x} + \mathbf{g}^T\mathbf{x} \tag{6.77}$$
$$\|\mathbf{x}\|_2 \leq \epsilon.$$

These algorithms are used within trust-region methods for nonlinear optimization, but they can also potentially be used to perform Tikhonov regularization by solving (5.3). To date, these methods have not been widely used in practice.

7

ADDITIONAL REGULARIZATION
TECHNIQUES

Synopsis: Three alternatives to Tikhonov regularization are introduced. Bounds constraint methods allow the use of prior knowledge regarding the permissible range of parameter values. Maximum entropy regularization maximizes a weighted entropy-like term that yields a non-negative solution while imposing a lesser penalty on sharply peaked solutions than Tikhonov regularization. Total variation (TV) uses a regularization term based on the 1-norm of the model gradient which does not penalize model discontinuities. A variant of TV allows for prescribing the number of discontinuities in a piecewise-constant solution.

7.1 USING BOUNDS AS CONSTRAINTS

In many physical situations, bounds exist on the maximum and/or minimum values of model parameters. For example, the model parameters may represent a physical quantity such as density that is inherently nonnegative, establishing a strict lower bound for model parameters of 0. The problem of solving for a least squares solution that includes this constraint can be written as

$$\min \quad \|\mathbf{Gm} - \mathbf{d}\|_2$$
$$\mathbf{m} \geq \mathbf{0} \tag{7.1}$$

where $\mathbf{m} \geq \mathbf{0}$ means that every element of the vector $\mathbf{m}$ must be greater than or equal to 0. This **nonnegative least squares** problem can be solved by an algorithm called NNLS that was originally developed by Lawson and Hanson [90]. MATLAB includes a command, **lsqnonneg**, that implements the NNLS algorithm.

We might also declare a strict upper bound, so that model parameters may not exceed some value, for example, a density 3.5 grams per cubic centimeter for crustal rocks in a particular region. Given the upper and lower bound vectors **l** and **u**, we can pose the **bounded variables least squares** (BVLS) problem

$$\min \quad \|\mathbf{Gm} - \mathbf{d}\|_2$$
$$\mathbf{m} \geq \mathbf{l} \tag{7.2}$$
$$\mathbf{m} \leq \mathbf{u}.$$

Stark and Parker developed an algorithm for solving the BVLS problem and implemented their algorithm in Fortran [153]. A similar algorithm is given in the 1995 edition of Lawson and Hanson's book [90]. Given the BVLS algorithm for (7.2), we can perform Tikhonov regularization with bounds.

A related optimization problem involves minimizing or maximizing a linear function of the model subject to bounds constraints and a constraint on the misfit. This problem can be formulated as

$$
\begin{aligned}
\min \quad & \mathbf{c}^T \mathbf{m} \\
& \|\mathbf{G}\mathbf{m} - \mathbf{d}\|_2 \le \delta \\
& \mathbf{m} \ge \mathbf{l} \\
& \mathbf{m} \le \mathbf{u}.
\end{aligned}
\tag{7.3}
$$

This problem can be solved by an algorithm given in Stark and Parker [153]. Solutions to this problem can be used to obtain bounds on the maximum and minimum possible values of model parameters.

■ **Example 7.1** Recall the source history reconstruction problem of Example 3.4, where data are taken in concentration units, u, at spatial positions, x, at a particular time, T. Figure 7.1 shows a hypothetical true source history, and Figure 7.2 shows the corresponding samples as a function of distance, x, at time $T = 300$, with $N(0, 0.001^2)$ noise added.

Figure 7.3 shows the least squares solution, which has the extremely large amplitudes and oscillatory behavior characteristic of an unregularized solution to an ill-posed problem. This solution is, furthermore, physically unrealistic in having negative concentrations. Figure 7.4 shows the nonnegative least squares solution, which, although certainly more realistic in having all of the concentration values nonnegative, does not accurately reconstruct the true source history, and is extremely rough. Suppose that the solubility limit of the contaminant in water is known to be 1.1. This provides a natural upper bound on model

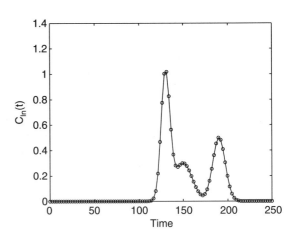

Figure 7.1   True source history.

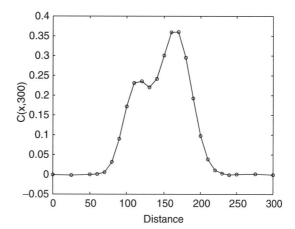

Figure 7.2 Concentration data as a function of position, x, taken at $T = 300$.

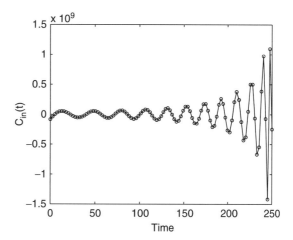

Figure 7.3 Least squares source history solution.

parameters. Figure 7.5 shows the corresponding BVLS solution. Further regularization is required.

Figure 7.6 shows the L-curve for a second-order Tikhonov regularization solution with bounds on the model parameter. Figure 7.7 shows the regularized solution for $\alpha = 0.0616$. This solution correctly shows the two major input concentration peaks. As is typical for cases of nonideal model resolution, the solution peaks are somewhat lower and broader than those of the true model. This solution does not resolve the smaller subsidiary peak near $t = 150$.

We can use (7.3) to establish bounds on the values of the model parameters. For example, we might want to establish bounds on the average concentration from $t = 125$ to $t = 150$. These concentrations appear in positions 51 through 60 of the model vector $\mathbf{m}$.

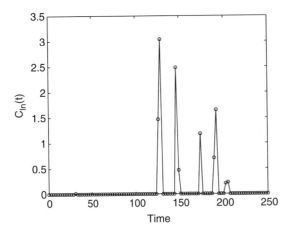

Figure 7.4   NNLS source history solution.

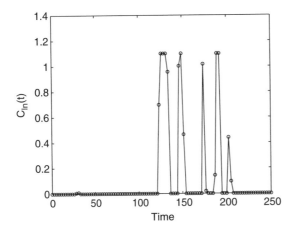

Figure 7.5   BVLS source history solution.

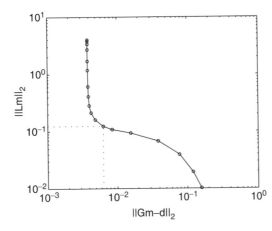

Figure 7.6   L-curve for the second-order Tikhonov solution with bounds, corner at $\alpha = 0.0616$.

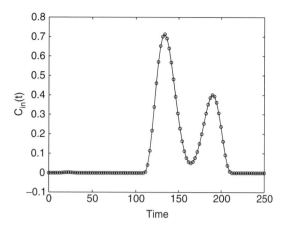

Figure 7.7   Second-order Tikhonov regularization source history solution, $\alpha = 0.0616$.

We let c_i be zero in positions 1 through 50 and 61 through 100, and let c_i be 0.1 in positions 51 through 60. The solution to (7.3) is then a bound on the average concentration from $t = 125$ to $t = 150$. After solving the optimization problem, we obtain a lower bound of 0.36 for the average concentration during this time period. Similarly, by minimizing $-\mathbf{c}^T\mathbf{M}$, we obtain an upper bound of 0.73 for the average concentration during this time period.   ∎

7.2   MAXIMUM ENTROPY REGULARIZATION

In **maximum entropy regularization**, we use a regularization function of the form $\sum_{i=1}^{n} m_i \ln(w_i m_i)$, where the positive weights w_i can be adjusted to favor particular types of solutions. Maximum entropy regularization is only used in problems where the model parameters are restricted to be positive, so logarithms are defined.

The term "maximum entropy" comes from a Bayesian approach to selecting a prior probability distribution, in which we select a discrete probability distribution that maximizes $-\sum_{i=1}^{n} p_i \ln p_i$, subject to the constraint $\sum_{i=1}^{n} p_i = 1$. The quantity $-\sum_{i=1}^{n} p_i \ln p_i$ has the same form as entropy in statistical physics. Because the model $\mathbf{m}$ is not itself a probability distribution, and because (7.4) incorporates the weights w_i, maximum entropy regularization is distinct from the maximum entropy approach to computing probability distributions discussed in Chapter 11.

In the maximum entropy regularization approach, we maximize the entropy of $\mathbf{m}$ subject to a constraint on the size of the misfit $\|\mathbf{Gm} - \mathbf{d}\|_2$:

$$\max \quad -\sum_{i=1}^{n} m_i \ln(w_i m_i)$$
$$\|\mathbf{Gm} - \mathbf{d}\|_2 \leq \delta$$
$$\mathbf{m} \geq \mathbf{0}. \tag{7.4}$$

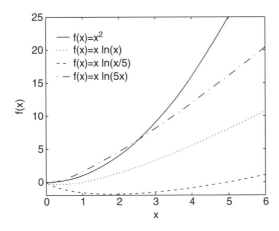

Figure 7.8 The maximum entropy regularization function compared with the 2-norm regularization function.

Using a Lagrange multiplier in the same way that we did with Tikhonov regularization, we can transform this problem into

$$\text{min} \quad \|\mathbf{Gm} - \mathbf{d}\|_2^2 + \alpha^2 \sum_{i=1}^{n} m_i \ln(w_i m_i) \tag{7.5}$$

$$\mathbf{m} \geq \mathbf{0}.$$

It can be shown that as long as $\alpha \geq 0$ the objective function is strictly convex, and thus (7.5) has a unique solution. See Exercise 7.2. However, the optimization problem can become badly conditioned as α approaches 0.

Figure 7.8 shows the maximum entropy regularization function $f(x) = x \ln(wx)$ for three different values of w, along with the Tikhonov regularization function, x^2. The function is zero at $x = 0$, decreases to a minimum, and then increases. For large values of x, x^2 grows faster than $x \ln(wx)$. Maximum entropy regularization thus penalizes solutions with large 2-norms, but not as heavily as zeroth-order Tikhonov regularization. The minimum of $f(x)$ occurs at $x = 1/(ew)$. Maximum entropy regularization thus favors solutions with parameters, m_i, that are close to $1/(ew_i)$, and penalizes parameters with smaller or especially much larger values. The choice of w_i can thus exert significant influence on the solution.

Maximum entropy regularization is widely applied in astronomical image processing [28, 107, 149]. Here the goal is to recover a solution which consists of bright spots (stars, galaxies, or other astronomical objects) on a dark background. The nonnegativity constraints in maximum entropy regularization ensure that the resulting image will not include features with negative intensities. While conventional Tikhonov regularization tends to broaden peaks in the solution, maximum entropy regularization may not penalize sharp peaks as much.

■ **Example 7.2** We will apply maximum entropy regularization to the Shaw problem. Our model, shown in Figure 7.9, consists of two spikes superimposed on a small random background intensity.

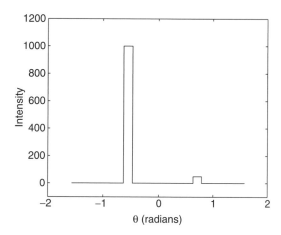

Figure 7.9 True model for maximum entropy regularization.

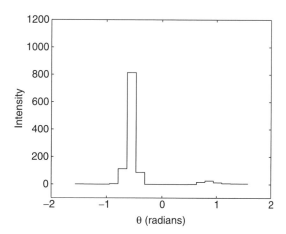

Figure 7.10   Maximum entropy solution.

We will assume that the data errors are independent and normally distributed with mean zero and standard deviation one. Following the discrepancy principle, we will seek a solution with $\|\mathbf{Gm} - \mathbf{d}\|_2$ around 4.4.

For this problem, default weights of $w_i = 1$ are appropriate, since the background noise level of about 0.5 is close to the minimum of the regularization term. We solved (7.5) for several values of the regularization parameter α. At $\alpha = 0.2$, the misfit is $\|\mathbf{Gm} - \mathbf{d}\|_2 = 4.4$. The corresponding solution is shown in Figure 7.10. The spike near $\theta = -0.5$ is visible, but the magnitude of the peak is incorrectly estimated. The second spike near $\theta = 0.7$ is poorly resolved.

For comparison, we applied zeroth-order Tikhonov regularization with a nonnegativity constraint to this problem. Figure 7.11 shows the Tikhonov solution. This solution is similar to the solution produced by maximum entropy regularization. This is consistent with the

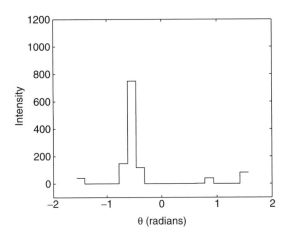

Figure 7.11   Zeroth-order Tikhonov regularization solution with nonnegativity constraints.

results from a number of sample problems in [1]. It was found that maximum entropy regularization was at best comparable to and often inferior to Tikhonov regularization with nonnegativity constraints.   ∎

7.3 TOTAL VARIATION

The total variation (TV) regularization function is appropriate for problems where we expect there to be discontinuous jumps in the model. In the one-dimensional case, the TV regularization function is

$$TV(\mathbf{m}) = \sum_{i=1}^{n-1} |m_{i+1} - m_i| \tag{7.6}$$

$$= \|\mathbf{Lm}\|_1 \tag{7.7}$$

where

$$\mathbf{L} = \begin{bmatrix} -1 & 1 & & & \\ & -1 & 1 & & \\ & & \ddots & & \\ & & & -1 & 1 \\ & & & & -1 & 1 \end{bmatrix}. \tag{7.8}$$

In higher-dimensional problems, $\mathbf{L}$ is a discretization of the gradient operator.

In first- and second-order Tikhonov regularization, discontinuities in the model are smoothed out and do not show up well in the inverse solution. This is because smooth transitions are penalized less by the regularization term than sharp transitions. The particular advantage of TV regularization is that the regularization term does not penalize discontinuous transitions in the model any more than smooth transitions.

This approach has seen wide use in the problem of "denoising" a model [118]. The denoising problem is a linear inverse problem in which $\mathbf{G} = \mathbf{I}$. In denoising, the general goal is to take a noisy data set and remove the noise while still retaining long term trends and even sharp discontinuities in the model.

We could insert the TV regularization term (7.7) in place of $\|\mathbf{Lm}\|_2^2$ in the Tikhonov regularization optimization problem to obtain

$$\min \|\mathbf{Gm} - \mathbf{d}\|_2^2 + \alpha \|\mathbf{Lm}\|_1. \tag{7.9}$$

However this is no longer a least squares problem, and the techniques for solving such problems such as the SVD will no longer be applicable. In fact, (7.9) is a nondifferentiable optimization problem because of the absolute values in $\|\mathbf{Lm}\|_1$.

One simple technique for dealing with this difficulty is to approximate the absolute value with a smooth function that removes the derivative discontinuity, such as

$$|x| \approx \sqrt{x^2 + \beta} \tag{7.10}$$

where β is a small positive parameter.

A simpler option is to switch to the 1-norm in the data misfit term of (7.9) as well, to obtain

$$\min \|\mathbf{Gm} - \mathbf{d}\|_1 + \alpha \|\mathbf{Lm}\|_1, \tag{7.11}$$

which can be rewritten as

$$\min \left\| \begin{bmatrix} \mathbf{G} \\ \alpha\mathbf{L} \end{bmatrix} \mathbf{m} - \begin{bmatrix} \mathbf{d} \\ 0 \end{bmatrix} \right\|_1. \tag{7.12}$$

A solution to this problem can be obtained using the iteratively reweighted least squares (IRLS) algorithm discussed in Chapter 2.

Hansen has suggested yet another approach which retains the 2-norm of the data misfit while incorporating the TV regularization term [59]. In the piecewise polynomial truncated singular value decomposition (PP-TSVD) method, the SVD and the k largest singular values of $\mathbf{G}$ are used to obtain a rank k approximation to $\mathbf{G}$:

$$\mathbf{G}_k = \sum_{i=1}^{k} s_i \mathbf{U}_{\cdot,i} \mathbf{V}_{\cdot,i}^T. \tag{7.13}$$

Note that the matrix $\mathbf{G}_k$ will be rank deficient. The point of the approximation is to obtain a matrix with a well-defined null space. The vectors $\mathbf{V}_{\cdot,k+1}, \ldots, \mathbf{V}_{\cdot,n}$ form a basis for the null space of $\mathbf{G}_k$. We will need this basis later, so let

$$\mathbf{B}_k = \begin{bmatrix} \mathbf{V}_{\cdot,k+1} \ldots \mathbf{V}_{\cdot,n} \end{bmatrix}. \tag{7.14}$$

Using the model basis set $\begin{bmatrix} \mathbf{V}_{\cdot,1} \dots \mathbf{V}_{\cdot,k} \end{bmatrix}$, the minimum length least squares solution is, from the SVD,

$$\mathbf{m}^k = \sum_{i=1}^{k} \frac{\mathbf{U}_{\cdot,i}^T \mathbf{d}}{s_i} \mathbf{V}_{\cdot,i}. \tag{7.15}$$

Adding any vector in the null space of $\mathbf{G}_k$ to this solution will increase the model norm $\|\mathbf{m}\|_2$, but have no effect on $\|\mathbf{G}_k\mathbf{m} - \mathbf{d}\|_2$.

We can use this formulation to find solutions that minimize some regularization function with minimum misfit. For example, in the modified truncated SVD (MTSVD) method, we seek a model that minimizes $\|\mathbf{Lm}\|_2$ among those models that minimize $\|\mathbf{G}_k\mathbf{m} - \mathbf{d}\|_2$. Because all models minimizing $\|\mathbf{G}_k\mathbf{m} - \mathbf{d}\|_2$ can be written as $\mathbf{m} = \mathbf{m}^k - \mathbf{B}_k\mathbf{z}$ for some vector $\mathbf{z}$, the MTSVD problem can be written as

$$\min \quad \|\mathbf{L}(\mathbf{m}^k - \mathbf{B}_k\mathbf{z})\|_2 \tag{7.16}$$

or

$$\min \quad \|\mathbf{L}\mathbf{B}_k\mathbf{z} - \mathbf{L}\mathbf{m}^k\|_2. \tag{7.17}$$

This is a least squares problem that can be solved with the SVD, by QR factorization, or by the normal equations.

The PP-TSVD algorithm uses a similar approach. First, we minimize $\|\mathbf{G}_k\mathbf{m} - \mathbf{d}\|_2$. Let β be the minimum value of $\|\mathbf{G}_k\mathbf{m} - \mathbf{d}\|_2$. Instead of minimizing the 2-norm of $\mathbf{Lm}$, we minimize the 1-norm of $\mathbf{Lm}$, subject to the constraint that $\mathbf{m}$ must be a least squares solution:

$$\min \quad \begin{aligned} &\|\mathbf{Lm}\|_1 \\ \|\mathbf{G}_k\mathbf{m} &- \mathbf{d}\|_2 = \beta. \end{aligned} \tag{7.18}$$

Any least squares solution can be expressed as $\mathbf{m} = \mathbf{m}^k - \mathbf{B}_k\mathbf{z}$, so the PP-TSVD problem can be reformulated as

$$\min \quad \|\mathbf{L}(\mathbf{m}^k - \mathbf{B}_k\mathbf{z})\|_1 \tag{7.19}$$

or

$$\min \|\mathbf{L}\mathbf{m}^k - \mathbf{L}\mathbf{B}_k\mathbf{z}\|_1 \tag{7.20}$$

which is a 1-norm minimization problem that can be solved by IRLS.

Note that $\mathbf{L}\mathbf{B}_k$ in (7.20) has $n - 1$ rows and $n - k$ columns. In general, when we solve this 1-norm minimization problem, we can find a solution for which $n - k$ of the equations are satisfied exactly. Thus, at most, $k - 1$ elements in the vector $\mathbf{L}(\mathbf{m}^k - \mathbf{B}_k\mathbf{z})$ will be nonzero. Since each of these nonzero entries corresponds to a model discontinuity, there will be at most $k - 1$ discontinuities in the solution. Furthermore, the zero elements of $\mathbf{L}(\mathbf{m}^k - \mathbf{B}_k\mathbf{z})$ correspond to points at which the model is constant. Thus, the solution will consist of k constant segments separated by $k-1$ discontinuities, some of which may be small. For example, if we use $k=1$, we will get a flat model with no discontinuities. For $k=2$, we can obtain a model with two flat sections and one discontinuity, and so on.

The PP-TSVD method can also be extended to piecewise linear functions and to piecewise higher-order polynomials by using a matrix **L** which approximates the second- or higher-order derivatives. The MATLAB function **pptsvd**, available from Hansen's web page, implements the PP-TSVD algorithm.

■ **Example 7.3**   In this example we consider the Shaw problem with a true model that consists of a step function. The true model is shown in Figure 7.12.

Figure 7.13 shows the zeroth-order Tikhonov regularization solution and Figure 7.14 shows the second-order Tikhonov regularization solution. Both solutions show the

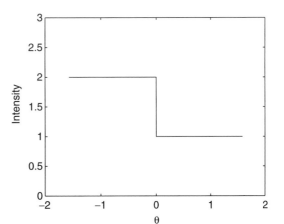

Figure 7.12   The true model.

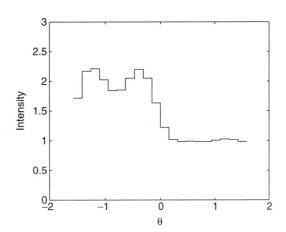

Figure 7.13   Zeroth-order Tikhonov regularization solution.

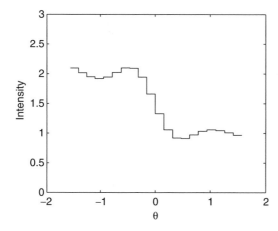

Figure 7.14 Second-order Tikhonov regularization solution.

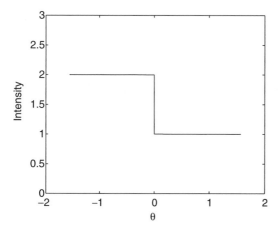

Figure 7.15 TV regularized solution.

discontinuity as a smooth transition because the regularization term penalizes model discontinuities. The relative error ($\|\mathbf{m}_{\text{true}} - \mathbf{m}\|_2 / \|\mathbf{m}_{\text{true}}\|_2$) is about 9% for the zeroth-order solution and about 7% for the second-order solution.

Next, we solved the problem by minimizing the 1-norm misfit with TV regularization. Because independent $N(0, 0.001^2)$ noise was added, we expected the 2-norm of the residual to be about 0.0045 for the 20 data points. Using $\alpha = 1.0$, we obtained a solution with $\|\mathbf{Gm} - \mathbf{d}\|_2 = 0.0039$. This solution is shown in Figure 7.15. The solution is extremely good, with $\|\mathbf{m}_{\text{true}} - \mathbf{m}\|_2 / \|\mathbf{m}_{\text{true}}\|_2 < 0.0005$.

Next, we solved the problem with **pptsvd**. The PP-TSVD solution with $k = 2$ is shown in Figure 7.16. Again, we get a very good solution, with $\|\mathbf{m}_{\text{true}} - \mathbf{m}\|_2 / \|\mathbf{m}_{\text{true}}\|_2 < 0.00007$. Of course, the PP-TSVD method with $k = 2$ is certain to produce a solution with a single

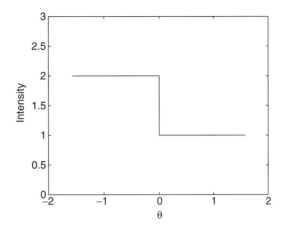

Figure 7.16   PP-TSVD solution, $k = 2$.

discontinuity. It is not clear how we can identify the appropriate number of discontinuities for the solution of a more general inverse problem.   ∎

7.4   EXERCISES

7.1   Using the method of Lagrange multipliers, develop a formula that can be used to solve

$$\min \qquad \mathbf{c}^T \mathbf{m}$$
$$\|\mathbf{Gm} - \mathbf{d}\|_2^2 \leq \delta^2. \tag{7.21}$$

7.2   In this exercise, you will show that (7.5) is a convex minimization problem.

   (a)   Compute the gradient of the function being minimized in (7.5).
   (b)   Compute the Hessian of the function in (7.5).
   (c)   Show that for $\alpha \neq 0$, the Hessian is positive definite for all $\mathbf{m} > 0$. Thus the function being minimized is strictly convex, and there is a unique minimum.

7.3   In applying maximum entropy regularization in image processing, it is common to add a constraint to (7.4) that

$$\sum_{i=1}^{n} m_i = c \tag{7.22}$$

for some constant c. Show how this constraint can be incorporated into (7.5) using a second Lagrange multiplier.

7.4   Returning to the problem in Exercise 4.5, solve for the density profile using total variation regularization. How does your solution compare to the solutions obtained in Exercise 4.5 and Exercise 5.4?

7.5 NOTES AND FURTHER READING

Methods for bounded variables least squares problems and minimizing a linear function subject to a bound on the misfit are given in [153]. Some applications of these techniques can be found in [67, 121, 124, 151, 152].

The maximum entropy method is widely applied in deconvolution and denoising of astronomical images from radio telescope data. Algorithms for maximum entropy regularization are discussed in [28, 107, 138, 149, 160]. More general discussions can be found in [1, 59]. Another widely used deconvolution algorithm in radio astronomy is the "CLEAN" algorithm [24, 68]. Briggs [19] compares the performance of CLEAN, maximum entropy regularization and NNLS. Methods for total variation regularization are discussed in [118, 175]. The PP-TSVD method is discussed in [61, 59].

8

FOURIER TECHNIQUES

Synopsis: The formulation of a general linear forward problem as a convolution is derived. The Fourier transform, Fourier basis functions, and the convolution theorem are introduced for continuous- and discrete-time systems. The inverse problem of deconvolution is explored in the context of the convolution theorem. Water level regularization is used to solve the deconvolution problem.

8.1 LINEAR SYSTEMS IN THE TIME AND FREQUENCY DOMAINS

A remarkable feature of linear time-invariant systems is that the forward problem can generally be described by a **convolution** (1.11),

$$d(t) = \int_{-\infty}^{\infty} m(\tau)g(t - \tau)\, d\tau. \tag{8.1}$$

Inverse problems involving such systems can be solved by **deconvolution**. Here, the independent variable t is time and the data d, model m, and system kernel g are all time functions. However, the results here are equally applicable to spatial problems (e.g., Example 8.1) and are also generalizable to higher dimensions. We will overview the essentials of Fourier theory in the context of performing convolutions and deconvolutions.

Consider a linear time-invariant operator, G, that converts an unknown model, $m(t)$ into an observable data function $d(t)$

$$d(t) = G[m(t)] \tag{8.2}$$

that follows the principles of superposition (1.5)

$$G[m_1(t) + m_2(t)] = G[m_1(t)] + G[m_2(t)] \tag{8.3}$$

and scaling (1.6)

$$G[\alpha m(t)] = \alpha G[m(t)], \tag{8.4}$$

where α is a scalar.

To show that any system satisfying (8.3) and (8.4) can be cast in the form of (8.1), we utilize the **sifting property** of the **impulse** or **delta function,** $\delta(t)$. The delta function can be conceptualized as the limiting case of a pulse as its width goes to zero, its height goes to infinity, and its area stays constant and equal to 1, e.g.,

$$\delta(t) = \lim_{\tau \to 0} \tau^{-1} \Pi(t/\tau) \tag{8.5}$$

where $\tau^{-1}\Pi(t/\tau)$ is a unit-area rectangle function of height τ^{-1} and width τ. The sifting property allows us to extract the value of a function at a particular point from within an integral,

$$\int_a^b f(t)\delta(t - t_0)\, dt = \begin{cases} f(t_0) & a \le t_0 \le b \\ 0 & \text{elsewhere} \end{cases}, \tag{8.6}$$

for any $f(t)$ continuous at finite $t = t_0$. The **impulse response,** or **Green's function** of a system, where the model and data are related by the operator G, is defined as the output produced when the input is a delta function

$$g(t) = G[\delta(t)]. \tag{8.7}$$

Note that any input signal, $m(t)$, can clearly be written as a summation of impulse functions by invoking (8.6):

$$m(t) = \int_{-\infty}^{\infty} m(\tau)\delta(t - \tau)\, d\tau. \tag{8.8}$$

Thus, a general linear system response $d(t)$ to an arbitrary input $m(t)$ can be written as

$$d(t) = G\left[\int_{-\infty}^{\infty} m(\tau)\delta(t - \tau)\, d\tau \right] \tag{8.9}$$

or, from the definition of the integral as a limit of a quadrature sum of Δt-width rectangular areas as Δt goes to zero,

$$d(t) = G\left[\lim_{\Delta\tau \to 0} \sum_{n=-\infty}^{\infty} m(\tau_n)\delta(t - \tau_n)\Delta\tau \right]. \tag{8.10}$$

Because G characterizes a linear process, we can apply (8.3) to move the operator inside of the summation in (8.10). Furthermore using the scaling relation (8.4), we can factor out the $m(\tau_n)$, to obtain

$$d(t) = \lim_{\Delta\tau \to 0} \sum_{n=-\infty}^{\infty} m(\tau_n)G[\delta(t - \tau_n)]\Delta\tau. \tag{8.11}$$

Taking the limit, (8.11) defines the integral

$$d(t) = \int_{-\infty}^{\infty} m(\tau) g(t - \tau) \, d\tau \tag{8.12}$$

which is identical to (8.1), the convolution of $m(t)$ and $g(t)$, often abbreviated as simply $d(t) = m(t) * g(t)$.

The convolution operation generally describes the transformation of models to data in any linear, time-invariant, physical process, including the output of any linear measuring instrument. For example, an unattainable perfect instrument that recorded some $m(t)$ with no distortion whatsoever would have a delta function impulse response, perhaps with a time delay t_0, in which case

$$d(t) = m(t) * \delta(t - t_0) \tag{8.13}$$

$$= \int_{-\infty}^{\infty} m(\tau)\delta(t - t_0 - \tau) \, d\tau \tag{8.14}$$

$$= m(t - t_0). \tag{8.15}$$

An important and useful relationship exists between convolution and the **Fourier transform**,

$$\mathcal{G}(f) = F[g(t)] \tag{8.16}$$

$$= \int_{-\infty}^{\infty} g(t) e^{-i2\pi ft} \, dt, \tag{8.17}$$

and its inverse operation

$$g(t) = F^{-1}[\mathcal{G}(f)] \tag{8.18}$$

$$= \int_{-\infty}^{\infty} \mathcal{G}(f) e^{i2\pi ft} \, df \tag{8.19}$$

where F denotes the Fourier transform operator and F^{-1} denotes the **inverse Fourier transform** operator. The impulse response $g(t)$ is called the **time-domain response**, when the independent variable characterizing the model is time, and its Fourier transform, $\mathcal{G}(f)$, is commonly called the **spectrum** of $g(t)$. $\mathcal{G}(f)$ is also referred to as the **frequency response** or **transfer function** of the system characterized by the impulse response $g(t)$. The Fourier transform (8.17) gives a formula for evaluating the spectrum, and the inverse Fourier transform (8.19) says that the time-domain function $g(t)$ can be exactly reconstructed by a complex weighted integration of functions of the form $e^{i2\pi ft}$, where the weighting is provided by $\mathcal{G}(f)$. The essence of Fourier analysis is the representation and analysis of functions using **Fourier basis functions** of the form $e^{i2\pi ft}$.

It is important to note that, for a real-valued function $g(t)$, the spectrum $\mathcal{G}(f)$ will be complex. $|\mathcal{G}(f)|$ is called the **spectral amplitude**, and the angle that $\mathcal{G}(f)$ makes in the complex plane

$$\theta = \tan^{-1}\left(\frac{\text{imag}(\mathcal{G}(f))}{\text{real}(\mathcal{G}(f))}\right) \tag{8.20}$$

is called the **spectral phase**.

It should be noted that in physics and geophysics applications the sign convention chosen for the complex exponentials in the Fourier transform and its inverse may be reversed, so that the forward transform (8.17) has a plus sign in the exponent and the inverse transform (8.19) has a minus sign in the exponent. This alternative sign convention merely causes complex conjugation in the spectrum that is reversed when the corresponding inverse transform is applied. An additional convention issue arises as to whether to express frequency in Hertz (f) or radians per second ($\omega = 2\pi f$). Alternative Fourier transform formulations using ω differ from (8.17) and (8.19) only by a simple change of variables and introduce scaling factors of 2π in the forward, reverse, or both transforms.

Consider the Fourier transform of the convolution of two functions

$$F[m(t) * g(t)] = \int_{-\infty}^{\infty}\left(\int_{-\infty}^{\infty} m(\tau)g(t-\tau)\,d\tau\right)e^{-\iota 2\pi ft}\,dt. \tag{8.21}$$

Reversing the order of integration and introducing a change of variables, $\xi = t - \tau$, gives

$$F[m(t) * g(t)] = \int_{-\infty}^{\infty} m(\tau)\left(\int_{-\infty}^{\infty} g(t-\tau)e^{-\iota 2\pi ft}dt\right)d\tau \tag{8.22}$$

$$= \int_{-\infty}^{\infty} m(\tau)\left(\int_{-\infty}^{\infty} g(\xi)e^{-\iota 2\pi f(\xi+\tau)}\,d\xi\right)d\tau \tag{8.23}$$

$$= \left(\int_{-\infty}^{\infty} m(\tau)e^{-\iota 2\pi f\tau}\,d\tau\right)\left(\int_{-\infty}^{\infty} g(\xi)e^{-\iota 2\pi f\xi}\,d\xi\right) \tag{8.24}$$

$$= \mathcal{M}(f)\mathcal{G}(f). \tag{8.25}$$

Equation (8.25) is called the **convolution theorem**. The convolution theorem states that convolution of two functions in the time domain has the simple effect of multiplying their Fourier transforms in the frequency domain. The Fourier transform of the impulse response, $\mathcal{G}(f) = F[g(t)]$, thus characterizes how $\mathcal{M}(f)$, the Fourier transform of the model, is altered in spectral amplitude and phase by the convolution.

To understand the implications of the convolution theorem more explicitly, consider the response of a linear system, characterized by an impulse response $g(t)$ in the time domain and the transfer function $\mathcal{G}(f)$ in the frequency domain, to a model Fourier basis function of frequency f_0,

$$m(t) = e^{\iota 2\pi f_0 t}. \tag{8.26}$$

The spectrum of $e^{i2\pi t f_0}$ can be shown to be $\delta(f - f_0)$ by examining the corresponding inverse Fourier transform (8.19) and invoking the sifting property of the delta function (8.6):

$$e^{i2\pi f_0} = \int_{-\infty}^{\infty} \delta(f - f_0) e^{i2\pi ft} \, df. \tag{8.27}$$

The response of a linear system to the basis function model (8.26) is thus, by (8.25),

$$F[e^{i2\pi f_0}]\mathcal{G}(f) = \delta(f - f_0)\mathcal{G}(f_0), \tag{8.28}$$

and the corresponding time-domain response is (8.19)

$$\int_{-\infty}^{\infty} \mathcal{G}(f_0)\delta(f - f_0) e^{i2\pi ft} \, df = \mathcal{G}(f_0) e^{i2\pi f_0 t}. \tag{8.29}$$

Linear time-invariant systems thus map model Fourier basis functions (8.26) to identical data functions and can only alter them in spectral amplitude and phase by the complex scalar $\mathcal{G}(f)$. Of particular interest is the result that model basis function amplitudes at frequencies that are weakly mapped to the data (frequencies where $|\mathcal{G}(f)|$ is small) and/or obscured by noise may be difficult or impossible to recover in an inverse problem.

The transfer function can be expressed in a particularly useful analytical form for the case where we can express the mathematical model in the form of a linear differential equation

$$a_n \frac{d^n y}{dt^n} + a_{n-1} \frac{d^{n-1} y}{dt^{n-1}} + \cdots + a_1 \frac{dy}{dt} + a_0 y = b_m \frac{d^m x}{dt^m} + b_{m-1} \frac{d^{m-1} x}{dt^{m-1}} + \cdots + b_1 \frac{dx}{dt} + b_0 x \tag{8.30}$$

where the a_i and b_i are constant coefficients. Because each term in (8.30) is linear (there are no powers or other nonlinear functions of x, y, or their derivatives), and because differentiation is itself a linear operation, (8.30) expresses a linear time-invariant system obeying superposition (8.3) and scaling (8.4).

If a system of the form expressed by (8.30) operates on a model of the form $m(t) = e^{i2\pi ft}$, (8.29) indicates that the corresponding output will be $d(t) = \mathcal{G}(f)e^{i2\pi ft}$. Inserting $m(t) = y$ and $d(t) = x$ into (8.30), differentiating each term (a time derivative, d/dt, simply generates a multiplier of $2\pi i f$ in this case), dividing the resulting equation on both sides by $e^{i2\pi ft}$, and finally solving for $\mathcal{G}(f)$ gives

$$\mathcal{G}(f) = \frac{\mathcal{D}(f)}{\mathcal{M}(f)} \tag{8.31}$$

$$= \frac{\sum_{j=0}^{m} b_j (2\pi i f)^j}{\sum_{k=0}^{n} a_k (2\pi i f)^k}. \tag{8.32}$$

The **transfer function** (8.32) is a ratio of two complex polynomials in f for any system expressible in the form of (8.30). The $m + 1$ complex frequencies, f_z, where the numerator (and transfer function) of (8.32) are zero are referred to as **zeros** of the transfer function $\mathcal{G}(f)$. The predicted data will be zero for inputs of the form $e^{\iota 2\pi f_z t}$, regardless of their amplitude. Any real-valued zero, f_r, corresponding to a Fourier model basis function, $e^{\iota 2\pi f_r t}$, will thus lie in the model null space and be unrecoverable by any inverse methodology. The $n + 1$ complex frequencies, f_p, for which the denominator of (8.32) is zero are called **poles**. The system will be unstable when excited by model basis functions of the form $e^{\iota 2\pi f_p t}$. Along with a scalar gain factor, the transfer function (and hence the response) of a general linear system can thus be completely characterized by the poles and zeros.

8.2 DECONVOLUTION FROM A FOURIER PERSPECTIVE

In recovering a model, $m(t)$, that has been convolved with some $g(t)$ using Fourier theory, we wish to recover the spectral amplitude and phase (the Fourier components) of $m(t)$ by reversing the changes in spectral amplitude and phase (8.25) caused by the convolution. As noted earlier, this may be difficult or impossible at and near frequencies where the spectral amplitude of the transfer function, $|\mathcal{G}(f)|$ is small (i.e., close in frequency to the zeros of $\mathcal{G}(f)$).

■ **Example 8.1** An illustrative and important physical example of an ill-posed inverse problem is the **downward continuation** of a vertical field. In downward continuation we wish to estimate a field at a surface using observations made at a height h above it. The corresponding forward problem is **upward continuation**, depicted in Figure 8.1.

   To understand the upward continuation problem, consider a point mass, M, located at the origin, that has a total gravitational field, expressed as a function of position, $\mathbf{r} = x\hat{x} + y\hat{y} + z\hat{z}$,

$$\mathbf{g}_t(\mathbf{r}) = (g_x, g_y, g_z) \tag{8.33}$$

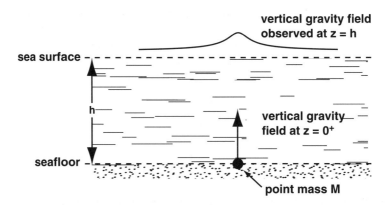

Figure 8.1 Upward continuation of a vertical gravitational field spatial impulse.

$$= \frac{-M\gamma \hat{r}}{\|\mathbf{r}\|_2^2} \tag{8.34}$$

$$= \frac{-M\gamma (x\hat{x} + y\hat{y} + z\hat{z})}{\|\mathbf{r}\|_2^3} \tag{8.35}$$

where γ is Newton's gravitational constant. The vertical component of (8.35) is

$$g_z = \hat{z} \cdot \mathbf{g}_t \tag{8.36}$$

$$= \frac{-M\gamma z}{\|\mathbf{r}\|_2^3}. \tag{8.37}$$

Integrating (8.37) across a horizontal plane at some height z, using the polar substitution of variables $\rho^2 = x^2 + y^2$, gives

$$\int_{-\infty}^{\infty} \int_{-\infty}^{\infty} g_z \, dx \, dy = -M\gamma z \int_0^{\infty} \int_0^{\infty} \frac{dx \, dy}{\|\mathbf{r}\|_2^3} \tag{8.38}$$

$$= -2\pi M\gamma z \int_0^{\infty} \frac{\rho \, d\rho}{(z^2 + \rho^2)^{3/2}} \tag{8.39}$$

$$= -2\pi M\gamma z \left(\frac{-1}{(z^2 + \rho^2)^{1/2}} \right) \Big|_0^{\infty} \tag{8.40}$$

$$= -2\pi M\gamma, \tag{8.41}$$

which is, remarkably, independent of the height, z, where the field is integrated.

The vertical field at an infinitesimal distance $z = 0^+$ above a point mass located at the origin will have zero vertical component unless we are directly above the origin. For $z = 0^+$ the vertical field is thus given by a delta function at the origin with a magnitude given by (8.41):

$$g_z|_{z=0^+} = -2\pi M\gamma \delta(x, y). \tag{8.42}$$

Now consider $z = 0$ to be the ocean floor (or the surface of the Earth) and suppose that gravity data are collected from a ship (or aircraft, or spacecraft) at a height h, where there are no appreciable mass variations contributing to field complexity on $z \geq 0$. If we normalize the response of (8.35) by the magnitude of the delta function at $z = 0^+$ (8.41), the vertical field observed at $z = h$ will be

$$g_z(h) = \frac{h}{2\pi (x^2 + y^2 + h^2)^{3/2}}. \tag{8.43}$$

Because the vertical gravitational field obeys superposition and linearity, upward continuation can be recognized as a linear problem where (8.43) is the impulse response for observations on a plane at height $z = h$. Vertical field measurements obtained at $z = h$ can consequently be expressed by a two-dimensional convolution operation in the (x, y) plane:

$$g_z(h, x, y) = \int_{-\infty}^{\infty} \int_{-\infty}^{\infty} g_z(0, \xi_1 - x, \xi_2 - y) \cdot \frac{h}{2\pi(\xi_1^2 + \xi_2^2 + h^2)^{3/2}} \, d\xi_1 \, d\xi_2. \qquad (8.44)$$

We can examine the effects of upward continuation from a Fourier perspective by evaluating the transfer function corresponding to (8.43) to see how sinusoidally varying field components at $z = 0^+$ are scaled when observed at $z = h$. Note that in characterizing the spatial variation of a field, the model and data Fourier basis functions are of the form $e^{i2\pi k_x x}$ and $e^{i2\pi k_y y}$, where $k_{x,y}$ are **spatial frequencies** analogous to the temporal frequency f encountered in time-domain problems. The transfer function is the Fourier transform of (8.43):

$$\mathcal{G}(k_x, k_y) = \int_{-\infty}^{\infty} \int_{-\infty}^{\infty} \frac{h \cdot e^{-i2\pi k_x x} e^{-i2\pi k_y y} \, dx \, dy}{2\pi(x^2 + y^2 + h^2)^{3/2}} \qquad (8.45)$$

$$= e^{-2\pi h(k_x^2 + k_y^2)^{1/2}}. \qquad (8.46)$$

The form of the upward continuation filter (8.46) shows that higher frequency (larger k) spatial basis functions representing components of the vertical field at $z = 0^+$ have their amplitudes attenuated exponentially to greater degrees. The field thus becomes smoother as it is observed at progressively greater heights. An operation that preferentially attenuates high frequencies relative to low ones is generally referred to as a **low-pass filter**, and upward continuation is thus a smoothing operator when applied to the field in the (x, y) plane.

In solving the inverse problem of downward continuation by inferring the field at $z = 0^+$ using measurements made at $z = h$, we seek an inverse filter to reverse the smoothing effect of upward continuation. An obvious candidate is the reciprocal of (8.46)

$$\left[\mathcal{G}(k_x, k_y)\right]^{-1} = e^{2\pi h(k_x^2 + k_y^2)^{1/2}}. \qquad (8.47)$$

Equation (8.47) is the transfer function of the deconvolution that undoes the convolution encountered in the forward problem (8.44). However, (8.47) grows without bound as spatial frequencies k_x and k_y become large. Thus, small components of field data acquired at high spatial frequencies in observations at $(h \gg 1/k)$ could lead to enormously amplified estimated amplitudes for those components in the field at $z = 0^+$. Downward continuation is thus a roughening operator when applied to the field in (x, y) space. Because of the exponential character of (8.47), if the data are noisy, the inverse problem will be severely ill-posed. ∎

The ill-posed character of downward continuation seen in Example 8.1 at high spatial frequencies is typical of forward problems characterized by smooth kernel functions and is recognizable as a manifestation of the Riemann–Lebesgue lemma. This instability means that solutions obtained using Fourier methodologies will frequently require regularization to produce stable and meaningful models.

8.3 LINEAR SYSTEMS IN DISCRETE TIME

We can readily approximate the continuous time transforms (8.17) and (8.19) in discretized characterizations of physical problems by using the **discrete Fourier transform**, or **DFT**. The DFT operates on a uniformly spaced (e.g., space or time) sequence. For example the DFT might operate on a vector **m** consisting of uniformly spaced samples of a continuous function $m(t)$. The frequency, f_s, at which the sampling occurs is called the **sampling rate**. The forward discrete Fourier transform is

$$\mathcal{M}_k = \{\text{DFT}[\mathbf{m}]\}_k \tag{8.48}$$

$$= \sum_{j=0}^{n-1} m_j e^{-\imath 2\pi jk/n} \tag{8.49}$$

and its inverse is

$$m_j = \left\{\text{DFT}^{-1}[\mathcal{M}]\right\}_j \tag{8.50}$$

$$= \frac{1}{n} \sum_{k=0}^{n-1} \mathcal{M}_k e^{\imath 2\pi jk/n}. \tag{8.51}$$

Equations (8.49) and (8.51) use the common convention that the indices range from 0 to $n-1$, and implement the same exponential sign convention as (8.17) and (8.19). Equation (8.51) states that a sequence m_j can be expressed as a linear combination of the n discrete basis functions $e^{\imath 2\pi jk/n}$, where the complex coefficients are the discrete spectral values $\mathcal{M}_k$. The DFT operations (8.49) and (8.51) are also widely referred to as the **FFT** and **IFFT** because a particularly efficient algorithm, the **fast Fourier transform**, is widely used to evaluate (8.49) and (8.51). They can be calculated in MATLAB using the **fft** and **ifft** commands. In using MATLAB routines to calculate DFTs, be sure to remember that MATLAB indexing begins at $k = 1$, rather than at $k = 0$.

DFT spectra, $\mathcal{M}_k$, are complex, discrete, and periodic, where the period is n. There is also an implicit assumption that the associated sequence, m_j, also has period n. Because of these periodicities, DFT results can be stored in complex vectors of length n without loss of information, although the DFT definitions (8.49) and (8.51) are valid for any integer index k. For a real-valued sequence, (8.49) exhibits **Hermitian symmetry** about $k = 0$ and $k = n/2$ where $\mathcal{M}_k = \mathcal{M}^*_{n-k}$. See Exercise 8.2. For n even, the positive frequencies, lf_s/n, where $l = 1, \ldots, n/2 - 1$, correspond to indices $k = 1, \ldots, n/2 - 1$, and the negative frequencies, $-lf_s/n$, correspond to indices $k = n/2 + 1, \ldots, n - 1$. The frequencies $\pm f_s/2$ have identical DFT values and correspond to index $k = n/2$. For n odd, there is no integer k corresponding to exactly half of the sampling rate. In this case positive frequencies correspond to indices 1 through $(n - 1)/2$ and negative frequencies correspond to indices $(n + 1)/2$ through $n - 1$. The $k = 0$ (zero frequency) term is the sum of the sequence elements m_j, or n times the average sequence value. Figure 8.2 displays the frequency mapping with respect to k for an $n = 16$-length DFT.

DFT Frequency Mapping (n = 16, Sampling Rate = f$_s$, Real Time Series)

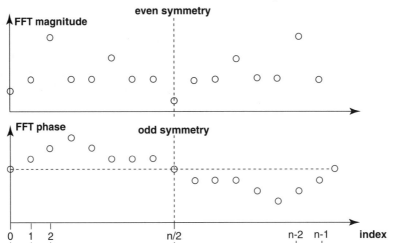

Figure 8.2 Frequency and index mapping describing the DFT of a real-valued sequence ($n = 16$) sampled at sampling rate f_s. For the DFT to adequately represent the spectrum of an assumed periodic sequence, f_s must be greater than or equal to the Nyquist frequency (8.52).

The Hermitian symmetry of the DFT means that, for a real-valued sequence, the spectral amplitude, $|\mathcal{M}|$, is symmetric and the spectral phase is antisymmetric with respect to $k = 0$ and $k = n/2$. See Figure 8.2. For this reason it is customary to only plot the positive frequency spectral amplitude and phase in depicting the spectrum of a real signal.

For a uniformly sampled sequence to accurately represent a continuous function containing appreciable spectral energy up to some maximum frequency f_{max}, the function must be sampled at a frequency at least as large as the **Nyquist frequency**

$$f_N = 2f_{max}. \tag{8.52}$$

Should (8.52) not be met, a nonlinear distortion called **aliasing** will occur. Generally speaking, aliasing causes spectral energy at frequencies $f > f_s/2$ to be folded and superimposed onto the DFT spectrum within the frequency range $-f_s/2 \leq f \leq f_s/2$. Aliasing is irreversible, in the sense that the original continuous time function cannot generally be reconstructed from its sampled time series via an inverse Fourier transform.

The discrete convolution of two sequences with equal sampling rates $f_s = 1/\Delta t$ can be performed in two ways. The first of these is **serial**,

$$d_j = \sum_{i=0}^{n-1} m_i g_{j-i} \Delta t \tag{8.53}$$

where it is assumed that the shorter of the two sequences is zero for all indices greater than m. In serial convolution the result has at most $n + m - 1$ nonzero terms. The second type of discrete convolution is **circular**. Circular convolution is applicable when the two series are of equal length. If the lengths differ, they may be equalized by padding the shorter of the two with zeros. The result of a circular convolution is as if each sequence had been joined to its tail and convolved in circular fashion. Equivalently, the same result can be obtained by making the sequences periodic, with period n, and calculating the convolution sums over a single period. A circular convolution results from the application of the discrete convolution theorem

$$d_i = \text{DFT}^{-1}[\text{DFT}[\mathbf{m}] \cdot \text{DFT}[\mathbf{g}]]_i \, \Delta t \qquad (8.54)$$

$$= \text{DFT}^{-1}[\mathcal{M} \cdot \mathcal{G}]_i \, \Delta t \qquad (8.55)$$

where $\mathcal{M}_i \cdot \mathcal{G}_i$ indicates element-by-element multiplication with no summation, *not* the dot product. To avoid **wrap-around** effects that may arise because of the assumed n-length periodicity of $\mathbf{m}$ and $\mathbf{g}$, and thus obtain a result that is indistinguishable from the serial convolution (8.53), it may be necessary to pad both series with up to n zeros and to apply (8.55) on sequences of length up to $2n$. Because of the factoring strategy used in the FFT algorithm, it is also desirable from the standpoint of computational efficiency to pad $\mathbf{m}$ and $\mathbf{g}$ to lengths that are powers of two.

Consider the case where we have a theoretically known, or accurately estimated, system impulse response, $g(t)$, convolved with an unknown model, $m(t)$. We note in passing that, although we will examine a one-dimensional deconvolution problem for simplicity, these results are generalizable to higher dimensions. The forward problem is

$$d(t) = \int_a^b g(t - \tau)m(\tau) \, d\tau. \qquad (8.56)$$

Uniformly discretizing this expression using simple collocation with a sampling interval, $\Delta t = 1/f_s$, that is short enough to avoid aliasing (8.52) gives

$$\mathbf{d} = \mathbf{Gm} \qquad (8.57)$$

where $\mathbf{d}$ and $\mathbf{m}$ are appropriate length sequences [sampled approximations of $d(t)$ and $m(t)$], and $\mathbf{G}$ is a matrix with rows that are padded, time-reversed, sampled impulse response vectors, so that

$$G_{i,j} = g(t_i - \tau_j)\Delta t. \qquad (8.58)$$

This time-domain representation of the forward problem was previously examined in Example 4.2.

An inverse solution using Fourier methodology can be obtained by first padding $\mathbf{d}$ and $\mathbf{g}$ appropriately with zeros so that they are of equal and sufficient length to render moot potential wrap-around artifacts associated with circular convolution. Applying the DFT and (8.25) allows us to cast the forward problem as a complex-valued linear system

$$\mathcal{D} = \mathbf{G}\mathcal{M}. \tag{8.59}$$

$\mathbf{G}$ in (8.59) is a diagonal matrix with

$$G_{i,i} = \mathcal{G}_i \tag{8.60}$$

where $\mathcal{G}$ is the discrete Fourier transform of the sampled impulse response, $\mathbf{g}$, $\mathcal{D}$ is the discrete Fourier transform of the data vector, $\mathbf{d}$, and $\mathcal{M}$ is the discrete Fourier transform of the model vector, $\mathbf{m}$.

Equation (8.59) suggests a straightforward solution by **spectral division**

$$\mathbf{m} = \text{DFT}^{-1}[\mathcal{M}] = \text{DFT}^{-1}\left[\mathbf{G}^{-1} \cdot \mathcal{D}\right]. \tag{8.61}$$

Equation (8.61) is appealing in its simplicity and efficiency. The application of (8.25), combined with the efficient FFT implementation of the DFT, reduces the necessary computational effort from solving a potentially very large linear system of time-domain equations (8.57) to just three n-length DFT operations and n complex divisions. If $\mathbf{d}$ and $\mathbf{g}$ are real, packing/unpacking algorithms exist that allow the DFT operations to be further reduced to operations involving complex vectors of just length $n/2$.

However, (8.61) does not avoid the instability potentially associated with deconvolution because the reciprocals of any very small diagonal elements (8.60) will become huge in the diagonal of $\mathbf{G}^{-1}$. Equation (8.61) will thus frequently require regularization to be useful.

8.4 WATER LEVEL REGULARIZATION

A straightforward and widely applied method of regularizing spectral division is **water level regularization**. The water level strategy employs a modified $\mathbf{G}$ matrix, $\mathbf{G}_w$, in the spectral division, where

$$G_{w,i,i} = \begin{cases} G_{i,i} & (|G_{i,i}| > w) \\ wG_{i,i}/|G_{i,i}| & (0 < |G_{i,i}| \leq w). \\ w & (G_{i,i} = 0) \end{cases} \tag{8.62}$$

The water level regularized model estimate is then

$$\mathbf{m}_w = \text{DFT}^{-1}\left(\mathbf{G}_w^{-1}\mathcal{D}\right). \tag{8.63}$$

The colorful name for this technique arises from the analogy of pouring water into the holes of the spectrum of $\mathbf{g}$ until the spectral amplitude levels there reach w. The effect in (8.63) is to prevent enormous amplifications, and attendant instability, from occurring at frequencies where the spectral amplitudes of the system transfer function are small.

An optimal water level value w will reduce the sensitivity to noise in the inverse solution while still recovering important model features. As is typical of the regularization process, it is possible to choose a "best" solution by assessing the trade-off between the norm of the residuals (fitting the data) and the model norm (smoothness of the model) as the regularization parameter w is varied. A useful property in evaluating data misfit and model length for calculations in the frequency domain is that the 2-norm of the Fourier transform vector (defined for complex vectors as the square root of the sum of the squared complex element amplitudes) is proportional to the 2-norm of the time-domain vector, where the constant of proportionality depends on DFT conventions. One can thus easily evaluate 2-norm trade-off metrics in the frequency domain without calculating inverse Fourier transforms. The 2-norm of the water level regularized solution, $\mathbf{m}_w$, will thus decrease monotonically as w increases because $|G_{w,i,i}| \geq |G_{i,i}|$.

■ **Example 8.2** In Example 4.2, we investigated time-domain seismometer deconvolution for uniformly sampled data with a sampling rate of $f_s = 2$ Hz using the truncated SVD. Here, we solve this problem using frequency-domain deconvolution regularized via the water level technique. The impulse response, true model, and noisy data for this example are plotted in Figures 4.9, 4.11, and 4.12, respectively. We first pad the 210-point data and impulse response vectors with 210 additional zeros to eliminate wrap-around artifacts, and apply the fast Fourier transform to both vectors to obtain corresponding discrete spectra. The spectral amplitudes of the impulse response, data, and noise are critical in assessing the stability of the spectral division solution. See Figure 8.3. The frequencies range from 0 to $f_s/2 = 1$ Hz. Because spectral amplitudes for real-valued sequences are symmetric about $k = 0$ and $k = n/2$ (Figure 8.2), only positive frequencies are shown.

Examining the impulse response spectral amplitude, $|\mathcal{G}_k|$, in Figure 8.3, we note that it decreases by approximately three orders of magnitude between very low frequencies and half of the sampling frequency ($f_s/2 = 1$ Hz). The convolution theorem (8.25) shows that

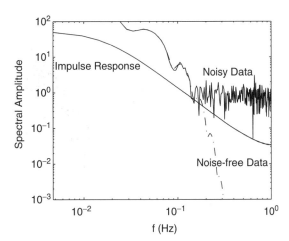

Figure 8.3 Impulse response, noise-free data, and noisy data spectral amplitudes for the seismometer deconvolution problem plotted as a function of frequency.

the forward problem convolution multiplies the spectrum of the model by $\mathcal{G}(f)$ in mapping it to the data. Thus, the convolution of a general signal with broad frequency content with this impulse response will strongly attenuate higher frequencies. Figure 8.3 also shows that the spectral amplitudes of the noise-free data fall off more quickly than the impulse response. This indicates that spectral division will be a stable process for noise-free data in this problem. Figure 8.3 also shows that the spectral amplitudes of the noisy data dominate the signal at frequencies higher than $f \approx 0.1$ Hz. Because of the small values of $\mathcal{G}_k$ at these frequencies, the spectral division solution using the noisy data will be dominated by noise (as was the case in the time-domain solution of Example 4.2; see Figure 4.14). Figure 8.4 shows the amplitude spectrum resulting from spectral division using the noisy data. The resulting spectrum, the Fourier transform amplitude of Figure 4.14, is dominated by noise at frequencies above about 0.1 Hz.

To regularize the spectral division solution, an optimal water level is sought. Because w has units of spectral amplitude, Figure 8.3 shows that the optimal value of w to deconvolve the portion of the data spectrum that is unobscured by noise (while suppressing the amplification of higher frequency noise) is of order 1. However, such a determination might be more difficult for real data with a more complex spectrum, or where the distinction between signal and noise is unclear. An adaptive way to select w is to examine the L-curve constructed using a range of water level values. Figure 8.5 shows the L-curve for this example, which suggests an optimal w close to 3. Figure 8.6 shows a corresponding range of solutions, and Figure 8.7 shows the solution for $w = 3.16$.

The solution shown in Figure 8.7, chosen from the corner of the trade-off curve of Figure 8.5, shows the familiar features of resolution reduction typical in regularized solutions. In this case, the reduction in resolution caused by regularization is manifested by reduced amplitude, oscillatory side lobes, and model spreading into adjacent elements relative to the true model.   ■

A significant new idea introduced by the Fourier methodology is that it provides a set of model and data basis functions of the form of (8.26), the complex exponentials, that have

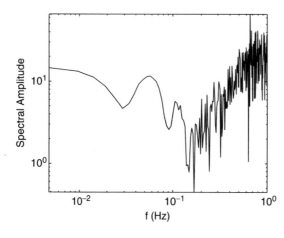

Figure 8.4 Spectral amplitudes resulting from the Fourier transform of the noisy data divided by the Fourier transform of the impulse response (the transfer function).

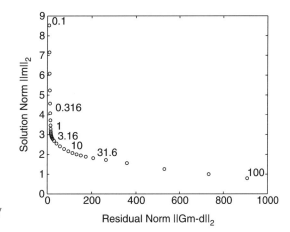

Figure 8.5   L-curve for a logarithmically distributed range of water level values.

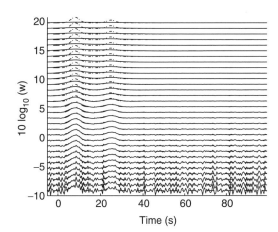

Figure 8.6   Models corresponding to the range of water level values used to construct Figure 8.5. Dashed curves show the true model.

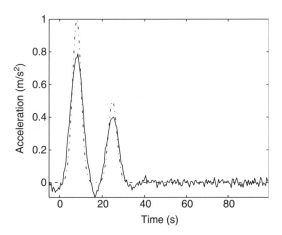

Figure 8.7   Model corresponding to $w = 3.16$. Dashed curve shows the true model.

the property of passing through a linear system altered in phase and amplitude, but not in functional character (8.29). This remarkable fact is the essence of the convolution theorem (8.25). The spectrum of the impulse response (such as in Figures 8.3 and 8.4) can thus be used to understand what frequency components may exhibit instability in an inverse solution. The information contained in the spectrum of Figure 8.3 is thus analogous to that obtained with a Picard plot in the context of the SVD in Chapter 5. The Fourier perspective also provides a link between linear inverse theory and the (vast) field of linear filtering. The deconvolution problem in this context is identical to finding an optimal inverse filter to recover the model while suppressing the influence of noise. Because of the FFT algorithm and the convolution theorem, Fourier deconvolution methods can also be spectacularly computationally efficient relative to time-domain methods. This efficiency can become critically important when larger and/or higher-dimensional models are of interest, a large number of deconvolutions must be performed, or computational speed is critical, such as in real-time applications.

8.5 EXERCISES

8.1   Given that the Fourier transform of a real-valued linear system, $g(t)$,

$$F[g(t)] = \mathcal{G}(f) = \text{real}(\mathcal{G}(f)) + \text{imag}(\mathcal{G}(f)) = \alpha(f) + \iota\beta(f), \qquad (8.64)$$

is Hermitian

$$\mathcal{G}(f) = \mathcal{G}^*(-f), \qquad (8.65)$$

show that convolving $g(t)$ with $\sin(2\pi f_0 t)$ and $\cos(2\pi f_0 t)$ produces the scaled and phase-shifted sinusoids

$$g(t) * \sin(2\pi f_0 t) = |\mathcal{G}(f_0)| \cdot \sin(2\pi f_0 t + \theta(f_0)) \qquad (8.66)$$

$$g(t) * \cos(2\pi f_0 t) = |\mathcal{G}(f_0)| \cdot \cos(2\pi f_0 t + \theta(f_0)) \qquad (8.67)$$

where the scale factor is the spectral amplitude

$$|\mathcal{G}(f_0)| = (\alpha^2(f_0) + \beta^2(f_0))^{1/2} \qquad (8.68)$$

and the phase-shift factor is the spectral phase

$$\theta(f_0) = \tan^{-1}\left(\frac{\beta(f_0)}{\alpha(f_0)}\right). \qquad (8.69)$$

8.2   (a)   Demonstrate using (8.49) that the DFT of an n-point, real-valued sequence, $\mathbf{x}$, is Hermitian, i.e.,

$$\mathcal{X}_{n-k} = \mathcal{X}_k^*. \qquad (8.70)$$

(b)   Demonstrate that the Hermitian symmetry shown in part (a) implies that the N independent elements in a time series $\mathbf{x}$ produce $N/2 + 1$ independent elements (N even) or $(N-1)/2 + 1$ independent elements (N odd) in the DFT $\mathcal{X}$. As the DFT has an inverse (8.51) that reproduces $\mathbf{x}$ from $\mathcal{X}$, clearly information has not been lost in taking the DFT, yet the number of independent elements in $\mathbf{x}$ and $\mathcal{X}$ differ. Explain this.

8.3   A linear damped vertical harmonic oscillator consisting of a mass suspended on a lossy spring is affixed to the surface of a terrestrial planet to function as a seismometer, where the recorded downward displacement $z(t)$ of the mass relative to its equilibrium position will depend on ground motion. For an upward ground displacement, $u(t)$, the system can be mathematically modeled in the form of (8.30) as

$$\frac{d^2 z}{dt^2} + \frac{D}{M} \frac{dz}{dt} + \frac{K}{M} z = \frac{d^2 u}{dt^2} \tag{8.71}$$

where the physical properties of the oscillator are defined by the mass M, the displacement-proportional spring force constant K, and the velocity-proportional damping force constant D.

(a)   Using $u(t) = e^{i 2\pi f t}$ and $z(t) = \mathcal{G}(f) e^{i 2\pi f t}$, obtain the transfer function $\mathcal{G}(f) = \mathcal{Z}(f)/\mathcal{U}(f)$, where $\mathcal{Z}(f)$ and $\mathcal{U}(f)$ are the Fourier transforms of $z(t)$ and $u(t)$, respectively.

In terms of M, K, and D:

(b)   For what general frequency range of ground motion will the response of this instrument be difficult to remove via a deconvolution?

(c)   For what general frequency range of ground motion will the output of this instrument be nearly identical to the true ground motion?

8.4   Consider a regularized deconvolution as the solution to

$$\mathbf{g} * \mathbf{m} = \mathbf{d} \tag{8.72}$$

subject to

$$w \mathbf{L} \mathbf{m} = \mathbf{0} \tag{8.73}$$

where $\mathbf{g}$ is an impulse response, $\mathbf{L}$ is an n by n roughening matrix, w is a water level parameter, and all vectors are time series of length n with unit sampling intervals.

(a)   Show, by taking the Fourier transforms of (8.72) and (8.73) and summing constraints, that a water level regularization-like solution incorporating the diagonal matrix

$$G_{w,i,i} = G_{i,i} + w \tag{8.74}$$

can be obtained when $\mathbf{L} = \mathbf{I}$.

(b) Show that, for **L** corresponding to a pth-order roughening matrix [e.g., (5.27) and (5.29)], higher-order water level solutions may be obtained where

$$G_{w,i,i} = G_{i,i} + (2\pi \iota f_i)^p w \qquad (8.75)$$

and f_i is the frequency associated with the ith element of the n-length discrete spectrum.

Hint: Apply the convolution theorem and note that the Fourier transform of $dg(t)/dt$ is $2\pi \iota f$ times the Fourier transform of $g(t)$.

8.5   A displacement seismogram is observed from a large earthquake at a far-field seismic station, from which the source region can be approximated as a point. A much smaller aftershock from the main shock region is used as an empirical Green's function for this event. It is supposed that the observed signal from the large event should be approximately equal to the convolution of the main shock's rupture history with this empirical Green's function. The 256-point seismogram is in the file **seis.mat**. The impulse response of the seismometer is in the file **impresp.mat**.

(a) Deconvolve the impulse response from the observed main shock seismogram using water level regularized deconvolution to solve for the source time function of the large earthquake. Note that the source time function is expected to consist of a nonnegative pulse or set of pulses. Estimate the source duration in samples and assess any evidence for subevents and their relative durations and amplitudes. Approximately what water level do you believe is best for this data set? Why?

(b) Perform the $p = 2$ order water level deconvolution of these data. See Exercise 8.3.

(c) Recast the problem as a discrete linear inverse problem, as described in the example for Chapter 4, and solve the system using second-order Tikhonov regularization.

(d) Are the results in (c) better or worse than in (a) or (b)? How and why? Compare the amount of time necessary to find the solution in each case on your computing platform.

8.6 NOTES AND FURTHER READING

Gubbins [54] also explores connections between Fourier and inverse theory in a geophysical context. Kak and Slaney [78] give an extensive treatment of Fourier-based methods for tomographic imaging. Vogel [175] discusses Fourier methods for image deblurring. Because of the tremendous utility of Fourier techniques, there are numerous resources on their use in the physical sciences, engineering, and pure mathematics. A basic text covering theory and some applications at the approximate level of this text is [18], and a recommended advanced text on the topic is [127].

9

NONLINEAR REGRESSION

Synopsis: Common approaches to solving nonlinear regression problems are introduced, extending the development of linear regression in Chapter 2. We begin with a discussion of Newton's method, which provides a general framework for solving nonlinear systems of equations and nonlinear optimization problems. Then we discuss the Gauss–Newton (GN) and Levenberg–Marquardt (LM) methods, which are versions of Newton's method specialized for nonlinear regression problems. The distinction between LM and Tikhonov regularization is also made. Statistical aspects and implementation issues are addressed, and examples of nonlinear regression are presented.

9.1 NEWTON'S METHOD

Consider a nonlinear system of m equations in m unknowns

$$\mathbf{F}(\mathbf{x}) = \mathbf{0}. \tag{9.1}$$

We will construct a sequence of vectors, $\mathbf{x}^0, \mathbf{x}^1, \ldots,$ that will converge to a solution $\mathbf{x}^*$. If $\mathbf{F}$ is continuously differentiable, we can construct a Taylor series approximation about $\mathbf{x}^0$

$$\mathbf{F}(\mathbf{x}^0 + \Delta\mathbf{x}) \approx \mathbf{F}(\mathbf{x}^0) + \nabla\mathbf{F}(\mathbf{x}^0)\Delta\mathbf{x} \tag{9.2}$$

where $\nabla\mathbf{F}(\mathbf{x}^0)$ is the Jacobian

$$\nabla\mathbf{F}(\mathbf{x}^0) = \begin{bmatrix} \dfrac{\partial F_1(\mathbf{x}^0)}{\partial x_1} & \cdots & \dfrac{\partial F_1(\mathbf{x}^0)}{\partial x_m} \\ \vdots & \ddots & \vdots \\ \dfrac{\partial F_m(\mathbf{x}^0)}{\partial x_1} & \cdots & \dfrac{\partial F_m(\mathbf{x}^0)}{\partial x_m} \end{bmatrix}. \tag{9.3}$$

Using (9.2) we can obtain an approximate equation for the difference between $\mathbf{x}^0$ and the unknown $\mathbf{x}^*$:

$$\mathbf{F}(\mathbf{x}^*) = \mathbf{0} \approx \mathbf{F}(\mathbf{x}^0) + \nabla \mathbf{F}(\mathbf{x}^0)\Delta\mathbf{x}. \tag{9.4}$$

Solving (9.4) for the difference, $\Delta\mathbf{x} = \mathbf{x}^* - \mathbf{x}^0$, between $\mathbf{x}^*$ and $\mathbf{x}^0$ gives

$$\nabla \mathbf{F}(\mathbf{x}^0)\Delta\mathbf{x} \approx -\mathbf{F}(\mathbf{x}^0) \tag{9.5}$$

which leads to **Newton's method**.

■ **Algorithm 9.1 Newton's Method** Given a system of equations $\mathbf{F}(\mathbf{x}) = \mathbf{0}$ and an initial solution $\mathbf{x}^0$, repeat the following steps to compute a sequence of solutions $\mathbf{x}^1, \mathbf{x}^2, \ldots$. Stop if and when the sequence converges to a solution with $\mathbf{F}(\mathbf{x}) = \mathbf{0}$.

1. Use Gaussian elimination to solve

$$\nabla \mathbf{F}(\mathbf{x}^k)\Delta\mathbf{x} = -\mathbf{F}(\mathbf{x}^k) \tag{9.6}$$

2. Let $\mathbf{x}^{k+1} = \mathbf{x}^k + \Delta\mathbf{x}$.
3. Let $k = k + 1$.   ■

The theoretical properties of Newton's method are summarized in the following theorem. For a proof, see [32].

■ **Theorem 9.1** If $\mathbf{x}^0$ is close enough to $\mathbf{x}^*$, $\mathbf{F}(\mathbf{x})$ is continuously differentiable in a neighborhood of $\mathbf{x}^*$, and $\nabla \mathbf{F}(\mathbf{x}^*)$ is nonsingular, then Newton's method will converge to $\mathbf{x}^*$. The convergence rate is quadratic in the sense that there is a constant c such that for large k,

$$\|\mathbf{x}^{k+1} - \mathbf{x}^*\|_2 \le c\|\mathbf{x}^k - \mathbf{x}^*\|_2^2. \tag{9.7}$$

■

In practical terms, quadratic convergence means that as we approach $\mathbf{x}^*$, the number of accurate digits in the solution doubles at each iteration. Unfortunately, if the hypotheses in the foregoing theorem are not satisfied, then Newton's method can converge very slowly or even fail altogether.

A simple modification to the basic Newton's method often helps with convergence problems. In the **damped Newton's method**, we use the Newton's method equations at each iteration to compute a direction in which to move. However, instead of simply taking the full step $\mathbf{x}^i + \Delta\mathbf{x}$, we search along the line between $\mathbf{x}^i$ and $\mathbf{x}^i + \Delta\mathbf{x}$ to find the point that minimizes $\|\mathbf{F}(\mathbf{x}^i + \alpha\Delta\mathbf{x})\|_2$, and take the step that minimizes the norm.

Now suppose that we wish to minimize a scalar-valued function $f(\mathbf{x})$. If we assume that $f(\mathbf{x})$ is twice continuously differentiable, we have a Taylor series approximation

$$f(\mathbf{x}^0 + \Delta\mathbf{x}) \approx f(\mathbf{x}^0) + \nabla f(\mathbf{x}^0)^T \Delta\mathbf{x} + \frac{1}{2}\Delta\mathbf{x}^T \nabla^2 f(\mathbf{x}^0)\Delta\mathbf{x} \tag{9.8}$$

where $\nabla f(\mathbf{x}^0)$ is the gradient

$$\nabla f(\mathbf{x}^0) = \begin{bmatrix} \dfrac{\partial f(\mathbf{x}^0)}{\partial x_1} \\ \vdots \\ \dfrac{\partial f(\mathbf{x}^0)}{\partial x_m} \end{bmatrix} \tag{9.9}$$

and $\nabla^2 f(\mathbf{x}^0)$ is the Hessian

$$\nabla^2 f(\mathbf{x}^0) = \begin{bmatrix} \dfrac{\partial^2 f(\mathbf{x}^0)}{\partial x_1^2} & \cdots & \dfrac{\partial^2 f(\mathbf{x}^0)}{\partial x_1 \partial x_m} \\ \vdots & \ddots & \vdots \\ \dfrac{\partial^2 f(\mathbf{x}^0)}{\partial x_m \partial x_1} & \cdots & \dfrac{\partial^2 f(\mathbf{x}^0)}{\partial x_m^2} \end{bmatrix}. \tag{9.10}$$

Note that we use ∇^2 to denote the Hessian here, *not* the Laplacian operator.

A necessary condition for $\mathbf{x}^*$ to be a minimum of $f(\mathbf{x})$ is that $\nabla f(\mathbf{x}^*) = \mathbf{0}$. We can approximate the gradient in the vicinity of x^0 by

$$\nabla f(\mathbf{x}^0 + \Delta \mathbf{x}) \approx \nabla f(\mathbf{x}^0) + \nabla^2 f(\mathbf{x}^0)\Delta \mathbf{x}. \tag{9.11}$$

Setting the approximate gradient (9.11) equal to zero gives

$$\nabla^2 f(\mathbf{x}^0)\Delta \mathbf{x} = -\nabla f(\mathbf{x}^0). \tag{9.12}$$

Solving (9.12) for successive solution steps leads to **Newton's method for minimizing** $f(\mathbf{x})$.

■ **Algorithm 9.2 Newton's Method for Minimizing** $f(\mathbf{x})$ Given a twice continuously differentiable function $f(\mathbf{x})$, and an initial solution $\mathbf{x}^0$, repeat the following steps to compute a sequence of solutions $\mathbf{x}^1, \mathbf{x}^2, \ldots$. Stop if and when the sequence converges to a solution with $\nabla f(\mathbf{x}) = \mathbf{0}$.

1. Solve $\nabla^2 f(\mathbf{x}^k)\Delta \mathbf{x} = -\nabla f(\mathbf{x}^k)$.
2. Let $\mathbf{x}^{k+1} = \mathbf{x}^k + \Delta \mathbf{x}$.
3. Let $k = k + 1$. ■

The theoretical properties of Newton's method for minimizing $f(\mathbf{x})$ are summarized in the following theorem. Since Newton's method for minimizing $f(\mathbf{x})$ is exactly Newton's method for solving a nonlinear system of equations applied to $\nabla f(\mathbf{x}) = \mathbf{0}$, the proof follows immediately from the proof of Theorem 9.1.

■ **Theorem 9.2** If $f(\mathbf{x})$ is twice continuously differentiable in a neighborhood of a local minimizer $\mathbf{x}^*$, and there is a constant λ such that $\|\nabla^2 f(\mathbf{x}) - \nabla^2 f(\mathbf{y})\|_2 \leq \lambda \|\mathbf{x} - \mathbf{y}\|_2$ for every vector $\mathbf{y}$ in the neighborhood, and $\nabla^2 f(\mathbf{x}^*)$ is positive definite, and $\mathbf{x}^0$ is close enough to $\mathbf{x}^*$, then Newton's method will converge quadratically to $\mathbf{x}^*$.   ■

Newton's method for minimizing $f(\mathbf{x})$ is very efficient when it works, but the method can also fail to converge. As with Newton's method for systems of equations, the convergence properties of the algorithm can be improved in practice by using a line search.

9.2 THE GAUSS–NEWTON AND LEVENBERG–MARQUARDT METHODS

Newton's method for systems of equations is not directly applicable to most nonlinear regression and inverse problems. This is because we may not have equal numbers of data points and model parameters and there may not be an exact solution to $\mathbf{G}(\mathbf{m}) = \mathbf{d}$. Instead, we will use Newton's method to minimize a nonlinear least squares problem.

Specifically, we consider the problem of fitting a vector of n parameters to a data vector $\mathbf{d}$. A vector of standard deviations σ for the measurements is also given. The parameters and data are related through a nonlinear system of equations $\mathbf{G}(\mathbf{m}) = \mathbf{d}$. Our goal is to find values of the parameters that best fit the data in the sense of minimizing the 2-norm of the residuals.

As with linear regression, if we assume that the measurement errors are normally distributed, then the maximum likelihood principle leads us to minimizing the sum of squared errors normalized by their respective standard deviations (2.13). We seek to minimize

$$f(\mathbf{m}) = \sum_{i=1}^{m} \left(\frac{G(\mathbf{m})_i - d_i}{\sigma_i} \right)^2 . \tag{9.13}$$

For convenience, we will let

$$f_i(\mathbf{m}) = \frac{G(\mathbf{m})_i - d_i}{\sigma_i} \qquad i = 1, 2, \ldots, m \tag{9.14}$$

and

$$\mathbf{F}(\mathbf{m}) = \begin{bmatrix} f_1(\mathbf{m}) \\ \vdots \\ f_m(\mathbf{m}) \end{bmatrix} . \tag{9.15}$$

Thus

$$f(\mathbf{m}) = \sum_{i=1}^{m} f_i(\mathbf{m})^2 . \tag{9.16}$$

The gradient of $f(\mathbf{m})$ can be written as the sum of the gradients of the individual terms:

$$\nabla f(\mathbf{m}) = \sum_{i=1}^{m} \nabla\left(f_i(\mathbf{m})^2\right). \tag{9.17}$$

The elements of the gradient are

$$\nabla f(\mathbf{m})_j = \sum_{i=1}^{m} 2\nabla f_i(\mathbf{m})_j \mathbf{F}(\mathbf{m})_j. \tag{9.18}$$

The gradient of $f(\mathbf{m})$ can be written in matrix notation as

$$\nabla f(\mathbf{m}) = 2\mathbf{J}(\mathbf{m})^T \mathbf{F}(\mathbf{m}) \tag{9.19}$$

where $\mathbf{J}(\mathbf{m})$ is the Jacobian

$$\mathbf{J}(\mathbf{m}) = \begin{bmatrix} \dfrac{\partial f_1(\mathbf{m})}{\partial m_1} & \cdots & \dfrac{\partial f_1(\mathbf{m})}{\partial m_n} \\ \vdots & \ddots & \vdots \\ \dfrac{\partial f_m(\mathbf{m})}{\partial m_1} & \cdots & \dfrac{\partial f_m(\mathbf{m})}{\partial m_n} \end{bmatrix}. \tag{9.20}$$

Similarly, we can express the Hessian of $f(\mathbf{m})$ using the $f_i(\mathbf{m})$ terms to obtain

$$\nabla^2 f(\mathbf{m}) = \sum_{i=1}^{m} \nabla^2\left(f_i(\mathbf{m})^2\right) \tag{9.21}$$

$$= \sum_{i=1}^{m} \mathbf{H}^i(\mathbf{m}) \tag{9.22}$$

where $\mathbf{H}^i(\mathbf{m})$ is the Hessian of $f_i(\mathbf{m})^2$. The j, k element of $\mathbf{H}^i(\mathbf{m})$ is

$$H^i_{j,k}(\mathbf{m}) = \frac{\partial^2(f_i(\mathbf{m})^2)}{\partial m_j \partial m_k} \tag{9.23}$$

$$= \frac{\partial}{\partial m_j}\left(2f_i(\mathbf{m})\frac{\partial f_i(\mathbf{m})}{\partial m_k}\right) \tag{9.24}$$

$$= 2\left(\frac{\partial f_i(\mathbf{m})}{\partial m_j}\frac{\partial f_i(\mathbf{m})}{\partial m_k} + f_i(\mathbf{m})\frac{\partial^2 f_i(\mathbf{m})}{\partial m_j \partial m_k}\right). \tag{9.25}$$

Thus

$$\nabla^2 f(\mathbf{m}) = 2\mathbf{J}(\mathbf{m})^T \mathbf{J}(\mathbf{m}) + \mathbf{Q}(\mathbf{m}) \tag{9.26}$$

where

$$\mathbf{Q}(\mathbf{m}) = 2 \sum_{i=1}^{m} f_i(\mathbf{m}) \nabla^2 f_i(\mathbf{m}). \tag{9.27}$$

In the **Gauss–Newton (GN) method**, we simply ignore the $\mathbf{Q}(\mathbf{m})$ term and approximate the Hessian by the first term of (9.26):

$$\nabla^2 f(\mathbf{m}) \approx 2\mathbf{J}(\mathbf{m})^T \mathbf{J}(\mathbf{m}). \tag{9.28}$$

In the context of nonlinear regression, we expect that the $f_i(\mathbf{m})$ terms will be reasonably small as we approach the optimal parameters $\mathbf{m}^*$, so that this is a reasonable approximation. This is not a reasonable approximation for nonlinear least squares problems in which the values of $f_i(\mathbf{m})$ are large.

Using (9.28) and dividing both sides by 2, the equations for successive iterations in the GN method become

$$\mathbf{J}(\mathbf{m}^k)^T \mathbf{J}(\mathbf{m}^k) \Delta \mathbf{m} = -\mathbf{J}(\mathbf{m}^k)^T \mathbf{F}(\mathbf{m}^k). \tag{9.29}$$

The left-hand side matrix in this system of equations is symmetric and positive semidefinite. If the matrix is actually positive definite then we can use the Cholesky factorization to solve the system of equations.

Although the GN method often works well in practice, it is based on Newton's method, and can thus fail for all of the same reasons. Furthermore, the method can fail when the matrix $\mathbf{J}(\mathbf{m}^k)^T \mathbf{J}(\mathbf{m}^k)$ is singular.

In the **Levenberg–Marquardt (LM) method**, the GN method equations (9.29) are modified to

$$(\mathbf{J}(\mathbf{m}^k)^T \mathbf{J}(\mathbf{m}^k) + \lambda \mathbf{I}) \Delta \mathbf{m} = -\mathbf{J}(\mathbf{m}^k)^T \mathbf{F}(\mathbf{m}^k). \tag{9.30}$$

Here the positive parameter λ is adjusted during the course of the algorithm to ensure convergence. One important reason for using a positive value of λ is that the $\lambda \mathbf{I}$ term ensures that the matrix is nonsingular. Since the matrix in this system of equations is symmetric and positive definite, we can use the Cholesky factorization to solve the system.

For very large values of λ,

$$\mathbf{J}(\mathbf{m}^k)^T \mathbf{J}(\mathbf{m}^k) + \lambda \mathbf{I} \approx \lambda \mathbf{I} \tag{9.31}$$

and

$$\Delta \mathbf{m} \approx -\frac{1}{\lambda} \nabla f(\mathbf{m}). \tag{9.32}$$

This is a **steepest-descent** step, meaning the algorithm simply moves down-gradient to most rapidly reduce $f(\mathbf{m})$. The steepest-descent approach provides very slow but certain convergence. Conversely, for very small values of λ, the LM method reverts to the GN method (9.29), which gives potentially fast but uncertain convergence.

One challenge associated with the LM method is determining the optimal value of λ. The general strategy is to use small values of λ in situations where the GN method is working well, but to switch to larger values of λ when the GN method fails to make progress. A simple approach is to start with a small value of λ, and then adjust it in every iteration. If the LM method leads to a reduction in $f(\mathbf{m})$, then take this step and decrease λ by a constant multiplicative factor (say 2) before the next iteration. Conversely, if the LM method does not lead to a reduction in $f(\mathbf{m})$, then do not take the step, but instead increase λ by a constant factor (say 2) and try again, repeating this process until a step is found which actually does decrease the value of $f(\mathbf{m})$. Robust implementations of the LM method use more sophisticated strategies for adjusting λ, but even this simple strategy works surprisingly well.

In practice, a careful LM implementation offers the good performance of the GN method as well as very good convergence properties. LM is the method of choice for small to medium sized nonlinear least squares problems.

The $\lambda\mathbf{I}$ term in the LM method (9.30) looks a lot like Tikhonov regularization, for example, (5.7). It is important to understand that this is not actually a case of Tikhonov regularization. The $\lambda\mathbf{I}$ term is used to stabilize the solution of the linear system of equations which determines the search direction to be used. Because the $\lambda\mathbf{I}$ term is only used as a way to improve the convergence of the algorithm and does *not* enter into the objective function that is being minimized, it does not regularize the nonlinear least squares problem. We discuss the regularization of nonlinear problems in Chapter 10.

9.3 STATISTICAL ASPECTS

Recall from Appendix B that if a vector $\mathbf{d}$ has a multivariate normal distribution, and $\mathbf{A}$ is an appropriately sized matrix, then $\mathbf{Ad}$ also has a multivariate normal distribution with an associated covariance matrix

$$\text{Cov}(\mathbf{Ad}) = \mathbf{A}\text{Cov}(\mathbf{d})\mathbf{A}^T. \tag{9.33}$$

We applied this formula to the least squares problem for $\mathbf{Gm} = \mathbf{d}$, which we solved by the normal equations. The resulting formula for $\text{Cov}(\mathbf{m})$ was

$$\text{Cov}(\mathbf{m}_{L_2}) = (\mathbf{G}^T\mathbf{G})^{-1}\mathbf{G}^T\text{Cov}(\mathbf{d})\mathbf{G}(\mathbf{G}^T\mathbf{G})^{-1}. \tag{9.34}$$

In the simplest case, where $\text{Cov}(\mathbf{d}) = \sigma^2\mathbf{I}$, (9.34) simplifies to

$$\text{Cov}(\mathbf{m}_{L_2}) = \sigma^2(\mathbf{G}^T\mathbf{G})^{-1}. \tag{9.35}$$

For the nonlinear regression problem we no longer have a linear relationship between the data and the estimated model parameters, so we cannot assume that the estimated model parameters have a multivariate normal distribution and cannot use the preceding formulas.

Since we are interested in how small data perturbations result in small model perturbations, we can consider a linearization of the misfit function $\mathbf{F}(\mathbf{m})$,

$$\mathbf{F}(\mathbf{m}^* + \Delta\mathbf{m}) \approx \mathbf{F}(\mathbf{m}^*) + \mathbf{J}(\mathbf{m}^*)\Delta\mathbf{m}. \tag{9.36}$$

Under this approximation, there is a linear relationship between changes in $\mathbf{F}$ and changes in the parameters $\mathbf{m}$:

$$\Delta\mathbf{F} \approx \mathbf{J}(\mathbf{m}^*)\Delta\mathbf{m}. \tag{9.37}$$

In the sense that the Hessian can be approximated by (9.28), $\mathbf{J}(\mathbf{m}^*)$ takes the place of $\mathbf{G}$ in an approximate estimate of the covariance of the model parameters. Because we have incorporated the σ_i into the formula for $f(\mathbf{m})$, $\text{Cov}(\mathbf{d})$ is the identity matrix, so

$$\text{Cov}(\mathbf{m}^*) \approx (\mathbf{J}(\mathbf{m}^*)^T\mathbf{J}(\mathbf{m}^*))^{-1}. \tag{9.38}$$

Unlike linear regression, the parameter covariance matrix in nonlinear regression is not exact. In nonlinear regression, the covariance matrix and confidence intervals depend critically on the accuracy of the linearization (9.37). If this linearization is not accurate over the range of likely models, then the resulting confidence intervals will not be correct. The Monte Carlo approach discussed in Section 2.5 provides a robust but computationally intensive alternative method for estimating $\text{Cov}(\mathbf{m}^*)$.

As in linear regression, we can perform a χ^2 test of goodness of fit (see Section 2.2). The appropriateness of this test also depends on how well the nonlinear model is approximated by the Jacobian linearization for points near the optimal parameter values. In practice, this approximation is typically adequate unless the data are extremely noisy.

As with linear regression, it is possible to apply nonlinear regression when the measurement errors are independent and normally distributed and the standard deviations are unknown but assumed to be equal. See Section 2.3. We set the σ_i to 1 and minimize the sum of squared errors. If we define a residual vector

$$r_i = G(\mathbf{m}^*)_i - d_i \qquad i = 1, 2, \ldots, m, \tag{9.39}$$

our estimate of the measurement standard deviation is

$$s = \sqrt{\frac{\sum_{i=1}^{m} r_i^2}{m - n}} \tag{9.40}$$

and the approximate covariance matrix for the estimated model parameters is

$$\text{Cov}(\mathbf{m}^*) = s^2(\mathbf{J}(\mathbf{m}^*)^T\mathbf{J}(\mathbf{m}^*))^{-1}. \tag{9.41}$$

Once we have $\mathbf{m}^*$ and $\text{Cov}(\mathbf{m}^*)$, we can establish confidence intervals for the model parameters exactly as we did in Chapter 2. Just as with linear regression, it is also important to examine the residuals for systematic patterns or deviations from normality. If we have not estimated the measurement standard deviation s, then it is also important to test the χ^2 value for goodness of fit.

■ **Example 9.1** A classic method in hydrology for determining the transmissivity and storage coefficient of an aquifer is called the "slug test" [42]. A known volume Q of water (the slug) is injected into a well, and the resulting effects on the head (water table elevation), h, at an observation well a distance d away from the injection well are observed at various times t. The head measured at the observation well typically increases rapidly and then decreases more slowly. We wish to determine the storage coefficient S, and the transmissivity, T.

The mathematical model for the slug test is

$$h = \frac{Q}{4\pi T t} e^{-d^2 S/(4Tt)}. \tag{9.42}$$

We know the parameters $Q = 50$ m^3 and $d = 60$ m, and the times t at which the head h is measured. Our data are given in Table 9.1, where head measurements were roughly accurate to 0.01 m ($\sigma_i = 0.01$ m).

The optimal parameter values were $S = 0.00207$ and $T = 0.585$ m^2/hr. The observed χ^2 value was 2.04, with a corresponding p-value of 73%. Thus this fit passes the χ^2 test. The data points and fitted curve are shown in Figure 9.1.

Using the Jacobian evaluated at the optimal parameter values, we computed an approximate covariance matrix for the fitted parameters. The resulting 95% confidence intervals for S and T were

$$S = 0.00207 \pm 0.00012 \tag{9.43}$$

$$T = 0.585 \pm 0.029 \text{ m}^2/\text{hr}. \tag{9.44}$$

A contour plot of the χ^2 surface obtained by varying S and T is shown in Figure 9.2. Note that, in contrast to our earlier linear regression problems, the contours are not ellipses. This is a consequence of the nonlinearity of the problem. If we zoom in to the region around the optimal parameter values, we find that near the optimal parameters, contours of the χ^2 surface are approximately elliptical. This indicates that the linear approximation of $G(m)$ around the optimal parameter values is a good approximation for small perturbations. Figure 9.3 shows the 95% confidence ellipse for the fitted parameters. For comparison, the individual confidence interval for S is shown with dashed lines. ■

Table 9.1 Slug test data.

t (hr)	5	10	20	30	40	50
h (m)	0.72	0.49	0.30	0.20	0.16	0.12

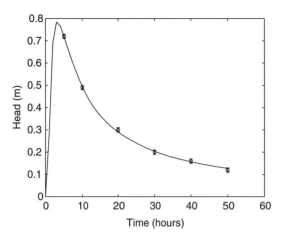

Figure 9.1 Data and fitted model for the slug test.

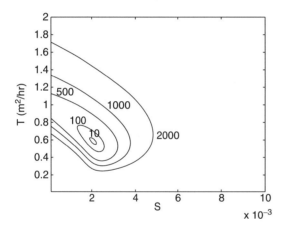

Figure 9.2 χ^2 contour plot for the slug test.

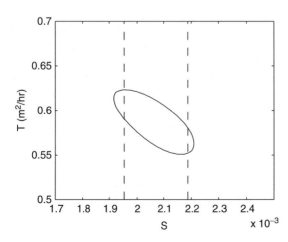

Figure 9.3 Closeup of the χ^2 contour plot for the slug test, showing the 95% confidence ellipsoid and the 95% confidence interval for S.

9.4 IMPLEMENTATION ISSUES

In this section we consider a number of important issues in the implementation of the GN and LM methods.

The most important difference between the linear regression problems that we solved in Chapter 2 and the nonlinear regression problems discussed in this chapter is that in nonlinear regression we have a nonlinear function $\mathbf{G}(\mathbf{m})$. Our iterative methods require the computation of the functions $f_i(\mathbf{m})$ and their partial derivatives with respect to the model parameters m_j. These partial derivatives in turn depend on the derivatives of $\mathbf{G}$

$$\frac{\partial f_i(\mathbf{m})}{\partial m_j} = \frac{1}{\sigma_i} \frac{\partial \mathbf{G}(\mathbf{m})_i}{\partial m_j}. \tag{9.45}$$

In some cases, we have explicit formulas for $\mathbf{G}(\mathbf{m})$ and its derivatives. In other cases, $\mathbf{G}(\mathbf{m})$ exists only as a **black box** subroutine that we can call as required to compute function values.

When an explicit formula for $\mathbf{G}(\mathbf{m})$ is available, and the number of parameters is relatively small, we can differentiate by hand or use a symbolic computation package. There are also **automatic differentiation** software packages that can translate the source code of a program that computes $\mathbf{G}(\mathbf{m})$ into a program that computes the derivatives of $\mathbf{G}(\mathbf{m})$.

Another approach is to use **finite differences** to approximate the first derivatives of $\mathbf{G}(\mathbf{m})_i$. A simple first-order scheme is

$$\frac{\partial G(\mathbf{m})_i}{\partial m_j} \approx \frac{G(\mathbf{m} + h\mathbf{e}_j)_i - G(\mathbf{m})_i}{h}. \tag{9.46}$$

Finite difference derivative approximations such as (9.46) are inevitably less accurate than exact formulas and can lead to numerical difficulties in the solution of a nonlinear regression problem. In particular, the parameter h must not be too large, because (9.46) arises from a Taylor series approximation that becomes inaccurate as h increases. On the other hand, as h becomes very small, significant round-off error in the numerator of (9.46) may occur. A good rule of thumb is to set $h = \sqrt{\epsilon}$, where ϵ is the accuracy of the evaluations of $\mathbf{G}(\mathbf{m})_i$. For example, if the function evaluations are accurate to 0.0001, then an appropriate choice of h would be about 0.01. Determining the actual accuracy of function evaluations can be difficult, especially when $\mathbf{G}$ is a black box routine. One useful assessment technique is to plot function values as a parameter of interest is varied over a small range. These plots should be smooth at the scale of h. When $\mathbf{G}$ is available only as a black box subroutine that can be called with particular values of $\mathbf{m}$, and the source code for the subroutine is not available, then the only possible approach is to use finite differences.

In practice many difficulties in solving nonlinear regression problems can be traced back to incorrect derivative computations. It is thus a good idea to cross-check any available analytical formulas for the derivative with finite difference approximations. Many software packages for nonlinear regression include options for checking the accuracy of derivative formulas.

A second important issue in the implementation of the GN and LM methods is deciding when to terminate the iterations. We would like to stop when the gradient $\nabla f(\mathbf{m})$ is approximately $\mathbf{0}$

and the values of **m** have stopped changing substantially from one iteration to the next. Because of scaling issues, it is not possible to set an absolute tolerance on $\|\nabla f(\mathbf{m})\|_2$ that would be appropriate for all problems. Similarly, it is difficult to pick a single absolute tolerance on $\|\mathbf{m}^{k+1} - \mathbf{m}^k\|_2$ or $|f(\mathbf{m}^{k+1}) - f(\mathbf{m}^k)|$.

The following convergence tests have been normalized so that they will work well on a wide variety of problems. We assume that values of $\mathbf{G}(\mathbf{m})$ can be calculated with an accuracy of ϵ. To ensure that the gradient of $f(\mathbf{m})$ is approximately $\mathbf{0}$, we require that

$$\|\nabla f(\mathbf{m}^k)\|_2 < \sqrt{\epsilon}(1 + |f(\mathbf{m}^k)|). \tag{9.47}$$

To ensure that successive values of **m** are close, we require

$$\|\mathbf{m}^k - \mathbf{m}^{k-1}\|_2 < \sqrt{\epsilon}(1 + \|\mathbf{m}^k\|_2). \tag{9.48}$$

Finally, to make sure that the values of $f(\mathbf{m})$ have stopped changing, we require that

$$|f(\mathbf{m}^k) - f(\mathbf{m}^{k-1})| < \epsilon(1 + |f(\mathbf{m}^k)|). \tag{9.49}$$

There are a number of additional problems that can arise during the solution of a nonlinear regression problem by the GN or LM methods related to the functional behavior of $f(\mathbf{m})$.

The first issue is that our methods assume that $f(\mathbf{m})$ is a smooth function. This means not only that $f(\mathbf{m})$ must be continuous, but also that its first and second partial derivatives with respect to the parameters must be continuous. Figure 9.4 shows a function which is itself continuous, but has discontinuities in the first derivative at $m = 0.2$ and the second derivative at $m = 0.5$. When $\mathbf{G}(\mathbf{m})$ is given by an explicit formula, it is usually easy to verify this assumption, but when $\mathbf{G}(\mathbf{m})$ is implemented as a black box routine, it can be very difficult.

A second issue is that the function $f(\mathbf{m})$ may have a "flat bottom." See Figure 9.5. In such cases, there are many values of **m** that come close to fitting the data, and it is difficult to determine the optimal $\mathbf{m}^*$. In practice, this condition is seen to occur when $\mathbf{J}(\mathbf{m}^*)^T \mathbf{J}(\mathbf{m}^*)$ is nearly singular. Because of this ill-conditioning, computing accurate confidence intervals for the model parameters can be effectively impossible. We will address this difficulty in the next chapter by applying Tikhonov regularization.

The final problem that we will consider is that $f(\mathbf{m})$ may be nonconvex and therefore have multiple local minimum points. See Figure 9.6.

The GN and LM methods are designed to converge to a local minimum, but depending on where we begin the search, there is no way to be certain that such a solution will be a global minimum. Depending on the particular problem, the optimization algorithm might well converge to a locally optimal solution.

Global optimization methods have been developed to deal with this issue [55, 70, 71, 143]. Deterministic global optimization procedures can be used on problems with a very small number of variables, whereas stochastic search procedures can be applied to large-scale problems.

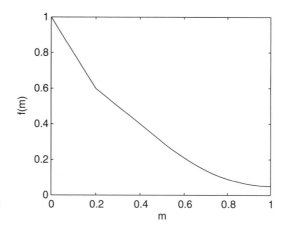

Figure 9.4 An example of a nonsmooth function.

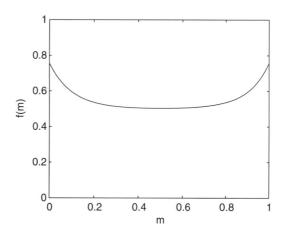

Figure 9.5 An example of a function with a flat bottom.

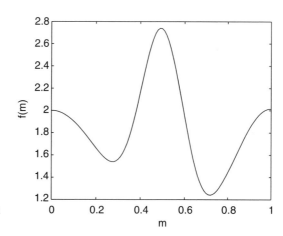

Figure 9.6 An example of a function with multiple local minima.

Stochastic search procedures can be quite effective in practice, even though they do not find a global optimum with certainty.

However, even a deterministic global optimization procedure is not a panacea. In the context of nonlinear regression, if the nonlinear least squares problem has multiple locally optimal solutions with similar objective function values, then each of these solutions will correspond to a statistically likely solution. We cannot simply report one globally optimal solution as our best estimate and construct confidence intervals using (9.38), because this would mean ignoring other likely solutions. However, if we could show that there is one globally optimal solution and other locally optimal solutions have very small p-values, then it would be appropriate to report the globally optimal solution and corresponding confidence intervals.

Although a thorough discussion of global optimization is beyond the scope of this book, we will discuss one simple global optimization procedure called the **multistart method**. In this procedure, we randomly generate a large number of initial solutions and perform the LM method starting with each of the random solutions. We then examine the local minimum solutions found by the procedure and select the one with the smallest value of $f(\mathbf{m})$. If other local minimum solutions are statistically unlikely, then it is appropriate to report the global optimum as our solution. The multistart approach has two important practical advantages. First, by finding many locally optimal solutions, we can determine whether there is more than one statistically likely solution. Second, we can make effective use of the fast convergence of the LM method to a locally optimal solution.

■ **Example 9.2** Consider the problem of fitting

$$y = m_1 e^{m_2 x} + m_3 x e^{m_4 x} \tag{9.50}$$

to a set of data. The true model parameters are $m_1 = 1.0$, $m_2 = -0.5$, $m_3 = 1.0$, and $m_4 = -0.75$, and the x values are 25 evenly spaced points between 1 and 7. We compute corresponding y values and add independent normally distributed noise with a standard deviation of 0.01 to obtain a synthetic data set.

We next use the LM method to solve the problem 20 times, using random initial solutions with each parameter uniformly distributed between -1 and 1. This produces a total of three different locally optimal solutions. See Table 9.2. Since solution number 1 has the best χ^2 value, and the other two solutions have unreasonably large χ^2 values, we will analyze only the first solution.

Table 9.2 Locally optimal solutions for the sample problem.

Solution Number	m_1	m_2	m_3	m_4	χ^2	p-value
1	0.9874	−0.5689	1.0477	−0.7181	17.3871	0.687
2	1.4368	0.1249	−0.5398	−0.0167	40.0649	0.007
3	1.5529	−0.1924	−0.1974	−0.1924	94.7845	< 0.001

The χ^2 value of 17.39 has an associated p-value (based on 21 degrees of freedom) of 0.69, so this regression fit passes the χ^2 test. Figure 9.7 shows the data points with 1-σ error bars and the fitted curve.

Figure 9.8 shows the normalized residuals for this regression fit. Note that the majority of the residuals are within 0.5 standard deviations, with a few residuals as large as 1.9 standard deviations. There is no obvious trend in the residuals as x ranges from 1 to 7.

Next, we compute the approximate covariance matrix for the model parameters using (9.38). The square roots of the diagonal elements of the covariance matrix are standard deviations for the individual model parameters. These are then used to compute 95% confidence intervals for model parameters. The solution parameters with 95% confidence intervals are

$$m_1 = +0.98 \pm 0.22 \tag{9.51}$$

$$m_2 = -0.57 \pm 0.77 \tag{9.52}$$

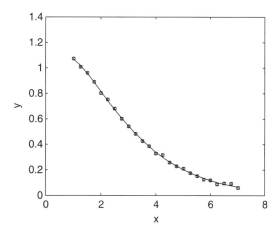

Figure 9.7   Data points and fitted curve.

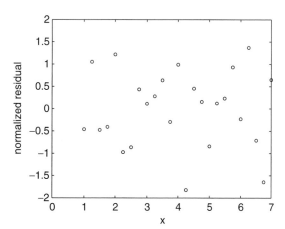

Figure 9.8   Normalized residuals.

$$m_3 = +1.05 \pm 0.50 \tag{9.53}$$

$$m_4 = -0.72 \pm 0.20. \tag{9.54}$$

The true parameters $(1, -0.5, 1, -0.75)$ are all covered by these confidence intervals. However, there is a large degree of uncertainty. This is an example of a poorly conditioned nonlinear regression problem in which the data do not constrain the parameter values very well.

The correlation matrix provides some insight into the nature of the ill-conditioning. For our preferred solution, the correlation matrix is

$$\rho = \begin{bmatrix} 1.00 & -0.84 & 0.68 & 0.89 \\ -0.84 & 1.00 & -0.96 & -0.99 \\ 0.68 & -0.96 & 1.00 & 0.93 \\ 0.89 & -0.99 & 0.93 & 1.00 \end{bmatrix}. \tag{9.55}$$

Note the strong positive and negative correlations between pairs of parameters. For example, the strong negative correlation between m_1 and m_2 tells us that by increasing m_1 and simultaneously decreasing m_2 we can obtain a solution that is very nearly as good as our optimal solution. There are also strong negative correlations between m_2 and m_3 and between m_2 and m_4.   ■

9.5 EXERCISES

9.1 A recording instrument sampling at 50 Hz records a noisy sinusoidal voltage signal in a 40-s-long record. The data are to be modeled using

$$y(t) = A \sin(2\pi f_0 t + \phi) + c + \sigma \eta(t) \text{ V} \tag{9.56}$$

where $\eta(t)$ is believed to be unit standard deviation, independent, and normally distributed noise, and σ is an unknown standard deviation. Using the data in the file **instdata**, solve for the parameters $(m_1, m_2, m_3, m_4) = (A, f_0, \phi, c)$, using the LM method. Show that it is critical to choose a good initial solution (suitable initial parameters can be found by examining a plot of the time series by eye). Once you are satisfied that you have found a good solution, use it to estimate the noise amplitude σ. Use your solution and estimate of σ to find corresponding covariance and correlation matrices and 95% parameter confidence intervals. Which pair of parameters is most strongly correlated? Are there multiple equally good solutions for this problem?

9.2 In hydrology, the van Genuchten model is often used to relate the volumetric water content in an unsaturated soil to the head [172]. The model is

$$\theta(h) = \theta_r + \frac{\theta_s - \theta_r}{(1 + (-\alpha h)^n)^{(1-1/n)}} \tag{9.57}$$

where θ_s is the volumetric water content at saturation, θ_r is the residual volumetric water content at a very large negative head, and α and n are two parameters which can be fit to laboratory measurements.

The file **vgdata.mat** contains measurements for a loam soil at the Bosque del Apache National Wildlife Refuge in New Mexico [62]. Fit the van Genuchten model to the data. The volumetric water content at saturation is $\theta_s = 0.44$, and the residual water content is $\theta_r = 0.09$. You may assume that the measurements of $\theta(h)$ are accurate to about 2% of the measured values.

You will need to determine appropriate values for σ_i, write functions to compute $\theta(h)$ and its derivatives, and then use the LM method to estimate the parameters. In doing so, you should consider whether or not this problem might have local minima. It will be helpful to know that typical values of α range from about 0.001 to 0.02, and typical values of n run from 1 to 10.

9.3 An alternative version of the LM method stabilizes the GN method by multiplicative damping. Instead of adding $\lambda \mathbf{I}$ to the diagonal of $\mathbf{J}(\mathbf{m}^k)^T \mathbf{J}(\mathbf{m}^k)$, this method multiplies the diagonal of $\mathbf{J}(\mathbf{m}^k)^T \mathbf{J}(\mathbf{m}^k)$ by a factor of $(1 + \lambda)$. Show that this method can fail by producing an example in which the modified $\mathbf{J}(\mathbf{m}^k)^T \mathbf{J}(\mathbf{m}^k)$ matrix is singular, no matter how large λ becomes.

9.4 A cluster of 10 small earthquakes occurs in a shallow geothermal reservoir. The field is instrumented with nine seismometers, eight of which are at the surface and one of which is 300 m down a borehole. The P-wave velocity of the fractured granite medium is thought to be an approximately uniform 2 km/s. The station locations (in meters relative to a central origin) are given in Table 9.3.

The arrival times of P-waves from the earthquakes are carefully measured at the stations, with an estimated error of approximately 1 ms. The arrival time estimates for each earthquake e_i at each station (in seconds, with the nearest clock second subtracted) are given in Table 9.4.

These data can be found in **eqdata.mat**.

(a) Apply the LM method to this data set to estimate least squares locations of the earthquakes.

Table 9.3 Station locations for the earthquake location problem.

Station	x (m)	y (m)	z (m)
1	500	−500	0
2	−500	−500	0
3	100	100	0
4	−100	0	0
5	0	100	0
6	0	−100	0
7	0	−50	0
8	0	200	0
9	10	50	−300

Table 9.4 Data for the earthquake location problem.

Station	e_1	e_2	e_3	e_4	e_5
1	0.8423	1.2729	0.8164	1.1745	1.1954
2	0.8680	1.2970	0.8429	1.2009	1.2238
3	0.5826	1.0095	0.5524	0.9177	0.9326
4	0.5975	1.0274	0.5677	0.9312	0.9496
5	0.5802	1.0093	0.5484	0.9145	0.9313
6	0.5988	1.0263	0.5693	0.9316	0.9480
7	0.5857	1.0141	0.5563	0.9195	0.9351
8	0.6017	1.0319	0.5748	0.9362	0.9555
9	0.5266	0.9553	0.5118	0.8533	0.8870

Station	e_6	e_7	e_8	e_9	e_{10}
1	0.5361	0.7633	0.8865	1.0838	0.9413
2	0.5640	0.7878	0.9120	1.1114	0.9654
3	0.2812	0.5078	0.6154	0.8164	0.6835
4	0.2953	0.5213	0.6360	0.8339	0.6982
5	0.2795	0.5045	0.6138	0.8144	0.6833
6	0.2967	0.5205	0.6347	0.8336	0.6958
7	0.2841	0.5095	0.6215	0.8211	0.6857
8	0.3025	0.5275	0.6394	0.8400	0.7020
9	0.2115	0.4448	0.5837	0.7792	0.6157

(b) Estimate the uncertainties in x, y, z (in m) and origin time (in s) for each earthquake using the diagonal elements of the appropriate covariance matrix. Do the earthquake locations follow any discernible trend?

9.5 The Lightning Mapping Array is a portable system that has recently been developed to locate the sources of lightning radiation in three spatial dimensions and time [129]. The system measures the arrival time of impulsive radiation events. The measurements are made at nine or more locations in a region 40 to 60 km in diameter. Each station records the peak radiation event in successive 100-μs time intervals; from this, several hundred to over a thousand radiation sources may be typically located per lightning discharge.

(a) Data from the LMA are shown in Table 9.5. Use the arrival times at stations 1, 2, 4, 6, 7, 8, 10, and 13 to find the time and location of the radio frequency source in the lightning flash. Assume that the radio waves travel along straight paths at the speed of light (2.997×10^8 m/s).

(b) During lightning storms we record the arrival of thousands of events each second. We use several methods to find which arrival times are due to the same event. The foregoing data were chosen as possible candidates to go together. We require any solution to use times from at least six stations. Find the largest subset of these data

Table 9.5   Data for the lightning mapping array problem.

Station	t (s)	x (km)	y (km)	z (km)
1	0.0922360280	−24.3471411	2.14673146	1.18923667
2	0.0921837940	−12.8746056	14.5005985	1.10808551
3	0.0922165500	16.0647214	−4.41975194	1.12675062
4	0.0921199690	0.450543748	30.0267473	1.06693166
6	0.0923199800	−17.3754105	−27.1991732	1.18526730
7	0.0922839580	−44.0424408	−4.95601205	1.13775547
8	0.0922030460	−34.6170855	17.4012873	1.14296361
9	0.0922797660	17.6625731	−24.1712580	1.09097830
10	0.0922497250	0.837203704	−10.7394229	1.18219520
11	0.0921672710	4.88218031	10.5960946	1.12031719
12	0.0921702350	16.9664920	9.64835135	1.09399160
13	0.0922357370	32.6468622	−13.2199767	1.01175261

that gives a good solution. See if it is the same subset suggested in part (a) of this
problem.

9.6   NOTES AND FURTHER READING

Newton's method is central to the field of optimization. Some references include [32, 82, 83,
108, 115]. Because of its speed, Newton's method is the basis for most methods of nonlinear
optimization. Various modifications to the basic method are used to ensure convergence to a
local minimum of $f(\mathbf{x})$ [108, 115]. One important difficulty in Newton's method is that, for
very large problems, it may be impractical to store the Hessian matrix. Specialized methods
have been developed for the solution of such large scale optimization problems [108, 115].

The GN and LM methods are discussed in [16, 108, 115]. Statistical aspects of nonlinear
regression are discussed in [8, 33, 106]. A more detailed discussion of the termination criteria
for the LM method that we describe in Section 9.4 can be found in [108]. There are a number of freely available and commercial software packages for nonlinear regression, including
GaussFit [77], MINPACK [103], and ODRPACK [17]. Automatic differentiation has applications in many areas of numerical computing, including optimization and numerical solution
of ordinary and partial differential equations. Two books that survey this topic are [27, 52].
Global optimization is a large field of research. Some basic references include [55, 70, 71]. A
survey of global optimization methods in geophysical inversion is [143].

10

NONLINEAR INVERSE PROBLEMS

Synopsis: The nonlinear regression approaches of Chapter 9 are generalized to problems requiring regularization. The Tikhonov regularization and Occam's inversion approaches are introduced. Seismic tomography and electrical conductivity inversion examples are used to illustrate the application of these methods.

10.1 REGULARIZING NONLINEAR LEAST SQUARES PROBLEMS

As with solution techniques for linear inverse problems, nonlinear least squares approaches can run into difficulty with ill-conditioned problems. This typically happens as the number of model parameters grows. Here, we will discuss regularization of nonlinear inverse problems and algorithms for computing a regularized solution to a nonlinear inverse problem.

The basic ideas of Tikhonov regularization can be extended to nonlinear problems. Suppose that we are given a nonlinear inverse problem involving a discrete n-point model $\mathbf{m}$ and discrete m-point data vector $\mathbf{d}$ that are related by a nonlinear system of equations $\mathbf{G}(\mathbf{m}) = \mathbf{d}$. For convenience, we will assume that the nonlinear equations have been scaled to incorporate the measurement standard deviations σ_i. We want to find the solution with smallest $\|\mathbf{Lm}\|_2$ that comes sufficiently close to fitting the data.

We can formulate this problem as

$$
\begin{aligned}
\min \quad & \|\mathbf{Lm}\|_2 \\
& \|\mathbf{G}(\mathbf{m}) - \mathbf{d}\|_2 \leq \delta.
\end{aligned}
\tag{10.1}
$$

Note that this is virtually identical to the problem considered in the linear case [e.g., (5.28)], with the only difference being that we now have a general function $\mathbf{G}(\mathbf{m})$ instead of a matrix–vector multiplication $\mathbf{Gm}$. As in the linear case, we can reformulate this problem in terms of minimizing the misfit subject to a constraint on $\|\mathbf{Lm}\|_2$

$$
\begin{aligned}
\min \quad & \|\mathbf{G}(\mathbf{m}) - \mathbf{d}\|_2 \\
& \|\mathbf{Lm}\|_2 \leq \epsilon
\end{aligned}
\tag{10.2}
$$

or as a damped least squares problem

$$\min \quad \|G(m) - d\|_2^2 + \alpha^2 \|Lm\|_2^2. \tag{10.3}$$

All three versions of the regularized least squares problem can be solved by applying standard nonlinear optimization software. In particular, (10.3) is a nonlinear least squares problem, so we could apply the LM or GN methods to it. Of course, any such approach will still have to deal with the possibility of local minima that are not global minimum points. In some cases, it is possible to show that the nonlinear least squares problem is convex and thus has only global minima. In other cases we will have to employ the multistart strategy or other methods to determine whether there are local minima.

To apply the GN method to (10.3), we rewrite it as

$$\min \left\| \begin{matrix} G(m) - d \\ \alpha Lm \end{matrix} \right\|_2^2. \tag{10.4}$$

The Jacobian for this damped least squares problem for the kth iteration is

$$K(m^k) = \begin{bmatrix} J(m^k) \\ \alpha L \end{bmatrix} \tag{10.5}$$

where $J(m^{(k)})$ is the Jacobian of $G(m^{(k)})$. A GN model step is obtained by solving

$$K(m^k)^T K(m^k) \Delta m = -K(m^k)^T \begin{bmatrix} G(m^k) - d \\ \alpha Lm^k \end{bmatrix}. \tag{10.6}$$

This can be simplified using (10.5) to

$$(J(m^k)^T J(m^k) + \alpha^2 L^T L) \Delta m = -J(m^k)^T (G(m^k) - d) - \alpha^2 L^T Lm^k. \tag{10.7}$$

Equation (10.7) resembles the LM method [e.g., (9.30)]. However, α in this formulation is fixed in the objective function (10.4) being minimized. To further stabilize the iterations as in the LM method, we could introduce a variable λI term to the left hand side. This is typically not necessary because the explicit regularization makes the system of equations nonsingular.

■ **Example 10.1** Consider a modified version of the cross-well tomography example from the Chapter 5 exercises, where we introduce nonlinearity by employing a more realistic forward model that incorporates ray path refraction. The two-dimensional velocity structure is parameterized using a matrix of uniformly spaced slowness nodes on an 8 by 8 grid that spans a 1600 m by 1600 m region.

We apply an approximate ray bending technique to predict travel times and to generate finite-difference estimates of the partial derivatives of travel times with respect to model parameters [170]. Figure 10.1 shows the true velocity model and the corresponding set of 64 ray paths. The true model consists of a background velocity of 2.9 km/s with embedded fast (+10%) and slow (−15%) Gaussian-shaped anomalies. The data set consists of the

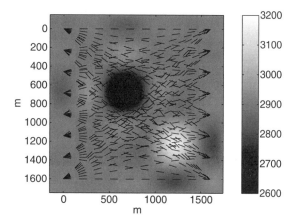

Figure 10.1   True velocity model (m/s) and the corresponding ray paths for the bent-ray cross-well tomography example.

64 travel times between each pair of opposing sources and receivers with $N(0, (0.001 \text{ s})^2)$ noise added.

Note that refracted ray paths tend to avoid low-velocity regions (dark shading) and are, conversely, "attracted" to high-velocity regions (light shading) in accordance with Fermat's least-time principle. In practice this effect makes low-velocity regions more difficult to resolve in such studies.

A discrete approximation of the two-dimensional Laplacian operator is used as the roughening matrix $\mathbf{L}$ for this problem. Iterative GN (10.7) solutions were obtained for a range of 16 values of α ranging logarithmically between approximately 4.9 and 367. Figure 10.2 shows

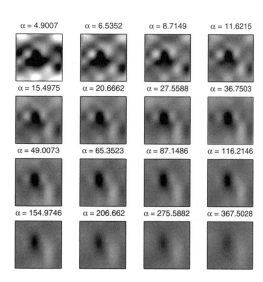

Figure 10.2   Suite of GN (10.7), second-order regularized solutions, ranging from least (upper left) to most (lower right) regularized, and associated α values. The physical dimensions and gray scale are identical to those of Figures 10.1 and 10.4. These solutions are shown in color on the cover of this book.

the suite of solutions after five iterations. An L-curve of seminorm versus data misfit is plotted in Figure 10.3, along with the discrepancy principle value ($\delta = 0.008$) expected for 64 data points with the assigned noise level. Note that the point for $\alpha = 4.9007$ is slightly out of the expected position because the GN method is unable to accurately solve the poorly conditioned least squares problem in this case. The solution best satisfying the discrepancy principle corresponds to $\alpha = 49$. See Figure 10.4. Because we know the true model in this example, it is instructive to examine how well the regularized solutions of Figure 10.2 compare to the true model. Figure 10.5 shows the 2-norm model misfit as a function of α and demonstrates that the discrepancy principle solution for this problem, and for this particular noise realization, is indeed close to the minimum in $\|\mathbf{m} - \mathbf{m}_{\text{true}}\|_2$. Note that the

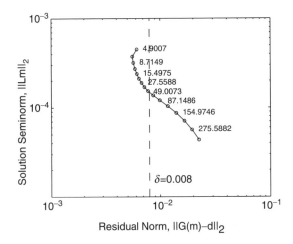

Figure 10.3   L-curve and corresponding α values for the solutions of Figure 10.2.

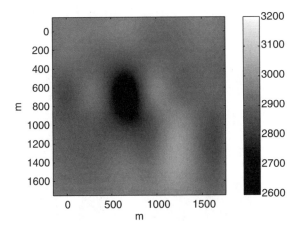

Figure 10.4   Best solution velocity structure (m/s), α selected using the discrepancy principle, $\alpha \approx 49$.

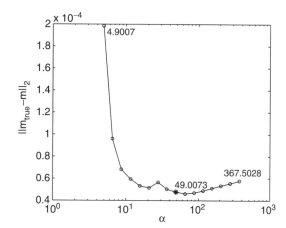

Figure 10.5 Model misfit 2-norm as a function of regularization parameter α, with preferred model highlighted.

solution shown in Figure 10.4 exhibits artifacts that are common in regularized solutions, such as streaking, side lobes, and underestimation of the true model variation. ∎

10.2 OCCAM'S INVERSION

Occam's inversion is a popular algorithm for nonlinear inversion introduced by Constable, Parker, and Constable [26]. The name refers to the 14th century philosopher William of Ockham, who argued that simpler explanations should always be preferred to more complicated explanations. A similar statement occurs as Rule 1 in Newton's "Rules for the Study of Natural Philosophy" [114]. This principle has become known as "Occam's razor."

Occam's inversion uses the discrepancy principle and searches for the solution that minimizes $\|\mathbf{Lm}\|_2$ subject to the constraint $\|\mathbf{G}(\mathbf{m}) - \mathbf{d}\|_2 \leq \delta$. The algorithm is straightforward to implement, requires only the nonlinear forward model $\mathbf{G}(\mathbf{m})$ and its Jacobian, and works well in practice.

We assume that our nonlinear inverse problem has been cast in the form of (10.1). The roughening matrix $\mathbf{L}$ can be equal to $\mathbf{I}$ to implement zeroth-order Tikhonov regularization, or it can be a finite-difference approximation of a first (5.27) or second (5.29) derivative for higher-order regularization. In practice, Occam's inversion is often used on two- or three-dimensional problems where $\mathbf{L}$ is a discrete approximation of the Laplacian operator.

As usual, we will assume that the measurement errors in $\mathbf{d}$ are independent and normally distributed. For convenience, we will also assume that the system of equations $\mathbf{G}(\mathbf{m}) = \mathbf{d}$ has been scaled so that the standard deviations σ_i are equal.

The basic idea behind Occam's inversion is an iteratively applied local linearization. Given a trial model $\mathbf{m}^k$, Taylor's theorem is applied to obtain the local approximation

$$\mathbf{G}(\mathbf{m}^k + \Delta\mathbf{m}) \approx \mathbf{G}(\mathbf{m}^k) + \mathbf{J}(\mathbf{m}^k)\Delta\mathbf{m} \qquad (10.8)$$

where $\mathbf{J}(\mathbf{m}^k)$ is the Jacobian

$$\mathbf{J}(\mathbf{m}^k) = \begin{bmatrix} \dfrac{\partial G_1(\mathbf{m}^k)}{\partial m_1} & \cdots & \dfrac{\partial G_1(\mathbf{m}^k)}{\partial m_n} \\ \vdots & \ddots & \vdots \\ \dfrac{\partial G_m(\mathbf{m}^k)}{\partial m_1} & \cdots & \dfrac{\partial G_m(\mathbf{m}^k)}{\partial m_n} \end{bmatrix}. \tag{10.9}$$

Using (10.8), the damped least squares problem (10.3) becomes

$$\min \quad \|\mathbf{G}(\mathbf{m}^k) + \mathbf{J}(\mathbf{m}^k)\Delta\mathbf{m} - \mathbf{d}\|_2^2 + \alpha^2\|\mathbf{L}(\mathbf{m}^k + \Delta\mathbf{m})\|_2^2 \tag{10.10}$$

where the variable is $\Delta\mathbf{m}$ and $\mathbf{m}^k$ is constant. Reformulating this as a problem in which the variable is $\mathbf{m}^{k+1} = \mathbf{m}^k + \Delta\mathbf{m}$ and letting

$$\hat{\mathbf{d}}(\mathbf{m}^k) = \mathbf{d} - \mathbf{G}(\mathbf{m}^k) + \mathbf{J}(\mathbf{m}^k)\mathbf{m}^k \tag{10.11}$$

gives

$$\min \quad \|\mathbf{J}(\mathbf{m}^k)(\mathbf{m}^k + \Delta\mathbf{m}) - (\mathbf{d} - \mathbf{G}(\mathbf{m}^k) + \mathbf{J}(\mathbf{m}^k)\mathbf{m}^k)\|_2^2 + \alpha^2\|\mathbf{L}(\mathbf{m}^k + \Delta\mathbf{m})\|_2^2 \tag{10.12}$$

or

$$\min \quad \|\mathbf{J}(\mathbf{m}^k)\mathbf{m}^{k+1} - \hat{\mathbf{d}}(\mathbf{m}^k)\|_2^2 + \alpha^2\|\mathbf{L}(\mathbf{m}^{k+1})\|_2^2. \tag{10.13}$$

Because $\mathbf{J}(\mathbf{m}^k)$ and $\hat{\mathbf{d}}(\mathbf{m}^k)$ are constant, (10.13) is in the form of a damped linear least squares problem that has the solution given by (5.7) :

$$\mathbf{m}^{k+1} = \mathbf{m}^k + \Delta\mathbf{m} = \left(\mathbf{J}(\mathbf{m}^k)^T\mathbf{J}(\mathbf{m}^k) + \alpha^2\mathbf{L}^T\mathbf{L}\right)^{-1}\mathbf{J}(\mathbf{m}^k)^T\hat{\mathbf{d}}(\mathbf{m}^k). \tag{10.14}$$

It is worth noting that this method is similar to the GN method applied to the damped least squares problem (10.3). See Exercise 10.1. The difference is that in Occam's inversion the parameter α is dynamically adjusted so that the solution will not exceed the allowable misfit. At each iteration we pick the largest value of α that keeps the χ^2 value of the solution from exceeding the bound on δ^2 specified in (10.1). If this is impossible, we instead pick the value of α that minimizes the χ^2 value. At the end of the procedure, we should have a solution with $\chi^2 = \delta^2$. We can now state the algorithm.

■ **Algorithm 10.1 Occam's inversion algorithm** Beginning with an initial solution $\mathbf{m}^0$, apply the formula

$$\mathbf{m}^{k+1} = \left(\mathbf{J}(\mathbf{m}^k)^T\mathbf{J}(\mathbf{m}^k) + \alpha^2\mathbf{L}^T\mathbf{L}\right)^{-1}\mathbf{J}(\mathbf{m}^k)^T\hat{\mathbf{d}}(\mathbf{m}^k). \tag{10.15}$$

In each iteration, pick the largest value of α such that $\chi^2(\mathbf{m}^{k+1}) \leq \delta^2$. If no such value exists, then pick a value of α that minimizes $\chi^2(\mathbf{m}^{k+1})$. Stop if and when the sequence converges to a solution with $\chi^2 = \delta^2$.   ∎

■ **Example 10.2** We will consider the problem of estimating subsurface electrical conductivities from above-ground EM induction measurements. The instrument used in this example is the Geonics EM-38 ground conductivity meter. A description of the instrument and the mathematical model of the response of the instrument can be found in [63]. The mathematical model is quite complicated, but we will treat it as a black box and concentrate on the inverse problem.

Measurements are taken at heights of 0, 10, 20, 30, 40, 50, 75, 100, and 150 cm above the surface, with the coils oriented in both the vertical and horizontal orientations. There are a total of 18 observations. The data are shown in Table 10.1. We will assume measurement standard deviations of 0.1 mS/m.

We discretize the subsurface electrical conductivity profile into 10 layers, each 20 cm thick, with a semiinfinite layer below 2 m. Thus we have 11 parameters to estimate.

The function $\mathbf{G}(\mathbf{m})$ is available to us in the form of a subroutine for computing predicted data. Since we do not have simple formulas for $\mathbf{G}(\mathbf{m})$, we cannot write down analytic expressions for the elements of the Jacobian. However, we can use finite-difference approximations [e.g., (9.46)] to estimate the necessary partial derivatives.

We first tried using the LM method to estimate the model parameters. After 50 iterations, this produced the model shown in Figure 10.6. The χ^2 value for this model is 9.62, with nine degrees of freedom, so the model actually fits the data adequately. Unfortunately, the least squares problem is very badly conditioned. The condition number of $\mathbf{J}^T\mathbf{J}$ is approximately 7.6×10^{17}. Furthermore, this model is unrealistic because it includes negative electrical conductivities and because it exhibits the high-amplitude and high-frequency oscillations that we have come to expect of underregularized solutions to inverse problems. Clearly a regularized solution is called for.

Table 10.1   Data for the EM-38 example.

Height (cm)	EMV (mS/m)	EMH (mS/m)
0	134.5	117.4
10	129.0	97.7
20	120.5	81.7
30	110.5	69.2
40	100.5	59.6
50	90.8	51.8
75	70.9	38.2
100	56.8	29.8
150	38.5	19.9

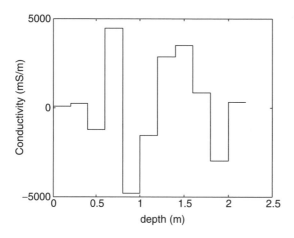

Figure 10.6   LM solution.

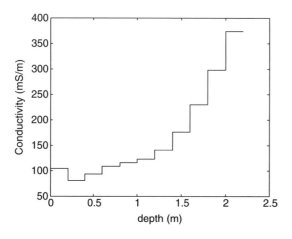

Figure 10.7   Occam's inversion solution.

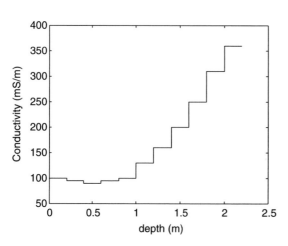

Figure 10.8   True model.

We next tried Occam's inversion with second-order regularization and $\delta = 0.4243$. The resulting model is shown in Figure 10.7. Figure 10.8 shows the true model. The Occam's inversion solution is a fairly good reproduction of the true model.   ∎

10.3 EXERCISES

10.1 Show that the GN step (10.7) and the Occam's inversion step (10.14) are identical.

10.2 Recall Example 1.5, in which we had gravity anomaly observations above a density perturbation of variable depth $m(x)$ and fixed density $\Delta\rho$. Use Occam's inversion to solve an instance of this inverse problem. Consider a gravity anomaly along a 1-km section, with observations taken every 50 m, and density perturbation of 200 kg/m^3 (0.2 g/cm^3). The perturbation is expected to be at a depth of roughly 200 m.

      The data file **gravprob.mat** contains a vector **x** of observation locations. Use the same coordinates for your discretization of the model. The vector **obs** contains the actual observations. Assume that the observations are accurate to about 1.0×10^{-12}.

  (a) Derive a formula for the elements of the Jacobian.
  (b) Write MATLAB routines to compute the model predictions and the Jacobian for this problem.
  (c) Use the supplied implementation of Occam's inversion to solve the inverse problem.
  (d) Discuss your results. What features in the inverse solution appear to be real? What is the resolution of your solution? Were there any difficulties with local minimum points?
  (e) What would happen if the true density perturbation was instead at about 1000 m depth?

10.3 Apply the GN method with explicit regularization to the EM inversion problem in Example 10.2. Compare your solution with the solution obtained by Occam's inversion. Which method required more computational effort?

10.4 Apply Occam's inversion to a cross-well bent-ray tomography problem with identical geometry to Example 10.1. Use the subroutine **getj.m** to forward model travel times and the Jacobian. Travel-time data and subroutine control parameters are contained in the file **benddata.mat**. Start with the uniform 2900 m/s velocity 8 by 8 node initial velocity model in **bendata.mat**, and assume independent and normally distributed data errors with $\sigma = 0.001$ ms.

      Hint: A search range of α^2 between 10 and 10^5 is appropriate for this problem.

      The following MATLAB code constructs a suitable roughening matrix **L** that approximates a two-dimensional Laplacian operator.

10.4 NOTES AND FURTHER READING

In inverse problems with a large number of parameters, the most difficult computational problem is often computing derivatives of $\mathbf{G}(\mathbf{m})$ with respect to the parameters. Computation of analytic formulas is commonly impractical. Finite-difference estimates require computational effort which increases with the number of parameters and may become impractical for large problems. A useful technique for problems in which the mathematical model is a differential equation is the adjoint equation approach [157, 39]. An alternative approach involves using the discretized differential equation as a set of constraints to be added to the nonlinear least squares problem [15].

For large-scale problems, it may be impractical to use direct factorization to solve the systems of equations (10.7) or (10.14) involved in computing the GN or Occam step. One approach in this case is to use an iterative method such as conjugate gradients to solve the linear systems of equations [108]. The conjugate gradient method can also be extended to minimize the nonlinear objective function directly [108, 145].

11

BAYESIAN METHODS

Synopsis: Following a review of the classical least squares approach to solving inverse problems, we introduce the Bayesian approach, which treats the model as a random variable with a probability distribution that we seek to estimate. A prior distribution for the model parameters is combined with the data to produce a posterior distribution for the model parameters. In special cases, the Bayesian approach produces solutions that are equivalent to the least squares, maximum likelihood, and Tikhonov regularization solutions. The maximum entropy method for selecting a prior *distribution is discussed. Several examples of the Bayesian approach are presented.*

11.1 REVIEW OF THE CLASSICAL APPROACH

In the classical approach to parameter estimation and inverse problems discussed in previous chapters, we begin with a mathematical model of the form $\mathbf{Gm} = \mathbf{d}$ in the linear case or $\mathbf{G(m)} = \mathbf{d}$ in the nonlinear case. We assume that there is a true model $\mathbf{m}_{\text{true}}$ and a true data set $\mathbf{d}_{\text{true}}$ such that $\mathbf{Gm}_{\text{true}} = \mathbf{d}_{\text{true}}$. We are given an actual data set $\mathbf{d}$, which is the sum of $\mathbf{d}_{\text{true}}$ and measurement noise. Our goal is to recover $\mathbf{m}_{\text{true}}$ from the noisy data.

For well-conditioned linear problems, under the assumption of independent and normally distributed data errors, the theory is well developed. In Chapter 2 it was shown that the maximum likelihood principle leads to the least squares solution. The least squares solution, $\mathbf{m}_{L_2}$, is found by minimizing the 2-norm of the residual, $\|\mathbf{Gm} - \mathbf{d}\|_2$.

Since there is noise in the data, we should expect some misfit between the data predictions of the forward model and the data, so that χ^2 will not typically be zero. We saw that the χ^2 distribution can be used to set a reasonable bound on $\|\mathbf{Gm} - \mathbf{d}\|_2$. This was used in the χ^2 goodness-of-fit test. We were also able to compute a covariance matrix for the estimated parameters

$$\text{Cov}(\mathbf{m}_{L_2}) = (\mathbf{G}^T\mathbf{G})^{-1}\mathbf{G}^T \text{Cov}(\mathbf{d})\mathbf{G}(\mathbf{G}^T\mathbf{G})^{-1} \tag{11.1}$$

and to use it to compute confidence intervals for the estimated parameters and correlations between the estimated parameters.

This approach works very well for linear regression problems in which the least squares problem is well-conditioned. We found, however, that in many cases the least squares problem is not well-conditioned. In such situations, the set of solutions that adequately fits the data is large and diverse, and commonly contains many physically unreasonable models.

In Chapters 4 through 8, we discussed a number of approaches to regularizing the least squares problem. These approaches pick one "best" solution out of the set of solutions that adequately fit the data. The different regularization methods differ in what constitutes the best solution. For example, zeroth-order Tikhonov regularization selects the model that minimizes the 2-norm $\|\mathbf{m}\|_2$ subject to the constraint $\|\mathbf{Gm} - \mathbf{d}\|_2 < \delta$, whereas higher-order Tikhonov regularization selects the model that minimizes $\|\mathbf{Lm}\|_2$ subject to $\|\mathbf{Gm} - \mathbf{d}\|_2 < \delta$.

Regularization can be applied to both linear and nonlinear problems. For relatively small linear problems the computation of the regularized solution is generally done with the help of the SVD. This process is straightforward, insightful, and robust. For large sparse linear problems iterative methods such as CGLS can be used.

For nonlinear problems things are more complicated. We saw in Chapters 9 and 10 that the GN and LM methods could be used to find a local minimum of the least squares problem. Unfortunately, nonlinear least squares problems may have a large number of local minimum solutions, and finding the global minimum can be extremely difficult. Furthermore, if there are several local minimum solutions with high likelihoods, then we cannot simply select a single "best" solution, and a more sophisticated approach is required.

How can we justify selecting one solution from the set of models which adequately fit the data? One justification is Occam's razor; when we have several different models to consider, we should select the simplest one. The solutions selected by regularization are in some sense the simplest models which fit the data. If fitting the data did not require a particular feature seen in the regularized solution, then that feature would have been smoothed out by the regularization. However, this answer is not entirely satisfactory because different choices of the roughening matrix $\mathbf{L}$ can result in very different solutions. Why should one regularized solution be preferred to another when the choice of the roughening matrix is subjective?

Recall from Chapter 4 that, once we have regularized a least squares problem, we lose the ability to obtain statistically useful confidence intervals for the parameters because regularization introduces bias in the solution. In particular this means that the expected value of a regularized solution is not the true solution. Bounds on the error in Tikhonov regularized solutions were discussed in Section 5.8, but these require model assumptions that are difficult if not impossible to justify in practice.

11.2 THE BAYESIAN APPROACH

The Bayesian approach is named after Thomas Bayes, an 18th-century pioneer in probability theory. The approach is based on philosophically different ideas than the classical approach. However, as we will see, it often results in similar solutions.

The most fundamental difference between the classical and Bayesian approaches is in the nature of the solution. In the classical approach, there is a specific but unknown model $\mathbf{m}_{\text{true}}$ that we would like to discover. In the Bayesian approach the model is a random variable, and the solution is a probability distribution for the model parameters. Once we have this probability distribution, we can use it to answer probabilistic questions about the model, such as, "What is the probability that m_5 is less than 1?" In the classical approach such questions do not make sense, since the true model that we seek is not a random variable.

A second very important difference between the classical and Bayesian approaches is that the Bayesian approach allows us to naturally incorporate prior information about the solution that comes from other data or from experience based intuition. This information is expressed as a **prior distribution** for **m**. If no other information is available, then under the **principle of indifference**, we may pick a prior distribution in which all model parameter values have equal likelihood. Such a prior distribution is said to be **uninformative**.

It should be pointed out that if the parameters **m** are contained in the range $(-\infty, \infty)$, then the uninformative prior is not a proper probability distribution. The problem is that there does not exist a probability distribution $f(x)$ such that

$$\int_{-\infty}^{\infty} f(x) \, dx = 1 \tag{11.2}$$

and $f(x)$ is constant. In practice, the use of this improper prior distribution can be justified, because the resulting posterior distribution for **m** is a proper distribution.

Once the data have been collected, they are combined with the prior distribution using Bayes' theorem to produce a **posterior distribution** for the model parameters.

One of the main objections to the Bayesian approach is that the method is "unscientific" because it allows the analyst to incorporate subjective judgments into the model that are not solely based on the data. Proponents of the approach reply that there are also subjective aspects to the classical approach, and that one is free to choose an uninformative prior distribution. Furthermore, it is possible to complete the Bayesian analysis with a variety of prior distributions and examine the effects of different prior distributions on the posterior distribution.

We will denote the prior distribution by $p(\mathbf{m})$ and assume that we can compute the likelihood that, given a particular model, a data vector, **d**, will be observed. We will use the notation $f(\mathbf{d}|\mathbf{m})$ for this conditional probability distribution. We seek the conditional distribution of the model parameter(s) given the data. We will denote the posterior probability distribution for the model parameters by $q(\mathbf{m}|\mathbf{d})$. Bayes' theorem relates the prior and posterior distributions in a way that makes the computation of $q(\mathbf{m}|\mathbf{d})$ possible. In the form that will be used in this chapter, Bayes' theorem can be stated as follows.

■ **Theorem 11.1**

$$q(\mathbf{m}|\mathbf{d}) = \frac{f(\mathbf{d}|\mathbf{m})p(\mathbf{m})}{\int_{\text{all models}} f(\mathbf{d}|\mathbf{m})p(\mathbf{m}) \, d\mathbf{m}} \tag{11.3}$$

$$= \frac{f(\mathbf{d}|\mathbf{m})p(\mathbf{m})}{c} \tag{11.4}$$

where

$$c = \int_{\text{all models}} f(\mathbf{d}|\mathbf{m})p(\mathbf{m}) \, d\mathbf{m}. \tag{11.5}$$

■

Note that the constant c in (11.4) simply normalizes the conditional distribution $q(\mathbf{m}|\mathbf{d})$ so that its integral is 1.

For many purposes, knowing c is not actually necessary. For example, we can compare two models $\hat{\mathbf{m}}$ and $\bar{\mathbf{m}}$ by computing the likelihood ratio

$$LR = \frac{q(\hat{\mathbf{m}}|\mathbf{d})}{q(\bar{\mathbf{m}}|\mathbf{d})} = \frac{f(\mathbf{d}|\hat{\mathbf{m}})p(\hat{\mathbf{m}})}{f(\mathbf{d}|\bar{\mathbf{m}})p(\bar{\mathbf{m}})}. \tag{11.6}$$

A very small likelihood ratio would indicate that the model $\bar{\mathbf{m}}$ is far more likely than the model $\hat{\mathbf{m}}$. Because c is not always needed, (11.4) is sometimes written as

$$q(\mathbf{m}|\mathbf{d}) \propto f(\mathbf{d}|\mathbf{m})p(\mathbf{m}). \tag{11.7}$$

Unfortunately, there are many other situations in which knowing c is required. In particular, c is needed to compute any posterior probabilities. Also, c is required to compute the expected value and variance of the posterior distribution.

It is important to emphasize that the probability distribution $q(\mathbf{m}|\mathbf{d})$ does not provide a single model that we can consider to be the "answer." In cases where we want to single out one model as the answer, it may be appropriate to use the model with the largest value of $q(\mathbf{m}|\mathbf{d})$, which is referred to as the **maximum *a posteriori* (MAP)** model. An alternative would be to use the mean of the posterior distribution. In situations where the posterior distribution is normal, the MAP model and the posterior mean model are identical.

In general, the computation of a posterior distribution can be problematic. The chief difficulty lies in evaluating the integrals in (11.5). These are often integrals in very high dimensions, for which numerical integration techniques are computationally expensive. Fortunately, there are a number of special cases in which computation of the posterior distribution is greatly simplified.

One simplification occurs when the prior distribution $p(\mathbf{m})$ is uninformative, in which case (11.7) becomes

$$q(\mathbf{m}|\mathbf{d}) \propto f(\mathbf{d}|\mathbf{m}) \tag{11.8}$$

and the posterior distribution is precisely the likelihood function, $L(\mathbf{m}|\mathbf{d})$. Under the maximum likelihood principle (see Section 2.2.) we would select the model $\mathbf{m}_{ML}$ that maximizes $L(\mathbf{m}|\mathbf{d})$. This is exactly the MAP model.

A further simplification occurs when the noise in the measured data is independent and normally distributed with standard deviation σ. Because the measurement errors are independent, we can write the likelihood function as the product of the likelihoods of the individual data points:

$$L(\mathbf{m}|\mathbf{d}) = f(\mathbf{d}|\mathbf{m}) = f(d_1|\mathbf{m}) \cdot f(d_2|\mathbf{m}) \cdots f(d_m|\mathbf{m}). \tag{11.9}$$

Because the individual data points d_i are normally distributed with expected values $(\mathbf{G(m)})_i$ and standard deviation σ, we can write

$$f(d_i|\mathbf{m}) = \frac{1}{\sigma\sqrt{2\pi}} e^{-\frac{((\mathbf{G(m)})_i - d_i)^2}{2\sigma^2}}. \tag{11.10}$$

Thus

$$L(\mathbf{m}|\mathbf{d}) = \left(\frac{1}{\sigma\sqrt{2\pi}}\right)^m e^{-\sum_{i=1}^m \frac{((\mathbf{G(m)})_i - d_i)^2}{2\sigma^2}}. \tag{11.11}$$

We can maximize (11.11) by maximizing the exponent or equivalently by minimizing the negative of the exponent:

$$\min \sum_{i=1}^m \frac{((\mathbf{G(m)})_i - d_i)^2}{2\sigma^2}. \tag{11.12}$$

This is a nonlinear least squares problem. Thus we have shown that when we have independent and normally distributed measurement errors and we use an uninformative prior, the MAP solution is the least squares solution.

■ **Example 11.1** Consider the very simple parameter estimation problem in which we perform repeated weighings of a very small object. The measurement errors are normally distributed with mean 0 and standard deviation $\sigma = 1$ μg. Our goal is to estimate the mass of the object.
   With this error model, we have

$$f(d|m) = \frac{1}{\sqrt{2\pi}} e^{-(m-d)^2/2}. \tag{11.13}$$

Suppose we weigh the mass and obtain a measurement of $d_1 = 10.3$ μg. What do we now know about m? An uninformative prior distribution (11.8) gives

$$q(m|d_1 = 10.3\ \mu g) \propto \frac{1}{\sqrt{2\pi}} e^{-(m-10.3)^2/2}. \tag{11.14}$$

Because this is itself a normal probability distribution, the constant of proportionality in (11.4) is 1, and the posterior distribution is

$$q(m|d_1 = 10.3\ \mu g) = \frac{1}{\sqrt{2\pi}} e^{-(m-10.3)^2/2}. \tag{11.15}$$

This posterior distribution is shown in Figure 11.1.

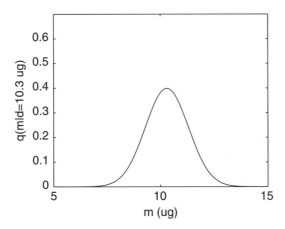

Figure 11.1   Posterior distribution $q(m|d_1 = 10.3\ \mu g)$, uninformative prior.

Next, suppose that we obtain a second measurement of $d_2 = 10.1\ \mu g$. We use the distribution (11.15) estimated from the first measurement as the prior distribution and compute a revised posterior distribution that incorporates information from both measurements,

$$q(m|d_1 = 10.3\ \mu g,\ d_2 = 10.1\ \mu g) \propto f(d_2 = 10.1\ \mu g|m)q(m|d_1 = 10.3\ \mu g). \quad (11.16)$$

$$q(m|d_1 = 10.3\ \mu g,\ d_2 = 10.1\ \mu g) \propto \frac{1}{\sqrt{2\pi}}e^{-(m-10.1)^2/2}\frac{1}{\sqrt{2\pi}}e^{-(m-10.3)^2/2}. \quad (11.17)$$

Adding the exponents and absorbing the $1/\sqrt{2\pi}$ factors into the constant of proportionality gives

$$q(m|d_1 = 10.3\ \mu g,\ d_2 = 10.1\ \mu g) \propto e^{-((m-10.3)^2+(m-10.1)^2)/2}. \quad (11.18)$$

Finally, we can simplify the exponent by combining terms and completing the square to obtain

$$(m - 10.3)^2 + (m - 10.1)^2 = 2(m - 10.2)^2 + 0.02. \quad (11.19)$$

Thus

$$q(m|d_1 = 10.3\ \mu g,\ d_2 = 10.1\ \mu g) \propto e^{-(2(m-10.2)^2+0.02)/2}. \quad (11.20)$$

The $e^{-0.02/2}$ factor can be absorbed into the constant of proportionality. We are left with

$$q(m|d_1 = 10.3\ \mu g,\ d_2 = 10.1\ \mu g) \propto e^{-(10.2-m)^2}. \quad (11.21)$$

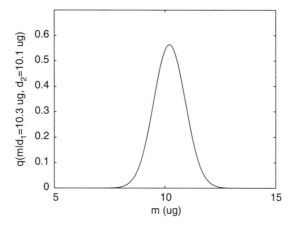

Figure 11.2   Posterior distribution $q(m|d_1 = 10.3\ \mu g,\ d_2 = 10.1\ \mu g)$, uninformative prior.

After normalization, we have

$$q(m|d_1 = 10.3\ \mu g,\ d_2 = 10.1\ \mu g) = \frac{1}{(1/\sqrt{2})\sqrt{2\pi}} e^{-\frac{(10.2-m)^2}{2(1/\sqrt{2})^2}}, \qquad (11.22)$$

which is a normal distribution with mean $10.2\ \mu g$ and a standard deviation of $1/\sqrt{2}\ \mu g$. This distribution is shown in Figure 11.2. Since we used an uninformative prior and the measurement errors were independent and normally distributed, the MAP solution is precisely the least squares solution for this problem.

It is remarkable that in this example we started with a normal prior distribution and took into account normally distributed data and obtained a normal posterior distribution. The property that the posterior distribution has the same form as the prior distribution is called **conjugacy**. There are other families of conjugate distributions for various parameter estimation problems, but in general this is a very rare property [45].

11.3   THE MULTIVARIATE NORMAL CASE

The idea that a normal prior distribution leads to a normal posterior distribution can be extended to situations with many model parameters. When we have a linear model $\mathbf{Gm} = \mathbf{d}$, the data errors have a multivariate normal distribution, and the prior distribution for the model parameters is also multivariate normal, the computation of the posterior distribution is relatively tractable [158].

Let $\mathbf{d}_{obs}$ be the observed data, and let $\mathbf{C}_D$ be the covariance matrix for the data. Let $\mathbf{m}_{prior}$ be the expected value of the prior distribution and let $\mathbf{C}_M$ be the covariance matrix for the prior distribution. The prior distribution is

$$p(\mathbf{m}) \propto e^{-\frac{1}{2}(\mathbf{m}-\mathbf{m}_{\text{prior}})^T \mathbf{C}_M^{-1}(\mathbf{m}-\mathbf{m}_{\text{prior}})}. \tag{11.23}$$

The conditional distribution of the data given $\mathbf{m}$ is

$$f(\mathbf{d}|\mathbf{m}) \propto e^{-\frac{1}{2}(\mathbf{Gm}-\mathbf{d})^T \mathbf{C}_D^{-1}(\mathbf{Gm}-\mathbf{d})}. \tag{11.24}$$

Thus (11.7) gives

$$q(\mathbf{m}|\mathbf{d}) \propto e^{-\frac{1}{2}((\mathbf{Gm}-\mathbf{d})^T \mathbf{C}_D^{-1}(\mathbf{Gm}-\mathbf{d})+(\mathbf{m}-\mathbf{m}_{\text{prior}})^T \mathbf{C}_M^{-1}(\mathbf{m}-\mathbf{m}_{\text{prior}}))}. \tag{11.25}$$

Tarantola shows that this can be simplified to

$$q(\mathbf{m}|\mathbf{d}) \propto e^{-\frac{1}{2}(\mathbf{m}-\mathbf{m}_{\text{MAP}})^T \mathbf{C}_{M'}^{-1}(\mathbf{m}-\mathbf{m}_{\text{MAP}})} \tag{11.26}$$

where $\mathbf{m}_{\text{MAP}}$ is the MAP solution, and

$$\mathbf{C}_{M'} = (\mathbf{G}^T \mathbf{C}_D^{-1} \mathbf{G} + \mathbf{C}_M^{-1})^{-1}. \tag{11.27}$$

The MAP solution can be found by maximizing the exponent in (11.25), or by minimizing its negative:

$$\min \ (\mathbf{Gm} - \mathbf{d})^T \mathbf{C}_D^{-1}(\mathbf{Gm} - \mathbf{d}) + (\mathbf{m} - \mathbf{m}_{\text{prior}})^T \mathbf{C}_M^{-1}(\mathbf{m} - \mathbf{m}_{\text{prior}}). \tag{11.28}$$

The key to minimizing this expression is to rewrite it in terms of the matrix square roots of $\mathbf{C}_M^{-1}$ and $\mathbf{C}_D^{-1}$. Note that every covariance matrix is positive definite and has a unique positive definite matrix square root. The matrix square root can be computed using the SVD. The MATLAB command **sqrtm** can also be used to find the square root of a matrix. This minimization problem can be reformulated as

$$\min \ (\mathbf{C}_D^{-1/2}(\mathbf{Gm} - \mathbf{d}))^T (\mathbf{C}_D^{-1/2}(\mathbf{Gm} - \mathbf{d}))$$
$$+ (\mathbf{C}_M^{-1/2}(\mathbf{m} - \mathbf{m}_{\text{prior}}))^T (\mathbf{C}_M^{-1/2}(\mathbf{m} - \mathbf{m}_{\text{prior}})) \tag{11.29}$$

or

$$\min \ \left\| \begin{bmatrix} \mathbf{C}_D^{-1/2}\mathbf{G} \\ \mathbf{C}_M^{-1/2} \end{bmatrix} \mathbf{m} - \begin{bmatrix} \mathbf{C}_D^{-1/2}\mathbf{d} \\ \mathbf{C}_M^{-1/2}\mathbf{m}_{\text{prior}} \end{bmatrix} \right\|_2^2. \tag{11.30}$$

This is a standard linear least squares problem.

In (11.30), notice that

$$\text{Cov}(\mathbf{C}_D^{-1/2}\mathbf{d}) = \mathbf{C}_D^{-1/2}\mathbf{C}_D(\mathbf{C}_D^{-1/2})^T. \tag{11.31}$$

This simplifies to

$$\text{Cov}(\mathbf{C}_D^{-1/2}\mathbf{d}) = \mathbf{I}. \tag{11.32}$$

The multiplication of $\mathbf{C}_D^{-1/2}$ times $\mathbf{d}$ in (11.30) can be thought of as a transformation of the data that effectively makes the data independent and normalizes the standard deviations (see Exercise 2.2). In the model space, multiplication by $\mathbf{C}_M^{-1/2}$ has a similar effect.

It is worthwhile to consider what happens to the posterior distribution in the extreme case in which the prior distribution provides essentially no information. Consider a prior distribution with a covariance matrix $\mathbf{C}_M = \alpha^2\mathbf{I}$ in the limit where α is extremely large. In this case, the diagonal elements of $\mathbf{C}_M^{-1}$ will be extremely small, and the posterior covariance matrix (11.27) will be well approximated by

$$\mathbf{C}_{M'} \approx (\mathbf{G}^T \; \mathbf{C}_D^{-1} \; \mathbf{G})^{-1}. \tag{11.33}$$

If the data covariance matrix is $\sigma^2\mathbf{I}$, then

$$\mathbf{C}_{M'} \approx \sigma^2(\mathbf{G}^T\mathbf{G})^{-1}. \tag{11.34}$$

This is precisely the covariance matrix for the model parameters in (11.1). Furthermore, when we solve (11.30) to obtain the MAP solution, we find that it simplifies to the least squares problem min $\|\mathbf{Gm} - \mathbf{d}\|_2^2$. Thus, under the common assumption of normally distributed and independent data errors with constant variance, a very broad prior distribution leads to the unregularized least squares solution.

It also worthwhile to consider what happens in the special case where $\mathbf{C}_D = \sigma^2\mathbf{I}$, and $\mathbf{C}_M = \alpha^2\mathbf{I}$. In this case, (11.30) simplifies to

$$\text{min} \;\; (1/\sigma)^2\|(\mathbf{Gm} - \mathbf{d})\|_2^2 + (1/\alpha)^2\|\mathbf{m} - \mathbf{m}_{\text{prior}}\|_2^2. \tag{11.35}$$

This is simply zeroth-order Tikhonov regularization. In other words, the MAP solution obtained by using a prior with independent and normally distributed model parameters is precisely the Tikhonov regularized solution. However, this does not mean that the Bayesian approach is entirely equivalent to Tikhonov regularization, since the Bayesian solution is a probability distribution, whereas the Tikhonov solution is a single model.

Once we have obtained the posterior distribution, it is straightforward to generate model realizations, for instance to help assess likely or unlikely model features. Following the method outlined in Example B.10, we compute the Cholesky factorization of the posterior distribution covariance matrix

$$\mathbf{C}_{M'} = \mathbf{R}^T\mathbf{R} \tag{11.36}$$

and then generate a random solution

$$\mathbf{m} = \mathbf{R}^T \mathbf{s} + \mathbf{m}_{\text{MAP}} \tag{11.37}$$

where the vector **s** consists of independent and normally distributed random numbers with zero mean and standard deviation one.

■ **Example 11.2** We return to the Shaw problem that was previously considered in Examples 3.2, 4.3, and 5.1.

For our first solution, we use a multivariate normal prior distribution with mean 0.5 and standard deviation 0.5 for each model parameter, with independent model parameters, so that $\mathbf{C}_M = 0.25\mathbf{I}$. As in the previous examples, the measurement noise has standard deviation 1.0×10^{-6}, so that $\mathbf{C}_D = 1.0 \times 10^{-12}\mathbf{I}$. Solving (11.30) produces the $\mathbf{m}_{\text{MAP}}$ solution shown in Figure 11.3. Figure 11.4 shows this same solution with error bars. These error bars are not classical 95% confidence intervals. Rather, they are 95% probability intervals calculated from the multivariate normal posterior distribution, so that there is 95% probability that each model parameter lies within the symmetric interval around $\mathbf{m}_{\text{MAP}}$.

Figure 11.5 shows a random model realization generated from the posterior distribution using (11.37). This solution varies considerably from the true model and demonstrates, along with the large confidence intervals in Figure 11.4, the great uncertainty in the inverse solution. The roughness of the solution is a consequence of the fact that the prior distribution $\mathbf{C}_M$ had zero covariances, and thus no preference for smoothness.

For our second solution, we considered a broader prior distribution. We used a prior mean of 0.5, but used variances of 25 in the prior distribution of the model parameters instead of variances of 0.25. Figure 11.6 shows the MAP model and error bars for this case. This solution is, not surprisingly, even worse than the model shown in Figure 11.3. With a very unrestrictive prior, we have depended mostly on the available data, which simply does not constrain the solution well in this severely ill-posed problem.

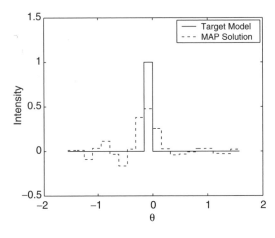

Figure 11.3 The MAP solution and the true model for the Shaw example.

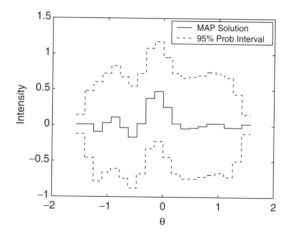

Figure 11.4   The MAP solution with error bars.

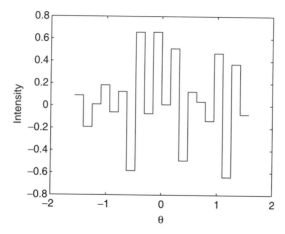

Figure 11.5   A model realization for the Shaw example.

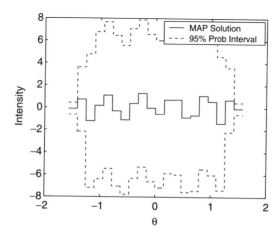

Figure 11.6   The MAP solution for the Shaw example using a broader prior distribution.

This result illustrates a major issue with applying the Bayesian approach to poorly conditioned problems. To obtain a reasonably tight posterior distribution, we have to make very strong assumptions in the prior distribution. Conversely, if these assumptions are not made, then we cannot obtain a useful solution to the inverse problem.

To be fair, Tikhonov regularization also requires strong assumptions on the model (see Section 5.8) to produce a solution with error bounds, and it is perhaps too easy to perform Tikhonov regularization without computing error bounds.   ∎

The approach described in this section can be extended to nonlinear problems. To find the MAP solution, we solve the nonlinear least squares problem

$$\min \ (\mathbf{G}(\mathbf{m}) - \mathbf{d})^T \mathbf{C}_D^{-1} (\mathbf{G}(\mathbf{m}) - \mathbf{d}) + (\mathbf{m} - \mathbf{m}_{\text{prior}})^T \mathbf{C}_M^{-1} (\mathbf{m} - \mathbf{m}_{\text{prior}}). \qquad (11.38)$$

We then linearize around the MAP solution to obtain the approximate posterior covariance

$$\mathbf{C}_{M'} = (\mathbf{J}(\mathbf{m}_{\text{MAP}})^T \mathbf{C}_D^{-1} \mathbf{J}(\mathbf{m}_{\text{MAP}}) + \mathbf{C}_M^{-1})^{-1} \qquad (11.39)$$

where $\mathbf{J}(\mathbf{m})$ is the Jacobian. As with other nonlinear optimization approaches, we must consider the possibility of multiple local minima. If (11.38) has multiple solutions with comparable likelihoods, then a single MAP solution and associated $\mathbf{C}_{M'}$ from (11.39) will not accurately characterize the posterior distribution.

11.4 MAXIMUM ENTROPY METHODS

We have seen that an essential issue in the Bayesian approach is the selection of the prior distribution. In this section we consider **maximum entropy methods** which can be used to select a prior distribution subject to available information such as bounds on a parameter or an average value of a parameter.

■ **Definition 11.1** The **entropy** of a discrete probability distribution

$$P(X = x_i) = p_i \qquad (11.40)$$

is given by

$$H(X) = -\sum p_i \ln p_i. \qquad (11.41)$$

The entropy of a continuous probability distribution

$$P(X \le a) = \int_{-\infty}^{a} f(x) \, dx \qquad (11.42)$$

is given by

$$H(X) = -\int_{-\infty}^{\infty} f(x) \ln f(x) \, dx.$$ (11.43)

∎

Under the **maximum entropy principle**, we select a prior distribution that has the largest possible entropy subject to constraints imposed by available information.

For continuous random variables, optimization problems resulting from the maximum entropy principle involve an unknown density function $f(x)$. Such problems can be solved using techniques from the calculus of variations [44]. Fortunately, maximum entropy distributions for a number of important cases have already been worked out [79].

∎ **Example 11.3** Suppose we know that X takes on only nonnegative values, and that the expected value of X is μ. It can be shown using the calculus of variations that the maximum entropy distribution is an exponential distribution [79]

$$f_X(x) = \frac{1}{\mu} e^{-x/\mu} \qquad x \geq 0.$$ (11.44)

∎

∎ **Definition 11.2** Given a discrete probability distribution

$$P(X = x_i) = p_i$$ (11.45)

and an alternative distribution

$$P(X = x_i) = q_i$$ (11.46)

the Kullback–Leibler cross-entropy [87] is given by

$$D(\mathbf{p}, \mathbf{q}) = \sum p_i \ln \frac{p_i}{q_i}.$$ (11.47)

Given continuous distributions $f(x)$ and $g(x)$, the cross-entropy is

$$D(f, g) = \int_{-\infty}^{\infty} f(x) \ln \frac{f(x)}{g(x)} \, dx.$$ (11.48)

∎

Cross-entropy is a measure of how close two distributions are. If the two distributions are identical, then the cross-entropy is zero.

Under the **minimum cross-entropy principle**, if we are given a prior distribution p and some additional constraints on the distribution, we should select a posterior distribution

q which minimizes the cross-entropy of p and q subject to the constraints. Note that this is not the same thing as using Bayes' theorem to update a prior distribution.

In the **minimum relative entropy (MRE)** method of Woodbury and Ulrych, the minimum cross-entropy principle is applied to linear inverse problems of the form $\mathbf{Gm} = \mathbf{d}$ subject to lower and upper bounds on the model elements [112, 113, 169, 179]. First, a maximum entropy prior distribution is computed using the lower and upper bounds and given mean values for the model parameters. Then, a posterior distribution is selected to minimize the cross-entropy subject to the constraint that the mean of the posterior distribution must satisfy the equations $\mathbf{Gm} = \mathbf{d}$.

■ **Example 11.4** Recall the source history reconstruction problem discussed in Examples 3.3 and 7.1. We have previously determined a lower bound of 0 for the input concentration and an upper bound of 1.1. Suppose we also have further reason to believe that most of the contamination occurred around $t = 150$. We thus select a prior distribution with a mean of

$$C(x, 0) = e^{-(t-150)^2/800}. \tag{11.49}$$

The true input concentration and mean of the prior distribution are shown in Figure 11.7.

Figure 11.8 shows the MRE solution with a 90% probability interval. Note that the information provided by the data was strong enough to overcome the prior distribution. In the time period between $t = 120$ and $t = 150$, the prior mean is significantly lower than the posterior mean, whereas this is reversed during the period from $t = 150$ to $t = 200$. ■

11.5 EPILOGUE

The main theme of this book has been obtaining, and statistically analyzing, solutions to discretized parameter estimation problems using classical and Bayesian approaches. We have

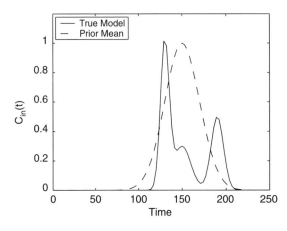

Figure 11.7 True model and mean of the prior distribution for the source history reconstruction example.

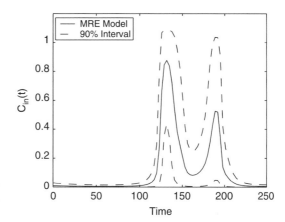

Figure 11.8 MRE solution and 90% probability intervals for the source history reconstruction example.

discussed computational procedures for both linear and nonlinear problems. Classical procedures produce estimates of the parameters and their associated uncertainties. In Bayesian methods, the model is a random variable, and the solution is its probability distribution. However, there are a number of crucial issues that need to be considered in applying these methods.

When we discretize a continuous problem, the choice of the discretization scheme, basis functions, and grid spacing can have large effects on the behavior of the discretized problem and its solutions, and these effects will not be reflected in the statistical analysis of the solution of the discretized problem. The discretization errors in the solution could potentially be far larger than any explicitly computed statistical uncertainty. Thus it is important to ensure that the discretization provides an adequate approximation to the continuous problem. If no formal analysis is performed, it is at least desirable to see whether varying the discretization has a significant effect on the solutions obtained.

For well-conditioned problems with normally distributed measurement errors, we can use the classical least squares approach. This results in unbiased parameter estimates and associated confidence intervals. For ill-conditioned problems, and for problems where we have good reason to prefer a specific bias in the character of the solution, Tikhonov or other regularization can be applied to obtain a solution. Although this is computationally tractable, the regularization introduces bias into the solution. We found that it is impossible to even bound this bias without making additional assumptions about the model.

Although the Bayesian approaches are also applicable to well-conditioned problems, they are particularly interesting in the context of ill-conditioned problems. By selecting a prior distribution we make our assumptions about the model explicit. The resulting posterior distribution is not affected by regularization bias. In the multivariate normal case for linear problems the Bayesian approach is no more difficult computationally than the least squares approach.

Various attempts have been made to avoid the use of subjective priors in the Bayesian approach. Principles such as maximum entropy can be used to derive prior distributions which have been claimed to be, in some sense, "objective." However, we do not find these arguments completely convincing.

Both the classical and Bayesian approaches can be extended to nonlinear inverse problems in a straightforward fashion. The computation of the estimated model parameters becomes substantially more difficult in that we must solve nonlinear optimization problems which may have multiple local minima. In both approaches, the statistical analysis is typically performed approximately by analyzing a linearization of the nonlinear model around the estimated parameters.

Inverse problems with nonnormally distributed measurement noise are more challenging. In such cases, the alternative to the least squares approach is maximum likelihood estimation. The Bayesian approach can in theory be applied when measurement errors are not normally distributed. However, in practice, the associated computations can be difficult.

11.6 EXERCISES

11.1   Reanalyze the data in Example 11.1 using a prior distribution that is uniform on the interval $[9, 11]$. Compute the posterior distribution after the first measurement of $10.3 \, \mu g$ and after the second measurement of $10.1 \, \mu g$. What is the posterior mean?

11.2   In writing (11.30) we made use of the matrix square root.

  (a)   Suppose that $\mathbf{A}$ is a symmetric and positive definite matrix. Using the SVD, find an explicit formula for the matrix square root. Your square root should itself be a symmetric and positive definite matrix.

  (b)   Show that instead of using the matrix square roots of $\mathbf{C}_D^{-1}$ and $\mathbf{C}_M^{-1}$, we could have used the Cholesky factorizations of $\mathbf{C}_D^{-1}$ and $\mathbf{C}_M^{-1}$ in formulating the least squares problem.

11.3   Consider the following coin tossing experiment. We repeatedly toss a coin, and each time record whether it comes up heads (0) or tails (1). The bias b of the coin is the probability that it comes up heads. We have reason to believe that this is not a fair coin, so we will not assume that $b = 1/2$. Instead, we will begin with a uniform prior distribution $p(b) = 1$, for $0 \le b \le 1$.

  (a)   What is $f(d|b)$? Note that the only possible data are 0 and 1, so this distribution will involve delta functions at $d = 0$, and $d = 1$.

  (b)   Suppose that on our first flip, the coin comes up heads. Compute the posterior distribution $q(b|d_1 = 0)$.

  (c)   The second, third, fourth, and fifth flips are 1, 1, 1, and 1. Find the posterior distribution $q(b|d_1 = 0, d_2 = 1, d_3 = 1, d_4 = 1, d_5 = 1)$. Plot the posterior distribution.

  (d)   What is your MAP estimate of the bias?

  (e)   Now, suppose that you initially felt that the coin was at least close to fair, with

$$p(b) \propto e^{-10(b-0.5)^2} \qquad 0 \le b \le 1. \qquad (11.50)$$

Repeat the analysis of the five coin flips described above.

11.4   Apply the Bayesian method to Exercise 5.2. Select what you consider to be a reasonable prior. How sensitive is your solution to the prior mean and covariance?

11.5   Consider a conventional six-sided die, with faces numbered 1, 2, 3, 4, 5, and 6. If each side were equally likely to come up, the mean value would be 7/2. Suppose instead that the mean is 9/2. Formulate an optimization problem that could be solved to find the maximum entropy distribution subject to the constraint that the mean is 9/2. Use the method of Lagrange multipliers to solve this optimization problem and obtain the maximum entropy distribution.

11.6   Let X be a discrete random variable that takes on the values $1, 2, \ldots, n$. Suppose that $E[X] = \mu$ is given. Find the maximum entropy distribution for X.

11.7 NOTES AND FURTHER READING

The arguments for and against the use of Bayesian methods in statistics and inverse problems have raged for decades. Some classical references that provide context for these arguments include [29, 35, 75, 76, 87, 139]. Sivia's book [146] is a good general introduction to Bayesian ideas for scientists and engineers. The textbook by Gelman *et al.* [45] provides a more comprehensive introduction to Bayesian statistics. An early paper by Tarantola and Valette on the application of the Bayesian approach was quite influential [159], and Tarantola's book is the standard reference work on Bayesian methods for inverse problems [158]. The book by Rodgers [131] focuses on application of the Bayesian approach to problems in atmospheric sounding. The paper of Gouveia and Scales [51] discusses the relative advantages and disadvantages of Bayesian and classical methods for inverse problems. The draft textbook by Scales and Smith [140] takes a Bayesian approach to inverse problems.

Sivia's book includes a brief introduction to the maximum entropy principle [146]. A complete survey of maximum entropy methods and results for particular distributions can be found in [79]. Other useful references on maximum entropy methods include [29, 75, 87].

In recent years there has been great interest in computationally tractable approaches to Bayesian inference with distributions that do not fit into the multivariate normal framework. One approach to such problems that has excited wide interest is the Markov chain Monte Carlo method (MCMC) [47, 130, 137, 105]. In MCMC, we are able to simulate samples from the posterior probability distribution without explicitly computing the distribution. The resulting samples can be statistically analyzed to estimate posterior probabilities of interest.

In many cases the solution to an inverse problem will be used in making a decision, with measurable consequences for making the "wrong" decision. Statistical decision theory can be helpful in determining the optimal decision. The paper by Evans and Stark provides a good introduction to the application of statistical decision theory to inverse problems [41].

Appendix A

REVIEW OF LINEAR ALGEBRA

Synopsis: A summary of essential concepts, definitions, and theorems in linear algebra used throughout this book.

A.1 SYSTEMS OF LINEAR EQUATIONS

A system of linear equations can be solved by the process of **Gaussian elimination**.

■ **Example A.1** Consider the system of equations

$$\begin{aligned} x + 2y + 3z &= 14 \\ x + 2y + 2z &= 11 \\ x + 3y + 4z &= 19. \end{aligned} \tag{A.1}$$

We eliminate x from the second and third equations by subtracting the first equation from the second and third equations to obtain

$$\begin{aligned} x + 2y + 3z &= 14 \\ -z &= -3 \\ y + z &= 5. \end{aligned} \tag{A.2}$$

We would like to eliminate y from the third equation, so we interchange the second and third equations:

$$\begin{aligned} x + 2y + 3z &= 14 \\ y + z &= 5 \\ -z &= -3. \end{aligned} \tag{A.3}$$

Next, we eliminate y from the first equation by subtracting two times the second equation from the first equation:

$$\begin{aligned} x \quad + z &= 4 \\ y + z &= 5 \\ -z &= -3. \end{aligned} \tag{A.4}$$

We then multiply the third equation by -1 to get an equation for z:

$$\begin{aligned} x \quad + z &= 4 \\ y + z &= 5 \\ z &= 3. \end{aligned}$$ (A.5)

Finally, we eliminate z from the first two equations:

$$\begin{aligned} x &= 1 \\ y &= 2 \\ z &= 3. \end{aligned}$$ (A.6)

The solution to the original system of equations is thus $x = 1$, $y = 2$, $z = 3$. Geometrically the constraints specified by the three equations of (A.1) describe three planes which, in this case, intersect at a single point. ∎

In solving (A.1), we used three **elementary row operations**: adding a multiple of one equation to another equation, multiplying an equation by a nonzero constant, and swapping two equations. This process can be extended to solve systems of equations with an arbitrary number of variables.

In performing the elimination process, the actual names of the variables are insignificant. We could have renamed the variables in the above example to a, b, and c without changing the solution in any significant way. Because the actual names of the variables are insignificant, we can save space by writing down the significant coefficients from the system of equations in **matrix** form as an **augmented matrix**. The augmented matrix form is also useful in solving a system of equations in computer algorithms, where the elements of the augmented matrix are stored in an array.

In augmented matrix form (A.1) becomes

$$\begin{bmatrix} 1 & 2 & 3 & | & 14 \\ 1 & 2 & 2 & | & 11 \\ 1 & 3 & 4 & | & 19 \end{bmatrix}.$$ (A.7)

In augmented notation, the elementary row operations become adding a multiple of one row to another row, multiplying a row by a nonzero constant, and interchanging two rows. The Gaussian elimination process is identical to the process used in Example A.1, with the final version of the augmented matrix given by

$$\begin{bmatrix} 1 & 0 & 0 & | & 1 \\ 0 & 1 & 0 & | & 2 \\ 0 & 0 & 1 & | & 3 \end{bmatrix}.$$ (A.8)

■ **Definition A.1** A matrix is said to be in **reduced row echelon form (RREF)** if it has the following properties:

1. The first nonzero element in each row is a one. The first nonzero row elements of the matrix are called **pivot elements**. A column in which a pivot element appears is called a **pivot column**.
2. Except for the pivot element, all elements in pivot columns are zero.
3. Any rows consisting entirely of zeros are at the bottom of the matrix. ■

In solving a system of equations in augmented matrix form, we apply elementary row operations to reduce the augmented matrix to RREF and then convert back to conventional notation to read off the solutions. The process of transforming a matrix into RREF can easily be automated. In MATLAB, this is done by the **rref** command.

It can be shown that any linear system of equations has no solutions, exactly one solution, or infinitely many solutions [91]. In a two-equation system in two dimensions, for example, the lines represented by the equations can fail to intersect (no solution), intersect at a point (one solution) or intersect in a line (many solutions). The following example shows how to determine the number of solutions from the RREF of the augmented matrix.

■ **Example A.2** Consider a system of two equations in three variables that has many solutions:

$$\begin{aligned} x_1 + \; x_2 + \; x_3 &= 0 \\ x_1 + 2x_2 + 2x_3 &= 0. \end{aligned} \tag{A.9}$$

We put this system of equations into augmented matrix form and then find the RREF, which is

$$\begin{bmatrix} 1 & 0 & 0 & | & 0 \\ 0 & 1 & 1 & | & 0 \end{bmatrix}. \tag{A.10}$$

We can translate this back into equation form as

$$\begin{aligned} x_1 \qquad\quad &= 0 \\ x_2 + x_3 &= 0. \end{aligned} \tag{A.11}$$

Clearly, x_1 must be 0 in any solution to the system of equations. However, x_2 and x_3 are not fixed. We could treat x_3 as a **free variable** and allow it to take on any value. Whatever value x_3 takes on, x_2 must be equal to $-x_3$. Geometrically, this system of equations describes the intersection of two planes, where the intersection consists of points on the line $x_2 = -x_3$ in the $x_1 = 0$ plane. ■

A linear system of equation may have more constraints than variables, in which case the system of equations is **overdetermined**. Although overdetermined systems often have no solutions, it is possible for an overdetermined system of equations to have either many solutions or exactly one solution.

Conversely, a system of equations with fewer equations than variables is **underdetermined**. Although in many cases underdetermined systems of equations have infinitely many solutions, it is also possible for such systems to have no solutions.

A system of equations with all zeros on the right-hand side is **homogeneous**. Every homogeneous system of equations has at least one solution, the trivial solution in which all of the variables are zero. A system of equations with a nonzero right-hand side is **nonhomogeneous**.

A.2 MATRIX AND VECTOR ALGEBRA

As we have seen in the previous section, a matrix is a table of numbers laid out in rows and columns. A **vector** is simply a matrix consisting of a single column of numbers.

There are several important notational conventions used here for matrices and vectors. Boldface capital letters such as $\mathbf{A}, \mathbf{B}, \ldots$ are used to denote matrices. Boldface lowercase letters such as $\mathbf{x}, \mathbf{y}, \ldots$ are used to denote vectors. Lowercase roman or Greek letters such as $m, n, \alpha, \beta, \ldots$ will be used to denote scalars.

At times we will need to refer to specific parts of a matrix. The notation $A_{i,j}$ denotes the element of the matrix $\mathbf{A}$ in row i and column j. We denote the jth element of the vector $\mathbf{x}$ by x_j. The notation $\mathbf{A}_{.,j}$ is used to refer to column j of the matrix $\mathbf{A}$, while $\mathbf{A}_{i,.}$ refers to row i of $\mathbf{A}$. See also Appendix D.

We can also build up larger matrices from smaller matrices. For example, the notation $\mathbf{A} = [\mathbf{B} \quad \mathbf{C}]$ means that the matrix $\mathbf{A}$ is composed of the matrices $\mathbf{B}$ and $\mathbf{C}$, with matrix $\mathbf{C}$ beside matrix $\mathbf{B}$.

If $\mathbf{A}$ and $\mathbf{B}$ are two matrices of the same size, we can add them by simply adding corresponding elements. Similarly, we can subtract $\mathbf{B}$ from $\mathbf{A}$ by subtracting the corresponding elements of $\mathbf{B}$ from those of $\mathbf{A}$. We can multiply a scalar times a matrix by multiplying the scalar times each vector element. Because vectors are just n by 1 matrices, we can perform the same arithmetic operations on vectors. A **zero matrix 0** is a matrix composed of all zero elements. A zero matrix plays the same role in matrix algebra as the scalar 0, with

$$\mathbf{A} + \mathbf{0} = \mathbf{A} \tag{A.12}$$

$$= \mathbf{0} + \mathbf{A}. \tag{A.13}$$

In general, matrices and vectors may contain complex numbers as well as real numbers.

Using vector notation, we can write a linear system of equations in **vector form**.

■ **Example A.3** Recall the system of equations (A.9)

$$\begin{aligned} x_1 + x_2 + x_3 &= 0 \\ x_1 + 2x_2 + 2x_3 &= 0 \end{aligned} \tag{A.14}$$

from Example A.2. We can write this in vector form as

$$x_1 \begin{bmatrix} 1 \\ 1 \end{bmatrix} + x_2 \begin{bmatrix} 1 \\ 2 \end{bmatrix} + x_3 \begin{bmatrix} 1 \\ 2 \end{bmatrix} = \begin{bmatrix} 0 \\ 0 \end{bmatrix}. \tag{A.15}$$

∎

The expression on the left-hand side of (A.15) where vectors are multiplied by scalars and the results are summed together is called a **linear combination**.

If $\mathbf{A}$ is an m by n matrix, and $\mathbf{x}$ is an n-element vector, we can multiply $\mathbf{A}$ times $\mathbf{x}$, where the product is defined by

$$\mathbf{A}\mathbf{x} = x_1 \mathbf{A}_{\cdot,1} + x_2 \mathbf{A}_{\cdot,2} + \cdots + x_n \mathbf{A}_{\cdot,n}. \tag{A.16}$$

∎ **Example A.4** Given

$$\mathbf{A} = \begin{bmatrix} 1 & 2 & 3 \\ 4 & 5 & 6 \end{bmatrix} \tag{A.17}$$

and

$$\mathbf{x} = \begin{bmatrix} 1 \\ 0 \\ 2 \end{bmatrix} \tag{A.18}$$

then

$$\mathbf{A}\mathbf{x} = 1 \begin{bmatrix} 1 \\ 4 \end{bmatrix} + 0 \begin{bmatrix} 2 \\ 5 \end{bmatrix} + 2 \begin{bmatrix} 3 \\ 6 \end{bmatrix} = \begin{bmatrix} 7 \\ 16 \end{bmatrix}. \tag{A.19}$$

∎

The formula (A.16) for $\mathbf{A}\mathbf{x}$ is a linear combination much like the one that occurred in the vector form of a system of equations. It is possible to write any linear system of equations in the form $\mathbf{A}\mathbf{x} = \mathbf{b}$, where $\mathbf{A}$ is a matrix containing the coefficients of the variables in the equations, $\mathbf{b}$ is a vector containing the numbers on the right-hand sides of the equations, and $\mathbf{x}$ is a vector of variables.

∎ **Definition A.2** If $\mathbf{A}$ is a matrix of size m by n, and $\mathbf{B}$ is a matrix of size n by r, then the product $\mathbf{C} = \mathbf{A}\mathbf{B}$ is obtained by multiplying $\mathbf{A}$ times each of the columns of $\mathbf{B}$ and assembling the matrix vector products in $\mathbf{C}$:

$$\mathbf{C} = \begin{bmatrix} \mathbf{A}\mathbf{B}_{\cdot,1} & \mathbf{A}\mathbf{B}_{\cdot,2} & \dots & \mathbf{A}\mathbf{B}_{\cdot,r} \end{bmatrix}. \tag{A.20}$$

This approach given in (A.20) for calculating a matrix–matrix product will be referred to as the **matrix–vector method**.   ∎

Note that the product (A.20) is only possible if the two matrices are of compatible sizes. If **A** has m rows and n columns, and **B** has n rows and r columns, then the product **AB** exists and is of size m by r. In some cases, it is thus possible to multiply **AB** but not **BA**. It is important to note that when both **AB** and **BA** exist, **AB** is not generally equal to **BA**!

An alternate way to compute the product of two matrices is the **row–column expansion method**, where the product element $C_{i,j}$ is calculated as the matrix product of row i of **A** and column j of **B**.

■ **Example A.5** Let

$$\mathbf{A} = \begin{bmatrix} 1 & 2 \\ 3 & 4 \\ 5 & 6 \end{bmatrix} \tag{A.21}$$

and

$$\mathbf{B} = \begin{bmatrix} 5 & 2 \\ 3 & 7 \end{bmatrix}. \tag{A.22}$$

The product matrix $\mathbf{C} = \mathbf{AB}$ will be of size 3 by 2. We compute the product using both methods. First, using the matrix–vector approach (A.20), we have

$$\mathbf{C} = \begin{bmatrix} \mathbf{AB}_{\cdot,1} & \mathbf{AB}_{\cdot,2} \end{bmatrix} \tag{A.23}$$

$$= \begin{bmatrix} 5\begin{bmatrix} 1 \\ 3 \\ 5 \end{bmatrix} + 3\begin{bmatrix} 2 \\ 4 \\ 6 \end{bmatrix} & 2\begin{bmatrix} 1 \\ 3 \\ 5 \end{bmatrix} + 7\begin{bmatrix} 2 \\ 4 \\ 6 \end{bmatrix} \end{bmatrix} \tag{A.24}$$

$$= \begin{bmatrix} 11 & 16 \\ 27 & 34 \\ 43 & 52 \end{bmatrix}. \tag{A.25}$$

Next, we use the row–column approach:

$$\mathbf{C} = \begin{bmatrix} 1 \cdot 5 + 2 \cdot 3 & 1 \cdot 2 + 2 \cdot 7 \\ 3 \cdot 5 + 4 \cdot 3 & 3 \cdot 2 + 4 \cdot 7 \\ 5 \cdot 5 + 6 \cdot 3 & 5 \cdot 2 + 6 \cdot 7 \end{bmatrix} \tag{A.26}$$

$$= \begin{bmatrix} 11 & 16 \\ 27 & 34 \\ 43 & 52 \end{bmatrix}. \tag{A.27}$$

■

■ **Definition A.3** The n by n **identity matrix** $\mathbf{I}_n$ is composed of ones in the diagonal and zeros in the off-diagonal elements.   ■

For example, the 3 by 3 identity matrix is

$$\mathbf{I}_3 = \begin{bmatrix} 1 & 0 & 0 \\ 0 & 1 & 0 \\ 0 & 0 & 1 \end{bmatrix}. \tag{A.28}$$

We often write $\mathbf{I}$ without specifying the size of the matrix in situations where the size of matrix is obvious from the context. It is easily shown that if $\mathbf{A}$ is an m by n matrix, then

$$\mathbf{AI}_n = \mathbf{A} \tag{A.29}$$

$$= \mathbf{I}_m\mathbf{A}. \tag{A.30}$$

Thus, multiplying by $\mathbf{I}$ in matrix algebra is similar to multiplying by 1 in conventional scalar algebra.

We have not defined matrix division, but it is possible at this point to define the matrix algebra equivalent of the reciprocal.

■ **Definition A.4** If $\mathbf{A}$ is an n by n matrix, and there is a matrix $\mathbf{B}$ such that

$$\mathbf{AB} = \mathbf{BA} = \mathbf{I}, \tag{A.31}$$

then $\mathbf{B}$ is the **inverse** of $\mathbf{A}$. We write $\mathbf{B} = \mathbf{A}^{-1}$.   ■

How do we compute the inverse of a matrix? If $\mathbf{AB} = \mathbf{I}$, then

$$\begin{bmatrix} \mathbf{AB}_{\cdot,1} & \mathbf{AB}_{\cdot,2} & \dots & \mathbf{AB}_{\cdot,n} \end{bmatrix} = \mathbf{I}. \tag{A.32}$$

Since the columns of the identity matrix and $\mathbf{A}$ are known, we can solve

$$\mathbf{AB}_{\cdot,1} = \begin{bmatrix} 1 & 0 & \dots & 0 \end{bmatrix}^T \tag{A.33}$$

to obtain $\mathbf{B}_{\cdot,1}$. We can find the remaining columns of the inverse in the same way. If any of these systems of equations are inconsistent, then $\mathbf{A}^{-1}$ does not exist.

The inverse matrix can be used to solve a system of linear equations with n equations and n variables. Given the system of equations $\mathbf{Ax} = \mathbf{b}$, and $\mathbf{A}^{-1}$, we can multiply $\mathbf{Ax} = \mathbf{b}$ on both sides by the inverse to obtain

$$\mathbf{A}^{-1}\mathbf{Ax} = \mathbf{A}^{-1}\mathbf{b}. \tag{A.34}$$

Because

$$\mathbf{A}^{-1}\mathbf{Ax} = \mathbf{Ix} \tag{A.35}$$

$$= \mathbf{x} \tag{A.36}$$

this gives the solution

$$\mathbf{x} = \mathbf{A}^{-1}\mathbf{b}. \tag{A.37}$$

This argument shows that if $\mathbf{A}^{-1}$ exists, then for any right-hand side $\mathbf{b}$, a system of equations has a unique solution. If $\mathbf{A}^{-1}$ does not exist, then the system of equations may either have many solutions or no solution.

■ **Definition A.5** When $\mathbf{A}$ is an n by n matrix, $\mathbf{A}^k$ is the product of k copies of $\mathbf{A}$. By convention, we define $\mathbf{A}^0 = \mathbf{I}$.   ■

■ **Definition A.6** The **transpose** of a matrix $\mathbf{A}$, denoted $\mathbf{A}^T$, is obtained by taking the columns of $\mathbf{A}$ and writing them as the rows of the transpose. We will also use the notation $\mathbf{A}^{-T}$ for $(\mathbf{A}^{-1})^T$.   ■

■ **Example A.6** Let

$$\mathbf{A} = \begin{bmatrix} 2 & 1 \\ 5 & 2 \end{bmatrix}. \tag{A.38}$$

Then

$$\mathbf{A}^T = \begin{bmatrix} 2 & 5 \\ 1 & 2 \end{bmatrix}. \tag{A.39}$$

■

■ **Definition A.7** A matrix is **symmetric** if $\mathbf{A} = \mathbf{A}^T$.   ■

Although many elementary textbooks on linear algebra consider only square diagonal matrices, we will have occasion to refer to m by n matrices that have nonzero elements only on the diagonal.

■ **Definition A.8** An m by n matrix $\mathbf{A}$ is **diagonal** if $A_{i,j} = 0$ whenever $i \neq j$.   ■

■ **Definition A.9** An m by n matrix $\mathbf{R}$ is **upper-triangular** if $R_{i,j} = 0$ whenever $i > j$. A matrix $\mathbf{L}$ is **lower-triangular** if $\mathbf{L}^T$ is upper-triangular.   ■

■ **Example A.7** The matrix

$$\mathbf{S} = \begin{bmatrix} 1 & 0 & 0 & 0 & 0 \\ 0 & 2 & 0 & 0 & 0 \\ 0 & 0 & 3 & 0 & 0 \end{bmatrix} \tag{A.40}$$

is diagonal, and the matrix

$$\mathbf{R} = \begin{bmatrix} 1 & 2 & 3 \\ 0 & 2 & 4 \\ 0 & 0 & 5 \\ 0 & 0 & 0 \end{bmatrix} \tag{A.41}$$

is upper-triangular.   ■

■ **Theorem A.1**   The following statements are true for any scalars s and t and matrices $\mathbf{A}$, $\mathbf{B}$, and $\mathbf{C}$. It is assumed that the matrices are of the appropriate size for the operations involved and that whenever an inverse occurs, the matrix is invertible.

1. $\mathbf{A} + \mathbf{0} = \mathbf{0} + \mathbf{A} = \mathbf{A}$.
2. $\mathbf{A} + \mathbf{B} = \mathbf{B} + \mathbf{A}$.
3. $(\mathbf{A} + \mathbf{B}) + \mathbf{C} = \mathbf{A} + (\mathbf{B} + \mathbf{C})$.
4. $\mathbf{A}(\mathbf{BC}) = (\mathbf{AB})\mathbf{C}$.
5. $\mathbf{A}(\mathbf{B} + \mathbf{C}) = \mathbf{AB} + \mathbf{AC}$.
6. $(\mathbf{A} + \mathbf{B})\mathbf{C} = \mathbf{AC} + \mathbf{BC}$.
7. $(st)\mathbf{A} = s(t\mathbf{A})$.
8. $s(\mathbf{AB}) = (s\mathbf{A})\mathbf{B} = \mathbf{A}(s\mathbf{B})$.
9. $(s + t)\mathbf{A} = s\mathbf{A} + t\mathbf{A}$.
10. $s(\mathbf{A} + \mathbf{B}) = s\mathbf{A} + s\mathbf{B}$.
11. $(\mathbf{A}^T)^T = \mathbf{A}$.
12. $(s\mathbf{A})^T = s(\mathbf{A}^T)$.
13. $(\mathbf{A} + \mathbf{B})^T = \mathbf{A}^T + \mathbf{B}^T$.
14. $(\mathbf{AB})^T = \mathbf{B}^T \mathbf{A}^T$.
15. $(\mathbf{AB})^{-1} = \mathbf{B}^{-1}\mathbf{A}^{-1}$.
16. $(\mathbf{A}^{-1})^{-1} = \mathbf{A}$.
17. $(\mathbf{A}^T)^{-1} = (\mathbf{A}^{-1})^T$.
18. If $\mathbf{A}$ and $\mathbf{B}$ are n by n matrices, and $\mathbf{AB} = \mathbf{I}$, then $\mathbf{A}^{-1} = \mathbf{B}$ and $\mathbf{B}^{-1} = \mathbf{A}$.   ■

The first 10 rules in this list are identical to rules of conventional algebra, and you should have little trouble in applying them. The rules involving transposes and inverses may be new, but they can be mastered without too much trouble.

Some students have difficulty with the following statements, which would appear to be true on the surface, but that are in fact **false** for at least some matrices.

1. $\mathbf{AB} = \mathbf{BA}$.
2. If $\mathbf{AB} = \mathbf{0}$, then $\mathbf{A} = \mathbf{0}$ or $\mathbf{B} = \mathbf{0}$.
3. If $\mathbf{AB} = \mathbf{AC}$ and $\mathbf{A} \neq \mathbf{0}$, then $\mathbf{B} = \mathbf{C}$.

It is a worthwhile exercise to construct examples of 2 by 2 matrices for which each of these statements is false.

A.3 LINEAR INDEPENDENCE

■ **Definition A.10** The vectors $\mathbf{v}_1, \mathbf{v}_2, \ldots, \mathbf{v}_n$ are **linearly independent** if the system of equations

$$c_1\mathbf{v}_1 + c_2\mathbf{v}_2 + \cdots + c_n\mathbf{v}_n = \mathbf{0} \qquad\qquad (A.42)$$

has only the trivial solution $\mathbf{c} = \mathbf{0}$. If there are multiple solutions, then the vectors are **linearly dependent**.   ■

Determining whether or not a set of vectors is linearly independent is simple. Just solve the system of equations (A.42).

■ **Example A.8** Let

$$\mathbf{A} = \begin{bmatrix} 1 & 2 & 3 \\ 4 & 5 & 6 \\ 7 & 8 & 9 \end{bmatrix}. \qquad\qquad (A.43)$$

Are the columns of $\mathbf{A}$ linearly independent vectors? To determine this we set up the system of equations $\mathbf{Ax} = \mathbf{0}$ in an augmented matrix, and then find the RREF

$$\left[\begin{array}{ccc|c} 1 & 0 & -1 & 0 \\ 0 & 1 & 2 & 0 \\ 0 & 0 & 0 & 0 \end{array}\right]. \qquad\qquad (A.44)$$

The solutions are

$$\mathbf{x} = x_3 \begin{bmatrix} 1 \\ -2 \\ 1 \end{bmatrix}. \qquad\qquad (A.45)$$

We can set $x_3 = 1$ and obtain the nonzero solution

$$\mathbf{x} = \begin{bmatrix} 1 \\ -2 \\ 1 \end{bmatrix}. \qquad\qquad (A.46)$$

Thus, the columns of $\mathbf{A}$ are linearly dependent.   ■

There are a number of important theoretical consequences of linear independence. For example, it can be shown that if the columns of an n by n matrix $\mathbf{A}$ are linearly independent, then $\mathbf{A}^{-1}$ exists, and the system of equations $\mathbf{Ax} = \mathbf{b}$ has a unique solution for every right-hand side $\mathbf{b}$ [91].

A.4 SUBSPACES OF R^n

So far, we have worked with vectors of real numbers in the n-dimensional space R^n. There are a number of properties of R^n that make it convenient to work with vectors. First, the operation of vector addition always works. We can take any two vectors in R^n and add them together and get another vector in R^n. Second, we can multiply any vector in R^n by a scalar and obtain another vector in R^n. Finally, we have the **0** vector, with the property that for any vector **x**, $\mathbf{x} + \mathbf{0} = \mathbf{0} + \mathbf{x} = \mathbf{x}$.

■ **Definition A.11** A **subspace** W of R^n is a subset of R^n which satisfies the following three properties:

1.   If **x** and **y** are vectors in W, then $\mathbf{x} + \mathbf{y}$ is also a vector in W.
2.   If **x** is a vector in W and s is any scalar, then $s\mathbf{x}$ is also a vector in W.
3.   The **0** vector is in W.   ■

■ **Example A.9** In R^3, the plane P defined by the equation

$$x_1 + x_2 + x_3 = 0 \tag{A.47}$$

is a subspace of R^n. To see this, note that if we take any two vectors in the plane and add them together, we get another vector in the plane. If we take a vector in this plane and multiply it by any scalar, we get another vector in the plane. Finally, **0** is a vector in the plane.   ■

Subspaces are important because they provide an environment within which all of the rules of matrix–vector algebra apply. An especially important subspace of R^n that we will work with is the **null space** of an m by n matrix.

■ **Definition A.12** Let **A** be an m by n matrix. The null space of **A**, written $N(\mathbf{A})$, is the set of all vectors **x** such that $\mathbf{A}\mathbf{x} = \mathbf{0}$.   ■

To show that $N(\mathbf{A})$ is actually a subspace of R^n, we need to show that:

1.   If **x** and **y** are in $N(\mathbf{A})$, then $\mathbf{A}\mathbf{x} = \mathbf{0}$ and $\mathbf{A}\mathbf{y} = \mathbf{0}$. By adding these equations, we find that $\mathbf{A}(\mathbf{x} + \mathbf{y}) = \mathbf{0}$. Thus $\mathbf{x} + \mathbf{y}$ is in $N(\mathbf{A})$.
2.   If **x** is in $N(\mathbf{A})$ and s is any scalar, then $\mathbf{A}\mathbf{x} = \mathbf{0}$. We can multiply this equation by s to get $s\mathbf{A}\mathbf{x} = \mathbf{0}$. Thus $\mathbf{A}(s\mathbf{x}) = \mathbf{0}$, and $s\mathbf{x}$ is in $N(\mathbf{A})$.
3. $\mathbf{A}\mathbf{0} = \mathbf{0}$, so **0** is in $N(\mathbf{A})$.

Computationally, the null space of a matrix can be determined by solving the system of equations $\mathbf{A}\mathbf{x} = \mathbf{0}$.

■ **Example A.10** Let

$$
\mathbf{A} = \begin{bmatrix} 3 & 1 & 9 & 4 \\ 2 & 1 & 7 & 3 \\ 5 & 2 & 16 & 7 \end{bmatrix}. \tag{A.48}
$$

To find the null space of $\mathbf{A}$, we solve the system of equations $\mathbf{Ax} = \mathbf{0}$. To solve the equations, we put the system of equations into an augmented matrix

$$
\begin{bmatrix} 3 & 1 & 9 & 4 & | & 0 \\ 2 & 1 & 7 & 3 & | & 0 \\ 5 & 2 & 16 & 7 & | & 0 \end{bmatrix} \tag{A.49}
$$

and find the RREF

$$
\begin{bmatrix} 1 & 0 & 2 & 1 & | & 0 \\ 0 & 1 & 3 & 1 & | & 0 \\ 0 & 0 & 0 & 0 & | & 0 \end{bmatrix}. \tag{A.50}
$$

From the augmented matrix, we find that

$$
\mathbf{x} = x_3 \begin{bmatrix} -2 \\ -3 \\ 1 \\ 0 \end{bmatrix} + x_4 \begin{bmatrix} -1 \\ -1 \\ 0 \\ 1 \end{bmatrix}. \tag{A.51}
$$

Any vector in the null space can be written as a linear combination of the above vectors, so the null space is a two-dimensional plane within R^4.

Now, consider the problem of solving $\mathbf{Ax} = \mathbf{b}$, where

$$
\mathbf{b} = \begin{bmatrix} 22 \\ 17 \\ 39 \end{bmatrix} \tag{A.52}
$$

and one particular solution is

$$
\mathbf{p} = \begin{bmatrix} 1 \\ 2 \\ 1 \\ 2 \end{bmatrix}. \tag{A.53}
$$

We can take any vector in the null space of $\mathbf{A}$ and add it to this solution to obtain another solution. Suppose that $\mathbf{x}$ is in $N(\mathbf{A})$. Then

$$\mathbf{A}(\mathbf{x} + \mathbf{p}) = \mathbf{A}\mathbf{x} + \mathbf{A}\mathbf{p}$$

$$\mathbf{A}(\mathbf{x} + \mathbf{p}) = \mathbf{0} + \mathbf{b}$$

$$\mathbf{A}(\mathbf{x} + \mathbf{p}) = \mathbf{b}.$$

For example,

$$\mathbf{x} = \begin{bmatrix} 1 \\ 2 \\ 1 \\ 2 \end{bmatrix} + 2 \begin{bmatrix} -2 \\ -3 \\ 1 \\ 0 \end{bmatrix} + 3 \begin{bmatrix} -1 \\ -1 \\ 0 \\ 1 \end{bmatrix} \tag{A.54}$$

is also a solution to $\mathbf{A}\mathbf{x} = \mathbf{b}$.   ■

In the context of inverse problems, the null space is critical because the presence of a nontrivial null space leads to nonuniqueness in the solution to a linear system of equations.

■ **Definition A.13** A **basis** for a subspace W is a set of vectors $\mathbf{v}_1, \ldots, \mathbf{v}_p$ such that

1. Any vector in W can be written as a linear combination of the basis vectors.
2. The basis vectors are linearly independent.   ■

A particularly simple and useful basis is the **standard basis**.

■ **Definition A.14** The **standard basis** for R^n is the set of vectors $\mathbf{e}_1, \ldots, \mathbf{e}_n$ such that the elements of $\mathbf{e}_i$ are all zero, except for the ith element, which is one.   ■

Any nontrivial subspace W of R^n will have an infinite number of different bases. For example, we can take any basis and multiply one of the basis vectors by 2 to obtain a new basis. It is possible to show that all bases for a subspace W have the same number of basis vectors [91].

■ **Theorem A.2** Let W be a subspace of R^n with basis $\mathbf{v}_1, \ldots, \mathbf{v}_p$. Then all bases for W have p basis vectors, and p is the **dimension** of W.   ■

It can be shown that the procedure used in the foregoing example always produces a basis for $N(\mathbf{A})$ [91].

■ **Definition A.15** Let $\mathbf{A}$ be an m by n matrix. The **column space** or **range** of $\mathbf{A}$ [written $R(\mathbf{A})$] is the set of all vectors $\mathbf{b}$ such that $\mathbf{A}\mathbf{x} = \mathbf{b}$ has at least one solution. In other words, the column space is the set of all vectors $\mathbf{b}$ that can be written as a linear combination of the columns of $\mathbf{A}$.   ■

The range is important in the context of discrete linear inverse problems, because $R(\mathbf{G})$ consists of all vectors $\mathbf{d}$ for which there is a model $\mathbf{m}$ such that $\mathbf{Gm} = \mathbf{d}$.

To find the column space of a matrix, we consider what happens when we compute the RREF of $[\mathbf{A} \mid \mathbf{b}]$. In the part of the augmented matrix corresponding to the left-hand side of the equations we always get the same result, namely the RREF of $\mathbf{A}$. The solution to the system of equations may involve some free variables, but we can always set these free variables to 0. Thus when we are able to solve $\mathbf{Ax} = \mathbf{b}$, we can solve the system of equations by using only variables corresponding to the pivot columns in the RREF of $\mathbf{A}$. In other words, if we can solve $\mathbf{Ax} = \mathbf{b}$, then we can write $\mathbf{b}$ as a linear combination of the pivot columns of $\mathbf{A}$. Note that these are columns from the original matrix $\mathbf{A}$, not columns from the RREF of $\mathbf{A}$.

■ **Example A.11** As in the previous example, let

$$\mathbf{A} = \begin{bmatrix} 3 & 1 & 9 & 4 \\ 2 & 1 & 7 & 3 \\ 5 & 2 & 16 & 7 \end{bmatrix}. \tag{A.55}$$

We want to find the column space of $\mathbf{A}$. We already know from Example A.10 that the RREF of $\mathbf{A}$ is

$$\begin{bmatrix} 1 & 0 & 2 & 1 \\ 0 & 1 & 3 & 1 \\ 0 & 0 & 0 & 0 \end{bmatrix}. \tag{A.56}$$

Thus whenever we can solve $\mathbf{Ax} = \mathbf{b}$, we can find a solution in which x_3 and x_4 are 0. In other words, whenever there is a solution to $\mathbf{Ax} = \mathbf{b}$, we can write $\mathbf{b}$ as a linear combination of the first two columns of $\mathbf{A}$:

$$\mathbf{b} = x_1 \begin{bmatrix} 3 \\ 2 \\ 5 \end{bmatrix} + x_2 \begin{bmatrix} 1 \\ 1 \\ 2 \end{bmatrix}. \tag{A.57}$$

Since these two vectors are linearly independent and span $R(\mathbf{A})$, they form a basis for $R(\mathbf{A})$. The dimension of $R(\mathbf{A})$ is two.   ■

In finding the null space and range of a matrix $\mathbf{A}$ we found that the basis vectors for $N(\mathbf{A})$ corresponded to nonpivot columns of $\mathbf{A}$, while the basis vectors for $R(\mathbf{A})$ corresponded to pivot columns of $\mathbf{A}$. Since the matrix $\mathbf{A}$ had n columns, we obtain the following theorem.

■ **Theorem A.3**

$$\dim N(\mathbf{A}) + \dim R(\mathbf{A}) = n. \tag{A.58}$$

■

In addition to the null space and range of a matrix $\mathbf{A}$, we will often work with the null space and range of the transpose of $\mathbf{A}$. Since the columns of $\mathbf{A}^T$ are rows of $\mathbf{A}$, the column space of $\mathbf{A}^T$ is also called the **row space** of $\mathbf{A}$. Since each row of $\mathbf{A}$ can be written as a linear combination of the nonzero rows of the RREF of $\mathbf{A}$, the nonzero rows of the RREF form a basis for the row space of $\mathbf{A}$. There are exactly as many nonzero rows in the RREF of $\mathbf{A}$ as there are pivot columns. Thus we have the following theorem.

■ **Theorem A.4**

$$\dim R(\mathbf{A}^T) = \dim R(\mathbf{A}). \tag{A.59}$$

■

■ **Definition A.16** The **rank** of an m by n matrix $\mathbf{A}$ is the dimension of $R(\mathbf{A})$. If $\text{rank}(\mathbf{A}) = \min(m, n)$, then $\mathbf{A}$ has **full rank**. If $\text{rank}(\mathbf{A}) = m$, then $\mathbf{A}$ has **full row rank**. If $\text{rank}(\mathbf{A}) = n$, then $\mathbf{A}$ has **full column rank**. If $\text{rank}(\mathbf{A}) < \min(m, n)$, then $\mathbf{A}$ is **rank-deficient**.   ■

The rank of a matrix is readily found in MATLAB by using the **rank** command.

A.5 ORTHOGONALITY AND THE DOT PRODUCT

■ **Definition A.17** Let $\mathbf{x}$ and $\mathbf{y}$ be two vectors in R^n. The **dot product** of $\mathbf{x}$ and $\mathbf{y}$ is

$$\mathbf{x} \cdot \mathbf{y} = \mathbf{x}^T \mathbf{y}$$
$$= x_1 y_1 + x_2 y_2 + \cdots + x_n y_n. \tag{A.60}$$

■

■ **Definition A.18** Let $\mathbf{x}$ be a vector in R^n. The **2-norm** or **Euclidean length** of $\mathbf{x}$ is

$$\|\mathbf{x}\|_2 = \sqrt{\mathbf{x}^T \mathbf{x}}$$
$$= \sqrt{x_1^2 + x_2^2 + \cdots + x_n^2}. \tag{A.61}$$

■

Later we will introduce two other ways of measuring the "length" of a vector. The subscript 2 is used to distinguish this 2-norm from the other norms.

You may be familiar with an alternative definition of the dot product in which $\mathbf{x} \cdot \mathbf{y} = \|\mathbf{x}\|_2 \|\mathbf{y}\|_2 \cos(\theta)$ where θ is the angle between the two vectors. The two definitions are equivalent. To see this, consider a triangle with sides $\mathbf{x}$, $\mathbf{y}$, and $\mathbf{x} - \mathbf{y}$. See Figure A.1. The angle

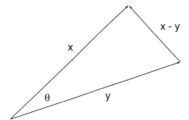

Figure A.1   Relationship between the dot product and the angle between two vectors.

between sides **x** and **y** is θ. By the law of cosines,

$$\|\mathbf{x} - \mathbf{y}\|_2^2 = \|\mathbf{x}\|_2^2 + \|\mathbf{y}\|_2^2 - 2\|\mathbf{x}\|_2\|\mathbf{y}\|_2 \cos(\theta)$$

$$(\mathbf{x} - \mathbf{y})^T (\mathbf{x} - \mathbf{y}) = \mathbf{x}^T\mathbf{x} + \mathbf{y}^T\mathbf{y} - 2\|\mathbf{x}\|_2\|\mathbf{y}\|_2 \cos(\theta)$$

$$\mathbf{x}^T\mathbf{x} - 2\mathbf{x}^T\mathbf{y} + \mathbf{y}^T\mathbf{y} = \mathbf{x}^T\mathbf{x} + \mathbf{y}^T\mathbf{y} - 2\|\mathbf{x}\|_2\|\mathbf{y}\|_2 \cos(\theta)$$

$$-2\mathbf{x}^T\mathbf{y} = -2\|\mathbf{x}\|_2\|\mathbf{y}\|_2 \cos(\theta)$$

$$\mathbf{x}^T\mathbf{y} = \|\mathbf{x}\|_2\|\mathbf{y}\|_2 \cos(\theta).$$

We can also use this formula to compute the angle between two vectors.

$$\theta = \cos^{-1}\left(\frac{\mathbf{x}^T\mathbf{y}}{\|\mathbf{x}\|_2\|\mathbf{y}\|_2}\right). \tag{A.62}$$

■ **Definition A.19** Two vectors **x** and **y** in R^n are **orthogonal**, or equivalently, **perpendicular** (written $\mathbf{x} \perp \mathbf{y}$), if $\mathbf{x}^T\mathbf{y} = 0$.   ■

■ **Definition A.20** A set of vectors $\mathbf{v}_1, \ldots, \mathbf{v}_p$ is **orthogonal** if each pair of vectors in the set is orthogonal.   ■

■ **Definition A.21** Two subspaces V and W of R^n are **orthogonal** if every vector in V is perpendicular to every vector in W.   ■

If **x** is in $N(\mathbf{A})$, then $\mathbf{A}\mathbf{x} = \mathbf{0}$. Since each element of the product $\mathbf{A}\mathbf{x}$ can be obtained by taking the dot product of a row of **A** and **x**, **x** is perpendicular to each row of **A**. Since **x** is perpendicular to all of the columns of $\mathbf{A}^T$, it is perpendicular to $R(\mathbf{A}^T)$. We have the following theorem.

■ **Theorem A.5** Let $\mathbf{A}$ be an m by n matrix. Then

$$N(\mathbf{A}) \perp R(\mathbf{A}^T). \tag{A.63}$$

Furthermore,

$$N(\mathbf{A}) + R(\mathbf{A}^T) = R^n. \tag{A.64}$$

That is, any vector $\mathbf{x}$ in R^n can be written uniquely as $\mathbf{x} = \mathbf{p} + \mathbf{q}$ where $\mathbf{p}$ is in $N(\mathbf{A})$ and $\mathbf{q}$ is in $R(\mathbf{A}^T)$. ■

■ **Definition A.22** A basis in which the basis vectors are orthogonal is an **orthogonal basis**. A basis in which the basis vectors are orthogonal and have length one is an **orthonormal basis**. ■

■ **Definition A.23** An n by n matrix $\mathbf{Q}$ is **orthogonal** if the columns of $\mathbf{Q}$ are orthogonal and each column of $\mathbf{Q}$ has length one. ■

With the requirement that the columns of an orthogonal matrix have length one, using the term "orthonormal" would make logical sense. However, the definition of "orthogonal" given here is standard.

Orthogonal matrices have a number of useful properties.

■ **Theorem A.6** If $\mathbf{Q}$ is an orthogonal matrix, then:

1. $\mathbf{Q}^T\mathbf{Q} = \mathbf{Q}\mathbf{Q}^T = \mathbf{I}$. In other words, $\mathbf{Q}^{-1} = \mathbf{Q}^T$.
2. For any vector $\mathbf{x}$ in R^n, $\|\mathbf{Q}\mathbf{x}\|_2 = \|\mathbf{x}\|_2$.
3. For any two vectors $\mathbf{x}$ and $\mathbf{y}$ in R^n, $\mathbf{x}^T\mathbf{y} = (\mathbf{Q}\mathbf{x})^T(\mathbf{Q}\mathbf{y})$. ■

A problem that we will often encounter in practice is projecting a vector $\mathbf{x}$ onto another vector $\mathbf{y}$ or onto a subspace W to obtain a projected vector $\mathbf{p}$. See Figure A.2. We know that

$$\mathbf{x}^T\mathbf{y} = \|\mathbf{x}\|_2 \|\mathbf{y}\|_2 \cos(\theta) \tag{A.65}$$

where θ is the angle between $\mathbf{x}$ and $\mathbf{y}$. Also,

$$\cos(\theta) = \frac{\|\mathbf{p}\|_2}{\|\mathbf{x}\|_2}. \tag{A.66}$$

Thus

$$\|\mathbf{p}\|_2 = \frac{\mathbf{x}^T\mathbf{y}}{\|\mathbf{y}\|_2}. \tag{A.67}$$

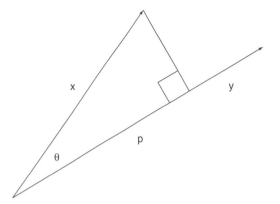

Figure A.2   The orthogonal projection of **x** onto **y**.

Since **p** points in the same direction as **y**,

$$\mathbf{p} = \left[\mathrm{proj}_\mathbf{y}\mathbf{x}\right] = \frac{\mathbf{x}^T\mathbf{y}}{\mathbf{y}^T\mathbf{y}}\mathbf{y}. \tag{A.68}$$

The vector **p** is called the **orthogonal projection** or simply the **projection** of **x** onto **y**, sometimes abbreviated as $\mathrm{proj}_\mathbf{y}\,\mathbf{x}$.

Similarly, if W is a subspace of R^n with an orthogonal basis $\mathbf{w}_1, \mathbf{w}_2, \ldots, \mathbf{w}_p$, then the **orthogonal projection of x onto** W is

$$\mathrm{proj}_W\mathbf{x} = \frac{\mathbf{x}^T\mathbf{w}_1}{\mathbf{w}_1^T\mathbf{w}_1}\mathbf{w}_1 + \frac{\mathbf{x}^T\mathbf{w}_2}{\mathbf{w}_2^T\mathbf{w}_2}\mathbf{w}_2 + \cdots + \frac{\mathbf{x}^T\mathbf{w}_p}{\mathbf{w}_p^T\mathbf{w}_p}\mathbf{w}_p. \tag{A.69}$$

It is inconvenient that the projection formula requires an orthogonal basis. The **Gram–Schmidt orthogonalization process** can be used to turn any basis for a subspace of R^n into an orthogonal basis. We begin with a basis $\mathbf{v}_1, \mathbf{v}_2, \ldots, \mathbf{v}_p$. The process recursively constructs an orthogonal basis by taking each vector in the original basis and then subtracting off its projection on the space spanned by the previous vectors. The formulas are

$$\mathbf{w}_1 = \mathbf{v}_1$$

$$\mathbf{w}_2 = \mathbf{v}_2 - \frac{\mathbf{v}_1^T\mathbf{v}_2}{\mathbf{v}_1^T\mathbf{v}_1}\mathbf{v}_1$$

$$\cdots$$

$$\mathbf{w}_p = \mathbf{v}_p - \frac{\mathbf{w}_1^T\mathbf{v}_p}{\mathbf{w}_1^T\mathbf{w}_1}\mathbf{w}_1 - \cdots - \frac{\mathbf{w}_p^T\mathbf{v}_p}{\mathbf{w}_p^T\mathbf{w}_p}\mathbf{w}_p. \tag{A.70}$$

Unfortunately, the Gram–Schmidt process is numerically unstable when applied to large bases. In MATLAB the command **orth** provides a numerically stable way to produce an orthogonal basis from a nonorthogonal basis. An important property of orthogonal projection is that the projection of $\mathbf{x}$ onto W is the point in W which is closest to $\mathbf{x}$. In the special case that $\mathbf{x}$ is in W, the projection of $\mathbf{x}$ onto W is $\mathbf{x}$.

Given an inconsistent system of equations $\mathbf{Ax} = \mathbf{b}$, it is often desirable to find an approximate solution. A natural measure of the quality of an approximate solution is the norm of the difference between $\mathbf{Ax}$ and $\mathbf{b}$, $\|\mathbf{Ax} - \mathbf{b}\|_2$. A solution which minimizes the 2-norm, $\|\mathbf{Ax} - \mathbf{b}\|_2$, is called a **least squares solution**, because it minimizes the sum of the squares of the errors.

The least squares solution can be obtained by projecting $\mathbf{b}$ onto the range of $\mathbf{A}$. This calculation requires us to first find an orthogonal basis for $R(\mathbf{A})$. There is an alternative approach that does not require the orthogonal basis. Let

$$\mathbf{Ax} = \mathbf{p}$$
$$= \text{proj}_{R(\mathbf{A})}\mathbf{b}. \tag{A.71}$$

Then $\mathbf{Ax} - \mathbf{b}$ is perpendicular to $R(\mathbf{A})$. In particular, each of the columns of $\mathbf{A}$ is orthogonal to $\mathbf{Ax} - \mathbf{b}$. Thus

$$\mathbf{A}^T(\mathbf{Ax} - \mathbf{b}) = \mathbf{0} \tag{A.72}$$

or

$$\mathbf{A}^T\mathbf{Ax} = \mathbf{A}^T\mathbf{b}. \tag{A.73}$$

This last system of equations is referred to as the **normal equations** for the least squares problem. It can be shown that if the columns of $\mathbf{A}$ are linearly independent, then the normal equations have exactly one solution [91], and this solution minimizes the sum of squared errors.

A.6 EIGENVALUES AND EIGENVECTORS

■ **Definition A.24** An n by n matrix $\mathbf{A}$ has an eigenvalue λ with an associated eigenvector $\mathbf{x}$ if $\mathbf{x}$ is not $\mathbf{0}$, and

$$\mathbf{Ax} = \lambda\mathbf{x}. \tag{A.74}$$

■

The eigenvalues and eigenvectors of a matrix $\mathbf{A}$ are important in analyzing difference equations of the form

$$\mathbf{x}_{k+1} = \mathbf{Ax}_k \tag{A.75}$$

and differential equations of the form

$$\mathbf{x}'(t) = \mathbf{A}\mathbf{x}(t). \tag{A.76}$$

To find eigenvalues and eigenvectors, we rewrite the eigenvector equation (A.74) as

$$(\mathbf{A} - \lambda\mathbf{I})\mathbf{x} = \mathbf{0}. \tag{A.77}$$

To find nonzero eigenvectors, the matrix $\mathbf{A} - \lambda\mathbf{I}$ must be singular. This leads to the **characteristic equation**

$$\det(\mathbf{A} - \lambda\mathbf{I}) = 0 \tag{A.78}$$

where det denotes the determinant. For small matrices (2 by 2 or 3 by 3), it is relatively simple to solve the characteristic equation (A.78) to find the eigenvalues. The eigenvalues can be subsequently substituted into (A.77) and the resulting system can be solved to find corresponding eigenvectors. Note that the eigenvalues can, in general, be complex. For larger matrices, solving the characteristic equation becomes impractical and more sophisticated numerical methods are used. The MATLAB command **eig** can be used to find eigenvalues and eigenvectors of a matrix.

Suppose that we can find a set of n linearly independent eigenvectors $\mathbf{v}_i$ of a matrix $\mathbf{A}$ with associated eigenvalues λ_i. These eigenvectors form a basis for R^n. We can use the eigenvectors to **diagonalize** the matrix as

$$\mathbf{A} = \mathbf{P}\mathbf{\Lambda}\mathbf{P}^{-1} \tag{A.79}$$

where

$$\mathbf{P} = \begin{bmatrix} \mathbf{v}_1 & \mathbf{v}_2 & \dots & \mathbf{v}_n \end{bmatrix} \tag{A.80}$$

and $\mathbf{\Lambda}$ is a diagonal matrix of eigenvalues

$$\Lambda_{ii} = \lambda_i. \tag{A.81}$$

To see that this works, simply compute $\mathbf{AP}$:

$$\mathbf{AP} = \mathbf{A}\begin{bmatrix} \mathbf{v}_1 & \mathbf{v}_2 & \dots & \mathbf{v}_n \end{bmatrix}$$
$$= \begin{bmatrix} \lambda_1\mathbf{v}_1 & \lambda_2\mathbf{v}_2 & \dots & \lambda_n\mathbf{v}_n \end{bmatrix}$$
$$= \mathbf{P}\mathbf{\Lambda}.$$

Thus $\mathbf{A} = \mathbf{P}\mathbf{\Lambda}\mathbf{P}^{-1}$. Not all matrices are diagonalizable, because not all matrices have n linearly independent eigenvectors. However, there is an important special case in which matrices can always be diagonalized.

■ **Theorem A.7** If **A** is a real symmetric matrix, then **A** can be written as

$$\mathbf{A} = \mathbf{Q}\boldsymbol{\Lambda}\mathbf{Q}^{-1} = \mathbf{Q}\boldsymbol{\Lambda}\mathbf{Q}^T \tag{A.82}$$

where **Q** is a real orthogonal matrix of eigenvectors of **A** and **Λ** is a real diagonal matrix of the eigenvalues of **A**.   ■

This **orthogonal diagonalization** of a real symmetric matrix **A** will be useful later on when we consider orthogonal factorizations of general matrices.

The eigenvalues of symmetric matrices are particularly important in the analysis of quadratic forms.

■ **Definition A.25** A **quadratic form** is a function of the form

$$f(\mathbf{x}) = \mathbf{x}^T \mathbf{A}\mathbf{x} \tag{A.83}$$

where **A** is a symmetric n by n matrix. The quadratic form $f(\mathbf{x})$ is **positive definite (PD)** if $f(\mathbf{x}) \geq 0$ for all **x** and $f(\mathbf{x}) = 0$ only when $\mathbf{x} = \mathbf{0}$. The quadratic form is **positive semidefinite** if $f(\mathbf{x}) \geq 0$ for all **x**.   Similarly, a symmetric matrix **A** is positive definite if the associated quadratic form $f(\mathbf{x}) = \mathbf{x}^T \mathbf{A}\mathbf{x}$ is positive definite. The quadratic form is **negative semidefinite** if $-f(\mathbf{x})$ is positive semidefinite. If $f(\mathbf{x})$ is neither positive semidefinite nor negative semidefinite, then $f(\mathbf{x})$ is **indefinite**.   ■

Positive definite quadratic forms have an important application in analytic geometry. Let **A** be a positive definite and symmetric matrix. Then the region defined by the inequality

$$(\mathbf{x} - \mathbf{c})^T \mathbf{A}(\mathbf{x} - \mathbf{c}) \leq \delta \tag{A.84}$$

is an ellipsoidal volume, with its center at **c**. We can diagonalize **A** as

$$\mathbf{A} = \mathbf{P}\boldsymbol{\Lambda}\mathbf{P}^{-1} \tag{A.85}$$

where the columns of **P** are normalized eigenvectors of **A**, and **Λ** is a diagonal matrix whose elements are the eigenvalues of **A**. It can be shown that the ith eigenvector of **A** points in the direction of the ith semimajor axis of the ellipsoid, and the length of the ith semimajor axis is given by $\sqrt{\delta/\lambda_i}$ [91].

An important connection between positive semidefinite matrices and eigenvalues is the following theorem.

■ **Theorem A.8** A symmetric matrix **A** is positive semidefinite if and only if its eigenvalues are greater than or equal to 0.   ■

This provides a convenient way to check whether or not a matrix is positive semidefinite.

The **Cholesky factorization** provides an another way to determine whether or not a symmetric matrix is positive definite.

■ **Theorem A.9** Let $\mathbf{A}$ be an n by n positive definite and symmetric matrix. Then $\mathbf{A}$ can be written uniquely as

$$\mathbf{A} = \mathbf{R}^T \mathbf{R} \qquad\qquad (A.86)$$

where $\mathbf{R}$ is a nonsingular upper-triangular matrix. Furthermore, $\mathbf{A}$ can be factored in this way only if it is positive definite [91].   ■

The MATLAB command **chol** can be used to compute the Cholesky factorization of a symmetric and positive definite matrix.

A.7 VECTOR AND MATRIX NORMS

Although the conventional Euclidean length (A.61) is most commonly used, there are alternative ways to measure the length of a vector.

■ **Definition A.26** Any measure of vector length satisfying the following four conditions is called a **norm**.

1. For any vector $\mathbf{x}$, $\|\mathbf{x}\| \geq 0$.
2. For any vector $\mathbf{x}$ and any scalar s, $\|s\mathbf{x}\| = |s| \|\mathbf{x}\|$.
3. For any vectors $\mathbf{x}$ and $\mathbf{y}$, $\|\mathbf{x} + \mathbf{y}\| \leq \|\mathbf{x}\| + \|\mathbf{y}\|$.
4. $\|\mathbf{x}\| = 0$ if and only if $\mathbf{x} = \mathbf{0}$.

If $\|\mathbf{x}\|$ satisfies conditions 1, 2, and 3, but does not satisfy condition 4, then $\|\mathbf{x}\|$ is called a **seminorm**.   ■

■ **Definition A.27** The p-**norm** of a vector in R^n is defined for $p \geq 1$ by

$$\|\mathbf{x}\|_p = (|x_1|^p + |x_2|^p + \cdots + |x_n|^p)^{1/p}. \qquad (A.87)$$

                                                                                                            ■

It can be shown that for any $p \geq 1$, (A.87) satisfies the conditions of Definition A.26 [49]. The conventional Euclidean length is just the 2-norm, but two other p-norms are also commonly used. The **1-norm** is the sum of the absolute values of the elements in $\mathbf{x}$. The ∞-**norm** is obtained by taking the limit as p goes to infinity. The ∞-norm is the maximum of the absolute values of the elements in $\mathbf{x}$. The MATLAB command **norm** can be used to compute the norm of a vector and has options for the 1, 2, and infinity norms.

The 2-norm is particularly important because of its natural connection with dot products and projections. The projection of a vector onto a subspace is the point in the subspace that is closest to the vector as measured by the 2-norm. We have also seen in (A.73) that the problem of minimizing $\|\mathbf{A}\mathbf{x} - \mathbf{b}\|_2$ can be solved by computing projections or by using the normal equations. In fact, the 2-norm can be tied directly to the dot product by the formula

$$\|\mathbf{x}\|_2 = \sqrt{\mathbf{x}^T \mathbf{x}}. \tag{A.88}$$

The 1- and ∞-norms can also be useful in finding approximate solutions to overdetermined linear systems of equations. To minimize the maximum of the errors, we minimize $\|\mathbf{A}\mathbf{x} - \mathbf{b}\|_\infty$. To minimize the sum of the absolute values of the errors, we minimize $\|\mathbf{A}\mathbf{x} - \mathbf{b}\|_1$. Unfortunately, these minimization problems are generally more difficult to solve than least squares problems.

■ **Definition A.28** Any measure of the size or length of an m by n matrix that satisfies the following five properties can be used as a **matrix norm**.

1. For any matrix $\mathbf{A}$, $\|\mathbf{A}\| \geq 0$.
2. For any matrix $\mathbf{A}$ and any scalar s, $\|s\mathbf{A}\| = |s| \|\mathbf{A}\|$.
3. For any matrices $\mathbf{A}$ and $\mathbf{B}$, $\|\mathbf{A} + \mathbf{B}\| \leq \|\mathbf{A}\| + \|\mathbf{B}\|$.
4. $\|\mathbf{A}\| = 0$ if and only if $\mathbf{A} = \mathbf{0}$.
5. For any two matrices $\mathbf{A}$ and $\mathbf{B}$ of compatible sizes, $\|\mathbf{A}\mathbf{B}\| \leq \|\mathbf{A}\| \|\mathbf{B}\|$. ■

■ **Definition A.29** The *p*-**norm** of a matrix $\mathbf{A}$ is

$$\|\mathbf{A}\|_p = \max_{\|\mathbf{x}\|_p=1} \|\mathbf{A}\mathbf{x}\|_p \tag{A.89}$$

where $\|\mathbf{x}\|_p$ and $\|\mathbf{A}\mathbf{x}\|_p$ are vector *p*-norms, while $\|\mathbf{A}\|_p$ is the matrix *p*-norm of $\mathbf{A}$. ■

Solving the maximization problem of (A.89) to determine a matrix *p*-norm could be extremely difficult. Fortunately, there are simpler formulas for the most commonly used matrix *p*-norms. See Exercises A.15, A.16, and C.4.

$$\|\mathbf{A}\|_1 = \max_j \sum_{i=1}^{m} |A_{i,j}| \tag{A.90}$$

$$\|\mathbf{A}\|_2 = \sqrt{\lambda_{\max}(\mathbf{A}^T \mathbf{A})} \tag{A.91}$$

$$\|\mathbf{A}\|_\infty = \max_i \sum_{j=1}^{n} |A_{i,j}| \tag{A.92}$$

where $\lambda_{\max}(\mathbf{A}^T \mathbf{A})$ denotes the largest eigenvalue of $\mathbf{A}^T \mathbf{A}$.

■ **Definition A.30** The **Frobenius norm** of an m by n matrix is given by

$$\|\mathbf{A}\|_F = \sqrt{\sum_{i=1}^{m}\sum_{j=1}^{n} A_{i,j}^2}. \tag{A.93}$$

■

■ **Definition A.31** A matrix norm and a vector norm are **compatible** if

$$\|\mathbf{Ax}\| \leq \|\mathbf{A}\|\|\mathbf{x}\|. \tag{A.94}$$

■

The matrix p-norm is by its definition compatible with the vector p-norm from which it was derived. It can also be shown that the Frobenius norm of a matrix is compatible with the vector 2-norm [101]. Thus the Frobenius norm is often used with the vector 2-norm.

In practice, the Frobenius norm, 1-norm, and ∞-norm of a matrix are easy to compute, while the 2-norm of a matrix can be difficult to compute for large matrices. The MATLAB command **norm** has options for computing the 1, 2, infinity, and Frobenius norms of a matrix.

A.8 THE CONDITION NUMBER OF A LINEAR SYSTEM

Suppose that we want to solve a system of n equations in n variables

$$\mathbf{Ax} = \mathbf{b}. \tag{A.95}$$

Suppose further that because of measurement errors in b, we actually solve

$$\mathbf{A\hat{x}} = \mathbf{\hat{b}}. \tag{A.96}$$

Can we get a bound on $\|\mathbf{x} - \mathbf{\hat{x}}\|$ in terms of $\|\mathbf{b} - \mathbf{\hat{b}}\|$? Starting with (A.95) and (A.96) we have

$$\mathbf{A}(\mathbf{x} - \mathbf{\hat{x}}) = \mathbf{b} - \mathbf{\hat{b}} \tag{A.97}$$

$$(\mathbf{x} - \mathbf{\hat{x}}) = \mathbf{A}^{-1}(\mathbf{b} - \mathbf{\hat{b}}) \tag{A.98}$$

$$\|\mathbf{x} - \mathbf{\hat{x}}\| = \|\mathbf{A}^{-1}(\mathbf{b} - \mathbf{\hat{b}})\| \tag{A.99}$$

$$\|\mathbf{x} - \mathbf{\hat{x}}\| \leq \|\mathbf{A}^{-1}\|\|\mathbf{b} - \mathbf{\hat{b}}\|. \tag{A.100}$$

This formula provides an absolute bound on the error in the solution. It is also worthwhile to compute a relative error bound.

$$\frac{\|\mathbf{x} - \mathbf{\hat{x}}\|}{\|\mathbf{b}\|} \leq \frac{\|\mathbf{A}^{-1}\|\|\mathbf{b} - \mathbf{\hat{b}}\|}{\|\mathbf{b}\|} \tag{A.101}$$

$$\frac{\|\mathbf{x} - \hat{\mathbf{x}}\|}{\|\mathbf{Ax}\|} \leq \frac{\|\mathbf{A}^{-1}\|\|\mathbf{b} - \hat{\mathbf{b}}\|}{\|\mathbf{b}\|} \qquad (A.102)$$

$$\|\mathbf{x} - \hat{\mathbf{x}}\| \leq \|\mathbf{Ax}\|\|\mathbf{A}^{-1}\|\frac{\|\mathbf{b} - \hat{\mathbf{b}}\|}{\|\mathbf{b}\|} \qquad (A.103)$$

$$\|\mathbf{x} - \hat{\mathbf{x}}\| \leq \|\mathbf{A}\|\|\mathbf{x}\|\|\mathbf{A}^{-1}\|\frac{\|\mathbf{b} - \hat{\mathbf{b}}\|}{\|\mathbf{b}\|} \qquad (A.104)$$

$$\frac{\|\mathbf{x} - \hat{\mathbf{x}}\|}{\|\mathbf{x}\|} \leq \|\mathbf{A}\|\|\mathbf{A}^{-1}\|\frac{\|\mathbf{b} - \hat{\mathbf{b}}\|}{\|\mathbf{b}\|}. \qquad (A.105)$$

The relative error in **b** is measured by

$$\frac{\|\mathbf{b} - \hat{\mathbf{b}}\|}{\|\mathbf{b}\|}. \qquad (A.106)$$

The relative error in **x** is measured by

$$\frac{\|\mathbf{x} - \hat{\mathbf{x}}\|}{\|\mathbf{x}\|}. \qquad (A.107)$$

The constant

$$\mathrm{cond}(\mathbf{A}) = \|\mathbf{A}\|\|\mathbf{A}^{-1}\| \qquad (A.108)$$

is called the **condition number** of **A**.

Note that nothing that we did in the calculation of the condition number depends on which norm we used. The condition number can be computed using the 1-norm, 2-norm, ∞-norm, or Frobenius norm. The MATLAB command **cond** can be used to find the condition number of a matrix. It has options for the 1, 2, infinity, and Frobenius norms.

The condition number provides an upper bound on how inaccurate the solution to a system of equations might be because of errors in the right-hand side. In some cases, the condition number greatly overestimates the error in the solution. As a practical matter, it is wise to assume that the error is of roughly the size predicted by the condition number. In practice, floating-point arithmetic only allows us to store numbers to about 16 digits of precision. If the condition number is greater than 10^{16}, then by the above inequality, there may be no accurate digits in the computer solution to the system of equations. Systems of equations with very large condition numbers are called **ill-conditioned**.

It is important to understand that ill-conditioning is a property of the system of equations and not of the algorithm used to solve the system of equations. Ill-conditioning cannot be fixed simply by using a better algorithm. Instead, we must either increase the precision of our arithmetic or find a different, better-conditioned system of equations to solve.

A.9 THE QR FACTORIZATION

Although the theory of linear algebra can be developed using the reduced row echelon form, there is an alternative computational approach that works better in practice. The basic idea is to compute factorizations of matrices that involve orthogonal, diagonal, and upper-triangular matrices. This alternative approach leads to algorithms which can quickly compute accurate solutions to linear systems of equations and least squares problems.

■ **Theorem A.10** Let $\mathbf{A}$ be an m by n matrix. $\mathbf{A}$ can be written as

$$\mathbf{A} = \mathbf{QR} \tag{A.109}$$

where $\mathbf{Q}$ is an m by m orthogonal matrix, and $\mathbf{R}$ is an m by n upper-triangular matrix. This is called the **QR factorization of A**.   ■

The MATLAB command **qr** can be used to compute the QR factorization of a matrix. In a common situation, $\mathbf{A}$ will be an m by n matrix with $m > n$ and the rank of $\mathbf{A}$ will be n. In this case, we can write

$$\mathbf{R} = \begin{bmatrix} \mathbf{R}_1 \\ \mathbf{0} \end{bmatrix} \tag{A.110}$$

where $\mathbf{R}_1$ is n by n, and

$$\mathbf{Q} = [\mathbf{Q}_1 \ \ \mathbf{Q}_2] \tag{A.111}$$

where $\mathbf{Q}_1$ is m by n and $\mathbf{Q}_2$ is m by $m - n$. In this case the QR factorization has some important properties.

■ **Theorem A.11** Let $\mathbf{Q}$ and $\mathbf{R}$ be the QR factorization of an m by n matrix $\mathbf{A}$ with $m > n$ and rank$(\mathbf{A}) = n$. Then

1. The columns of $\mathbf{Q}_1$ are an orthonormal basis for $R(\mathbf{A})$.
2. The columns of $\mathbf{Q}_2$ are an orthonormal basis for $N(\mathbf{A}^T)$.
3. The matrix $\mathbf{R}_1$ is nonsingular.   ■

Now, suppose that we want to solve the least squares problem

$$\min \|\mathbf{Ax} - \mathbf{b}\|_2. \tag{A.112}$$

Since multiplying a vector by an orthogonal matrix does not change its length, this is equivalent to

$$\min \|\mathbf{Q}^T (\mathbf{Ax} - \mathbf{b})\|_2. \tag{A.113}$$

But

$$\mathbf{Q}^T\mathbf{A} = \mathbf{Q}^T\mathbf{Q}\mathbf{R}$$
$$= \mathbf{R}. \tag{A.114}$$

So, we have

$$\min \|\mathbf{R}\mathbf{x} - \mathbf{Q}^T\mathbf{b}\|_2 \tag{A.115}$$

or

$$\min \left\| \begin{matrix} \mathbf{R}_1\mathbf{x} - \mathbf{Q}_1^T\mathbf{b} \\ \mathbf{0}\mathbf{x} - \mathbf{Q}_2^T\mathbf{b} \end{matrix} \right\|_2. \tag{A.116}$$

Whatever value of $\mathbf{x}$ we pick, we will probably end up with nonzero error because of the $\mathbf{0}\mathbf{x} - \mathbf{Q}_2^T\mathbf{b}$ part of the least squares problem. We cannot minimize the norm of this part of the vector. However, we can find an $\mathbf{x}$ that exactly solves $\mathbf{R}_1\mathbf{x} = \mathbf{Q}_1^T\mathbf{b}$. Thus we can minimize the least squares problem by solving the square system of equations

$$\mathbf{R}_1\mathbf{x} = \mathbf{Q}_1^T\mathbf{b}. \tag{A.117}$$

The advantage of solving this system of equations instead of the normal equations (A.73) is that it is typically much better conditioned.

A.10 LINEAR ALGEBRA IN SPACES OF FUNCTIONS

So far, we have considered only vectors in R^n. The concepts of linear algebra can be extended to other contexts. In general, as long as the objects that we want to consider can be multiplied by scalars and added together, and as long as they obey the laws of vector algebra, then we have a **vector space** in which we can practice linear algebra. If we can also define a vector product similar to the dot product, then we have what is called an **inner product space**, and we can define orthogonality, projections, and the 2-norm.

There are many different vector spaces used in various areas of science and mathematics. For our work in inverse problems, a very commonly used vector space is the space of functions defined on an interval $[a, b]$.

Multiplying a scalar times a function or adding two functions together clearly produces another function. In this space, the function $z(x) = 0$ takes the place of the $\mathbf{0}$ vector, since $f(x) + z(x) = f(x)$. Two functions $f(x)$ and $g(x)$ are linearly independent if the only solution to

$$c_1 f(x) + c_2 g(x) = z(x) \tag{A.118}$$

is $c_1 = c_2 = 0$.

We can define the dot product of two functions f and g to be

$$f \cdot g = \int_a^b f(x)g(x)\,dx. \tag{A.119}$$

Another commonly used notation for this dot product or **inner product** of f and g is

$$f \cdot g = \langle f, g \rangle. \tag{A.120}$$

It is easy to show that this inner product has all of the algebraic properties of the dot product of two vectors in R^n. A more important motivation for defining the dot product in this way is that it leads to a useful definition of the 2-norm of a function. Following our earlier formula that $\|\mathbf{x}\|_2 = \sqrt{\mathbf{x}^T \mathbf{x}}$, we have

$$\|f\|_2 = \sqrt{\int_a^b f(x)^2\,dx}. \tag{A.121}$$

Using this definition, the distance between two functions f and g is

$$\|f - g\|_2 = \sqrt{\int_a^b (f(x) - g(x))^2\,dx}. \tag{A.122}$$

This measure is obviously zero when $f(x) = g(x)$ everywhere, and is only zero when $f(x) = g(x)$ except possibly at some isolated points.

Using this inner product and norm, we can reconstruct the theory of linear algebra from R^n in our space of functions. This includes the concepts of orthogonality, projections, norms, and least squares solutions.

■ **Definition A.32** Given a collection of functions $f_1(x), \ldots, f_m(x)$ in an inner product space, the **Gram matrix** of the functions is the m by m matrix $\boldsymbol{\Gamma}$, whose elements are given by

$$\boldsymbol{\Gamma}_{i,j} = f_i \cdot f_j. \tag{A.123}$$

■

The Gram matrix has several important properties. It is symmetric and positive semidefinite. If the functions are linearly independent, then the Gram matrix is also positive definite. Furthermore, the rank of $\boldsymbol{\Gamma}$ is equal to the size of the largest linearly independent subset of the functions $f_1(x), \ldots, f_m(x)$.

A.11 EXERCISES

A.1 Is it possible for an underdetermined system of equations to have exactly one solution? If so, construct an example. If not, then explain why it is not possible.

A.2 Let $\mathbf{A}$ be an m by n matrix with n pivot columns in its RREF. Can the system of equations $\mathbf{Ax} = \mathbf{b}$ have infinitely many solutions?

A.3 If $\mathbf{C} = \mathbf{AB}$ is a 5 by 4 matrix, then how many rows does $\mathbf{A}$ have? How many columns does $\mathbf{B}$ have? Can you say anything about the number of columns in $\mathbf{A}$?

A.4 Suppose that $\mathbf{v}_1$, $\mathbf{v}_2$, and $\mathbf{v}_3$ are three vectors in R^3 and that $\mathbf{v}_3 = -2\mathbf{v}_1 + 3\mathbf{v}_2$. Are the vectors linearly dependent or linearly independent?

A.5 Let

$$\mathbf{A} = \begin{bmatrix} 1 & 2 & 3 & 4 \\ 2 & 2 & 1 & 3 \\ 4 & 6 & 7 & 11 \end{bmatrix}. \tag{A.124}$$

Find bases for $N(\mathbf{A})$, $R(\mathbf{A})$, $N(\mathbf{A}^T)$, and $R(\mathbf{A}^T)$. What are the dimensions of the four subspaces?

A.6 Let $\mathbf{A}$ be an n by n matrix such that $\mathbf{A}^{-1}$ exists. What are $N(\mathbf{A})$, $R(\mathbf{A})$, $N(\mathbf{A}^T)$, and $R(\mathbf{A}^T)$?

A.7 Let $\mathbf{A}$ be any 9 by 6 matrix. If the dimension of the null space of $\mathbf{A}$ is 5, then what is the dimension of $R(\mathbf{A})$? What is the dimension of $R(\mathbf{A}^T)$? What is the rank of $\mathbf{A}$?

A.8 Suppose that a nonhomogeneous system of equations with four equations and six unknowns has a solution with two free variables. Is it possible to change the right-hand side of the system of equations so that the modified system of equations has no solutions?

A.9 Let W be the set of vectors $\mathbf{x}$ in R^4 such that $x_1 x_2 = 0$. Is W a subspace of R^4?

A.10 Let $\mathbf{v}_1$, $\mathbf{v}_2$, and $\mathbf{v}_3$ be a set of three nonzero orthogonal vectors. Show that the vectors are also linearly independent.

A.11 Show that if $\mathbf{x} \perp \mathbf{y}$, then

$$\|\mathbf{x} + \mathbf{y}\|_2^2 = \|\mathbf{x}\|_2^2 + \|\mathbf{y}\|_2^2. \tag{A.125}$$

A.12 Prove the **parallelogram law**

$$\|\mathbf{u} + \mathbf{v}\|_2^2 + \|\mathbf{u} - \mathbf{v}\|_2^2 = 2\|\mathbf{u}\|_2^2 + 2\|\mathbf{v}\|_2^2. \tag{A.126}$$

A.13 Suppose that a nonsingular matrix $\mathbf{A}$ can be diagonalized as

$$\mathbf{A} = \mathbf{P}\mathbf{\Lambda}\mathbf{P}^{-1}. \tag{A.127}$$

Find a diagonalization of $\mathbf{A}^{-1}$. What are the eigenvalues of $\mathbf{A}^{-1}$?

A.14 Suppose that $\mathbf{A}$ is diagonalizable and that all eigenvalues of $\mathbf{A}$ have absolute value less than one. What is the limit as k goes to infinity of $\mathbf{A}^k$?

A.15   In this exercise, we will derive the formula (A.90) for the 1-norm of a matrix. Begin
with the optimization problem

$$\|\mathbf{A}\|_1 = \max_{\|\mathbf{x}\|_1=1} \|\mathbf{A}\mathbf{x}\|_1. \tag{A.128}$$

(a)   Show that if $\|\mathbf{x}\|_1 = 1$, then

$$\|\mathbf{A}\mathbf{x}\|_1 \leq \max_j \sum_{i=1}^{m} |A_{i,j}|. \tag{A.129}$$

(b)   Find a vector $\mathbf{x}$ such that $\|\mathbf{x}\|_1 = 1$, and

$$\|\mathbf{A}\mathbf{x}\|_1 = \max_j \sum_{i=1}^{m} |A_{i,j}|. \tag{A.130}$$

(c)   Conclude that

$$\|\mathbf{A}\|_1 = \max_{\|\mathbf{x}\|_1=1} \|\mathbf{A}\mathbf{x}\|_1 = \max_j \sum_{i=1}^{m} |A_{i,j}|. \tag{A.131}$$

A.16   Derive the formula (A.92) for the infinity norm of a matrix.

A.17   Let $\mathbf{A}$ be an m by n matrix.

(a)   Show that $\mathbf{A}^T\mathbf{A}$ is symmetric.
(b)   Show that $\mathbf{A}^T\mathbf{A}$ is positive semidefinite. Hint: Use the definition of positive
semidefinite rather than trying to compute eigenvalues.
(c)   Show that if $\text{rank}(\mathbf{A}) = n$, then the only solution to $\mathbf{A}\mathbf{x} = \mathbf{0}$ is $\mathbf{x} = \mathbf{0}$.
(d)   Use part c to show that if $\text{rank}(\mathbf{A}) = n$, then $\mathbf{A}^T\mathbf{A}$ is positive definite.
(e)   Use part d to show that if $\text{rank}(\mathbf{A}) = n$, then $\mathbf{A}^T\mathbf{A}$ is nonsingular.
(f)   Show that $N(\mathbf{A}^T\mathbf{A}) = N(\mathbf{A})$.

A.18   Show that

$$\text{cond}(\mathbf{A}\mathbf{B}) \leq \text{cond}(\mathbf{A})\text{cond}(\mathbf{B}). \tag{A.132}$$

A.19   Let $\mathbf{A}$ be a symmetric and positive definite matrix with Cholesky factorization

$$\mathbf{A} = \mathbf{R}^T\mathbf{R}. \tag{A.133}$$

Show how the Cholesky factorization can be used to solve $\mathbf{A}\mathbf{x} = \mathbf{b}$ by solving two
systems of equations, each of which has $\mathbf{R}$ or $\mathbf{R}^T$ as its matrix.

A.20   Let $P_3[0, 1]$ be the space of polynomials of degree less than or equal to 3 on the interval [0, 1]. The polynomials $p_1(x) = 1$, $p_2(x) = x$, $p_3(x) = x^2$, and $p_4(x) = x^3$ form a basis for $P_3[0, 1]$, but they are not orthogonal with respect to the inner product

$$f \cdot g = \int_0^1 f(x)g(x) \, dx. \qquad (A.134)$$

Use the Gram–Schmidt orthogonalization process to construct an orthogonal basis for $P_3[0, 1]$. Once you have your basis, use it to find the third-degree polynomial that best approximates $f(x) = e^{-x}$ on the interval [0, 1].

A.12 NOTES AND FURTHER READING

Much of this and associated material is typically covered in sophomore-level linear algebra courses, and there are an enormous number of textbooks at this level. One good introductory linear algebra textbook is [91]. At a slightly more advanced level, [155] and [101] are both excellent. The book by Strang and Borre [156] reviews linear algebra in the context of geodetic problems.

Fast and accurate algorithms for linear algebra computations are a somewhat more advanced and specialized topic. A classic reference is [49]. Other good books on this topic include [167] and [31].

The extension of linear algebra to spaces of functions is a topic in the subject of functional analysis. Unfortunately, most textbooks on functional analysis assume that the reader has considerable mathematical background. One book that is reasonably accessible to readers with limited mathematical backgrounds is [97].

Appendix B

REVIEW OF PROBABILITY AND STATISTICS

Synopsis: A brief review is given of the topics in classical probability and statistics that are used in this book. Connections between probability theory and its application to the analysis of data with random measurement errors are highlighted. Note that some very different philosophical interpretations of probability theory are discussed in Chapter 11.

B.1 PROBABILITY AND RANDOM VARIABLES

The mathematical theory of probability begins with an **experiment**, which has a set S of possible outcomes. We will be interested in **events** which are subsets A of S.

■ **Definition B.1** The **probability function** P is a function defined on subsets of S with the following properties:

1. $P(S) = 1$
2. For every event $A \subseteq S$, $P(A) \geq 0$
3. If events $A_1, A_2, \ldots$ are pairwise mutually exclusive (that is, if $A_i \cap A_j$ is empty for all pairs i, j), then

$$P(\cup_{i=1}^{\infty} A_i) = \sum_{i=1}^{\infty} P(A_i). \tag{B.1}$$

■

The probability properties just given are fundamental to developing the mathematics of probability theory. However, applying this definition of probability to real-world situations frequently requires ingenuity.

■ **Example B.1** Consider the experiment of throwing a dart at a dart board. We will assume that our dart thrower is an expert who always hits the dart board. The sample space S consists of the points on the dart board. We can define an event A that consists of the points in the bullseye, so that $P(A)$ is the probability that the thrower hits the bullseye. ■

In practice, the outcome of an experiment is often a number rather than an event. Random variables are a useful generalization of the basic concept of probability.

■ **Definition B.2** A **random variable** X is a function $X(s)$ that assigns a value to each outcome s in the sample space S.

Each time we perform an experiment, we obtain a particular value of the random variable. These values are called **realizations** of the random variable. ■

■ **Example B.2** To continue our previous example, let X be the function that takes a point on the dart board and returns the associated score. Suppose that throwing the dart in the bullseye scores 50 points. Then for each point s in the bullseye, $X(s) = 50$. ■

In this book we deal frequently with experimental measurements that can include some random measurement error.

■ **Example B.3** Suppose we measure the mass of an object five times to obtain the realizations $m_1 = 10.1$ kg, $m_2 = 10.0$ kg, $m_3 = 10.0$ kg, $m_4 = 9.9$ kg, and $m_5 = 10.1$ kg. We will assume that there is one true mass m, and that the measurements we obtained varied because of random measurement errors e_i, so that

$$m_1 = m + e_1, \quad m_2 = m + e_2, \quad m_3 = m + e_3, \quad m_4 = m + e_4, \quad m_5 = m + e_5. \quad \text{(B.2)}$$

We can treat the measurement errors as realizations of a random variable E. Equivalently, since the true mass m is just a constant, we could treat the measurements $m_1, m_2, \ldots, m_5$ as realizations of a random variable M. In practice it makes little difference whether we treat the measurements or the measurement errors as random variables.

Note that, in a Bayesian approach, the mass m of the object would itself be a random variable. This is a viewpoint that we consider in Chapter 11. ■

The relative probability of realization values for a random variable can be characterized by a nonnegative **probability density function (PDF)**, $f_X(x)$, with

$$P(X \leq a) = \int_{-\infty}^{a} f_X(x)\, dx. \quad \text{(B.3)}$$

Because the random variable always has some value,

$$\int_{-\infty}^{\infty} f_X(x)\, dx = 1. \quad \text{(B.4)}$$

The following definitions give some useful random variables that frequently arise in inverse problems.

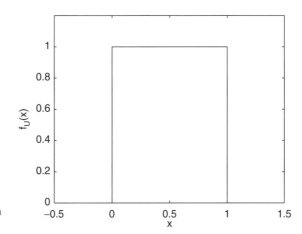

Figure B.1   The PDF for the uniform random variable on [0, 1].

■ **Definition B.3** The **uniform** random variable on the interval $[a, b]$ has the probability density function

$$f_U(x) = \begin{cases} \dfrac{1}{b-a} & a \le x \le b \\ 0 & x < a \\ 0 & x > b \end{cases}. \tag{B.5}$$

See Figure B.1.   ■

■ **Definition B.4** The **normal** or **Gaussian** random variable has the probability density function

$$f_N(x) = \frac{1}{\sigma\sqrt{2\pi}} e^{-\frac{1}{2}(x-\mu)^2/\sigma^2}. \tag{B.6}$$

See Figure B.2. The notation $N(\mu, \sigma^2)$ is used to denote a normal distribution with parameters μ and σ. The **standard normal** random variable, $N(0, 1)$, has $\mu = 0$ and $\sigma = 1$.   ■

■ **Definition B.5** The **exponential** random variable has the probability density function

$$f_{exp}(x) = \begin{cases} \lambda e^{-\lambda x} & x \ge 0 \\ 0 & x < 0 \end{cases}. \tag{B.7}$$

See Figure B.3.   ■

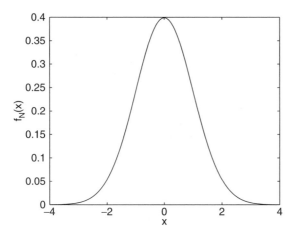

Figure B.2   The PDF of the standard normal random variable.

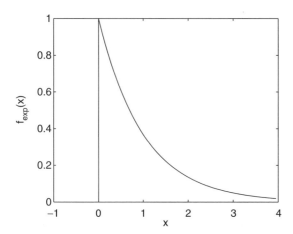

Figure B.3   The exponential probability density function ($\lambda = 1$).

■ **Definition B.6**   The **double–sided exponential** random variable has the probability density function

$$f_{\mathrm{dexp}}(x) = \frac{1}{2^{3/2}\sigma} e^{-\sqrt{2}|x-\mu|/\sigma}. \tag{B.8}$$

See Figure B.4.   ■

■ **Definition B.7**   The χ^2 random variable has the probability density function

$$f_{\chi^2}(x) = \frac{1}{2^{\nu/2}\Gamma(\nu/2)} x^{\frac{1}{2}\nu-1} e^{-x/2} \tag{B.9}$$

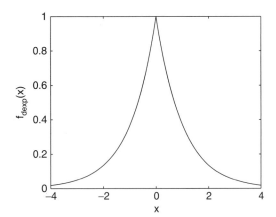

Figure B.4   The double-sided exponential probability density function ($\mu = 0$, $\lambda = 1$).

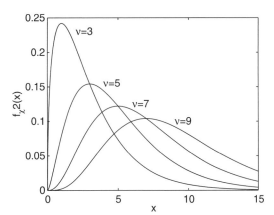

Figure B.5   The χ^2 probability density function for several values of ν.

where the **gamma function** is

$$\Gamma(x) = \int_0^\infty \xi^{x-1} e^{-\xi} d\xi \tag{B.10}$$

and the parameter ν is called the **number of degrees of freedom**. See Figure B.5.

It can be shown that for n independent random variables, X_i with standard normal distributions, the random variable

$$Z = \sum_{i=1}^n X_i^2 \tag{B.11}$$

is a χ^2 random variable with $\nu = n$ degrees of freedom [40].   ■

■ **Definition B.8** The **Student's t distribution** with ν degrees of freedom has the probability density function

$$f_t(x) = \frac{\Gamma((\nu+1)/2)}{\Gamma(\nu/2)} \frac{1}{\sqrt{\nu\pi}} \left(1 + \frac{x^2}{\nu}\right)^{-(\nu+1)/2}. \tag{B.12}$$

■

See Figure B.6. The Student's t distribution is so named because W. S. Gosset used the pseudonym "Student" in publishing the first paper in which the distribution appeared. In the limit as ν goes to infinity, Student's t distribution approaches a standard normal distribution.

The **cumulative distribution function (CDF)** $F_X(a)$ of a one–dimensional random variable X is given by the definite integral of the associated PDF:

$$F_X(a) = P(X \le a) = \int_{-\infty}^{a} f_X(x)\,dx. \tag{B.13}$$

Note that $F_X(a)$ must lie in the interval $[0, 1]$ for all a, and is a nondecreasing function of a because of the unit area and nonnegativity of the PDF.

For the uniform PDF on the unit interval, for example, the CDF is a ramp function

$$F_U(a) = \int_{-\infty}^{a} f_u(z)\,dz \tag{B.14}$$

$$F_U(a) = \begin{cases} 0 & a \le 0 \\ a & 0 \le a \le 1 \\ 1 & a > 1 \end{cases}. \tag{B.15}$$

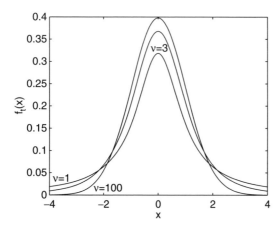

Figure B.6 The Student's t probability density function for $\nu = 1, 3, 100$.

The PDF, $f_X(x)$, or CDF, $F_X(a)$, completely determine the probabilistic properties of a random variable. The probability that a particular realization of X will lie within a general interval $[a, b]$ is

$$P(a \leq X \leq b) = P(X \leq b) - P(X \leq a) = F(b) - F(a) \tag{B.16}$$

$$= \int_{-\infty}^{b} f(x)\, dx - \int_{-\infty}^{a} f(x)\, dx = \int_{a}^{b} f(x)\, dx. \tag{B.17}$$

B.2 EXPECTED VALUE AND VARIANCE

■ **Definition B.9** The **expected value** of a random variable X, denoted by $E[X]$ or μ_X, is

$$E[X] = \int_{-\infty}^{\infty} x f_X(x)\, dx. \tag{B.18}$$

In general, if $g(X)$ is some function of a random variable X, then

$$E[g(X)] = \int_{-\infty}^{\infty} g(x) f_X(x)\, dx. \tag{B.19}$$

■

Some authors use the term "mean" for the expected value of a random variable. We will reserve this term for the average of a set of data. Note that the expected value of a random variable is not necessarily identical to the **mode** [the value with the largest value of $f(x)$] nor is it necessarily identical to the **median**, the value of x for which the value of the CDF is $F(x) = 1/2$.

■ **Example B.4** The expected value of an $N(\mu, \sigma)$ random variable X is

$$E[X] = \int_{-\infty}^{\infty} x \frac{1}{\sigma\sqrt{2\pi}} e^{-\frac{(x-\mu)^2}{2\sigma^2}}\, dx \tag{B.20}$$

$$= \int_{-\infty}^{\infty} \frac{1}{\sigma\sqrt{2\pi}} (x + \mu) e^{-\frac{x^2}{2\sigma^2}}\, dx \tag{B.21}$$

$$= \mu \int_{-\infty}^{\infty} \frac{1}{\sigma\sqrt{2\pi}} e^{-\frac{x^2}{2\sigma^2}}\, dx + \int_{-\infty}^{\infty} \frac{1}{\sigma\sqrt{2\pi}} x e^{-\frac{x^2}{2\sigma^2}}\, dx. \tag{B.22}$$

The first integral term is μ because the integral of the entire PDF is 1, and the second term is zero because it is an odd function integrated over a symmetric interval. Thus

$$E[X] = \mu. \tag{B.23}$$

■

■ **Definition B.10** The **variance** of a random variable X, denoted by $\text{Var}(X)$ or σ_X^2, is given by

$$\text{Var}(X) = \sigma_X^2$$
$$= E[(X - \mu_X)^2]$$
$$= E[X^2] - \mu_X^2$$
$$= \int_{-\infty}^{\infty} (x - \mu_X)^2 f_X(x)\, dx. \tag{B.24}$$

The **standard deviation** of X, often denoted σ_X, is

$$\sigma_X = \sqrt{\text{Var}(X)}. \tag{B.25}$$

■

The variance and standard deviation serve as measures of the spread of the random variable about its expected value. Since the units of σ are the same as the units of μ, the standard deviation is generally more practical as a measure of the spread of the random variable. However, the variance has many properties that make it more useful for certain calculations.

B.3 JOINT DISTRIBUTIONS

■ **Definition B.11** If we have two random variables X and Y, they *may* have a **joint probability density function (JDF)**, $f(x, y)$ with

$$P(X \leq a \quad \text{and} \quad Y \leq b) = \int_{-\infty}^{a} \int_{-\infty}^{b} f(x, y)\, dy\, dx. \tag{B.26}$$

■

If X and Y have a joint probability density function, then we can use it to evaluate the expected value of a function of X and Y. The expected value of $g(X, Y)$ is

$$E[g(X, Y)] = \int_{-\infty}^{\infty} \int_{-\infty}^{\infty} g(x, y) f(x, y)\, dy\, dx. \tag{B.27}$$

■ **Definition B.12** Two random variables X and Y are **independent** if a JDF exists and is defined by

$$f(x, y) = f_X(x) f_Y(y). \tag{B.28}$$

■

■ **Definition B.13** If X and Y have a JDF, then the **covariance** of X and Y is

$$\text{Cov}(X, Y) = E[(X - E[X])(Y - E[Y])] = E[XY] - E[X]E[Y]. \qquad \text{(B.29)}$$

■

If X and Y are independent, then $E[XY] = E[X]E[Y]$, and $\text{Cov}(X, Y) = 0$. However if X and Y are dependent, it is still possible, given some particular distributions, for X and Y to have $\text{Cov}(X, Y) = 0$. If $\text{Cov}(X, Y) = 0$, X and Y are called **uncorrelated**.

■ **Definition B.14** The **correlation** of X and Y is

$$\rho(X, Y) = \frac{\text{Cov}(X, Y)}{\sqrt{\text{Var}(X)\text{Var}(Y)}}. \qquad \text{(B.30)}$$

Correlation is thus a scaled covariance.   ■

■ **Theorem B.1** The following properties of Var, Cov, and correlation hold for any random variables X and Y and scalars s and a.

1. $\text{Var}(X) \geq 0$
2. $\text{Var}(X + a) = \text{Var}(X)$
3. $\text{Var}(sX) = s^2\text{Var}(X)$
4. $\text{Var}(X + Y) = \text{Var}(X) + \text{Var}(Y) + 2\text{Cov}(X, Y)$
5. $\text{Cov}(X, Y) = \text{Cov}(Y, X)$
6. $\rho(X, Y) = \rho(Y, X)$
7. $-1 \leq \rho_{XY} \leq 1$ ■

The following example demonstrates the use of some of these properties.

■ **Example B.5** Suppose that Z is a standard normal random variable. Let

$$X = \mu + \sigma Z. \qquad \text{(B.31)}$$

Then

$$E[X] = E[\mu] + \sigma E[Z] \qquad \text{(B.32)}$$

so

$$E[X] = \mu. \qquad \text{(B.33)}$$

Also,

$$\text{Var}(X) = \text{Var}(\mu) + \sigma^2\text{Var}(Z)$$

$$= \sigma^2. \qquad \text{(B.34)}$$

Thus if we have a program to generate random numbers with the standard normal distribution, we can use it to generate normal random numbers with any desired expected value and standard deviation. The MATLAB command **randn** generates independent realizations of an $N(0, 1)$ random variable.   ∎

■ **Example B.6** What is the CDF (or PDF) of the sum of two independent random variables $X + Y$? To see this, we write the desired CDF in terms of an appropriate integral over the JDF, $f(x, y)$, which gives

$$F_{X+Y}(z) = P(X + Y \leq z) \tag{B.35}$$

$$= \iint_{x+y \leq z} f(x, y)\, dx\, dy \tag{B.36}$$

$$= \iint_{x+y \leq z} f_X(x) f_Y(y)\, dx\, dy \tag{B.37}$$

$$= \int_{-\infty}^{\infty} \int_{-\infty}^{z-y} f_X(x) f_Y(y)\, dx\, dy \tag{B.38}$$

$$= \int_{-\infty}^{\infty} \int_{-\infty}^{z-y} f_X(x)\, dx\ f_Y(y)\, dy \tag{B.39}$$

$$= \int_{-\infty}^{\infty} F_X(z - y) f_Y(y)\, dy. \tag{B.40}$$

See Figure B.7. The associated PDF is

$$f_{X+Y}(z) = \frac{d}{dz} \int_{-\infty}^{\infty} F_X(z - y) f_Y(y)\, dy \tag{B.41}$$

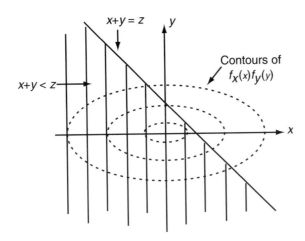

Figure B.7 Integration of a joint probability density function for two independent random variables, X, and Y, to evaluate the CDF of $Z = X + Y$.

$$= \int_{-\infty}^{\infty} \frac{d}{dz} F_X(z - y) f_Y(y) \, dy \tag{B.42}$$

$$= \int_{-\infty}^{\infty} f_X(z - y) f_Y(y) \, dy \tag{B.43}$$

$$= f_X(z) * f_Y(z). \tag{B.44}$$

Adding two independent random variables thus produces a new random variable that has a PDF given by the convolution of the PDF's of the two individual variables. ■

The JDF can be used to evaluate the CDF or PDF arising from a general function of jointly distributed random variables. The process is identical to the previous example except that the specific form of the integral limits is determined by the specific function.

■ **Example B.7** Consider the product of two independent, identically distributed, standard normal random variables,

$$Z = XY \tag{B.45}$$

with a JDF given by

$$f(x, y) = f(x) f(y) = \frac{1}{2\pi\sigma^2} e^{-(x^2 + y^2)/2\sigma^2}. \tag{B.46}$$

The CDF of Z is

$$F(z) = P(Z \leq z) = P(XY \leq z). \tag{B.47}$$

For $z \leq 0$, this is the integral of the JDF over the exterior of the hyperbolas defined by $xy \leq z \leq 0$, whereas for $z \geq 0$, we integrate over the interior of the complementary hyperbolas $xy \leq z \geq 0$. At $z = 0$, the integral covers exactly half of the (x, y) plane (the second and fourth quadrants) and, because of the symmetry of the JDF, has accumulated half of the probability, or 1/2.

The integral is thus

$$F(z) = 2 \int_{-\infty}^{0} \int_{z/x}^{\infty} \frac{1}{2\pi\sigma^2} e^{-(x^2 + y^2)/2\sigma^2} \, dy \, dx \quad (z \leq 0) \tag{B.48}$$

and

$$F(z) = 1/2 + 2 \int_{-\infty}^{0} \int_{0}^{z/x} \frac{1}{2\pi\sigma^2} e^{-(x^2 + y^2)/2\sigma^2} \, dy \, dx \quad (z \geq 0). \tag{B.49}$$

As in the previous example for the sum of two random variables, the PDF may be obtained from the CDF by differentiating with respect to z. ■

B.4 CONDITIONAL PROBABILITY

In some situations we will be interested in the probability of an event happening given that some other event has also happened.

■ **Definition B.15** The **conditional probability** of A given that B has occurred is given by

$$P(A|B) = \frac{P(A \cap B)}{P(B)}. \tag{B.50}$$

■

Arguments based on conditional probabilities are often very helpful in computing probabilities. The key to such arguments is the **law of total probability**.

■ **Theorem B.2** Suppose that $B_1, B_2, \ldots, B_n$ are mutually disjoint and exhaustive events. That is, $B_i \cap B_j = \emptyset$ (the empty set) for $i \neq j$, and

$$\bigcup_{i=1}^{n} B_i = S. \tag{B.51}$$

Then

$$P(A) = \sum_{i=1}^{n} P(A|B_i)P(B_i). \tag{B.52}$$

■

It is often necessary to reverse the order of conditioning in a conditional probability. Bayes' theorem provides a way to do this.

■ **Theorem B.3 Bayes' Theorem**

$$P(B|A) = \frac{P(A|B)P(B)}{P(A)}. \tag{B.53}$$

■

■ **Example B.8** A screening test has been developed for a very serious but rare disease. If a person has the disease, then the test will detect the disease with probability 99%. If a person does not have the disease, then the test will give a false positive detection with probability 1%. The probability that any individual in the population has the disease is 0.01%. Suppose that a randomly selected individual tests positive for the disease. What is the probability that this individual actually has the disease?

Let A be the event "the person tests positive." Let B be the event "the person has the disease." We then want to compute $P(B|A)$. By Bayes' theorem,

$$P(B|A) = \frac{P(A|B)P(B)}{P(A)}. \tag{B.54}$$

We have that $P(A|B)$ is 0.99, and that $P(B)$ is 0.0001. To compute $P(A)$, we apply the law of total probability, considering separately the probability of a diseased individual testing positive and the probability of someone without the disease testing positive.

$$P(A) = 0.99 \times 0.0001 + 0.01 \times 0.9999 = 0.010098. \tag{B.55}$$

Thus

$$P(B|A) = \frac{0.99 \times 0.0001}{0.010098} = 0.0098. \tag{B.56}$$

In other words, even after a positive screening test, it is still unlikely that the individual will have the disease. The vast majority of those individuals who test positive will in fact not have the disease. ∎

The concept of conditioning can be extended from simple events to distributions and expected values of random variables. If the distribution of X depends on the value of Y, then we can work with the **conditional PDF** $f_{X|Y}(x)$, the **conditional CDF** $F_{X|Y}(a)$, and the **conditional expected value** $E[X|Y]$.

In this notation, we can also specify a particular value of Y by using the notation $f_{X|Y=y}$, $F_{X|Y=y}$, or $E[X|Y = y]$. In working with conditional distributions and expected values, the following versions of the law of total probability can be very useful.

∎ **Theorem B.4** Given two random variables X and Y, with the distribution of X depending on Y, we can compute

$$P(X \le a) = \int_{-\infty}^{\infty} P(X \le a|Y = y)f_Y(y)\,dy \tag{B.57}$$

and

$$E[X] = \int_{-\infty}^{\infty} E[X|Y = y]f_Y(y)\,dy. \tag{B.58}$$

∎

∎ **Example B.9** Let U be a random variable uniformly distributed on $(1, 2)$. Let X be an exponential random variable with parameter $\lambda = U$. We will find the expected value of X:

$$E[X] = \int_{1}^{2} E[X|U = u]f_U(u)\,du. \tag{B.59}$$

Since the expected value of an exponential random variable with parameter λ is $1/\lambda$, and the PDF of a uniform random variable on $(1, 2)$ is $f_U(u) = 1$,

$$E[X] = \int_1^2 \frac{1}{u}\, du$$

$$= \ln 2. \tag{B.60}$$

■

B.5 THE MULTIVARIATE NORMAL DISTRIBUTION

■ **Definition B.16** If the random variables $X_1, \ldots, X_n$ have a **multivariate normal** (MVN) **distribution**, then the joint probability density function is

$$f(\mathbf{x}) = \frac{1}{(2\pi)^{n/2}} \frac{1}{\sqrt{\det(\mathbf{C})}} e^{-(\mathbf{x}-\boldsymbol{\mu})^T \mathbf{C}^{-1}(\mathbf{x}-\boldsymbol{\mu})/2} \tag{B.61}$$

where $\boldsymbol{\mu} = [\mu_1, \mu_2, \ldots, \mu_n]^T$ is a vector containing the expected values along each of the coordinate directions of $X_1, \ldots, X_n$, and $\mathbf{C}$ contains the covariances between the random variables:

$$C_{i,j} = \mathrm{Cov}(X_i, X_j). \tag{B.62}$$

Notice that if $\mathbf{C}$ is singular, then the joint probability density function involves a division by zero and is simply not defined.   ■

The vector $\boldsymbol{\mu}$ and the covariance matrix $\mathbf{C}$ completely characterize the MVN distribution. There are other multivariate distributions that are not completely characterized by the expected values and covariance matrix.

■ **Theorem B.5** Let $\mathbf{X}$ be a multivariate normal random vector with expected values defined by the vector $\boldsymbol{\mu}$ and covariance matrix $\mathbf{C}$, and let $\mathbf{Y} = \mathbf{AX}$. Then $\mathbf{Y}$ is also multivariate normal, with

$$E[\mathbf{Y}] = \mathbf{A}\boldsymbol{\mu} \tag{B.63}$$

and

$$\mathrm{Cov}(\mathbf{Y}) = \mathbf{ACA}^T. \tag{B.64}$$

■

■ **Theorem B.6** If we have an n–dimensional MVN distribution with covariance matrix $\mathbf{C}$ and expected value $\boldsymbol{\mu}$, and the covariance matrix is of full rank, then the quantity

$$Z = (\mathbf{X} - \boldsymbol{\mu})^T \mathbf{C}^{-1}(\mathbf{X} - \boldsymbol{\mu}) \tag{B.65}$$

has a χ^2 distribution with n degrees of freedom. ■

■ **Example B.10** We can generate vectors of random numbers according to an MVN distribution with known mean and covariance matrix by using the following process, which is very similar to the process for generating random normal scalars.

1. Find the Cholesky factorization $\mathbf{C} = \mathbf{LL}^T$.
2. Let $\mathbf{Z}$ be a vector of n independent $N(0, 1)$ random numbers.
3. Let $\mathbf{X} = \boldsymbol{\mu} + \mathbf{LZ}$.

Because $E[\mathbf{Z}] = \mathbf{0}$, $E[\mathbf{X}] = \boldsymbol{\mu} + \mathbf{L0} = \boldsymbol{\mu}$. Also, since $\text{Cov}(\mathbf{Z}) = \mathbf{I}$ and $\text{Cov}(\boldsymbol{\mu}) = \mathbf{0}$, $\text{Cov}(\mathbf{X}) = \text{Cov}(\boldsymbol{\mu} + \mathbf{LZ}) = \mathbf{LIL}^T = \mathbf{C}$. ■

B.6 THE CENTRAL LIMIT THEOREM

■ **Theorem B.7** Let $X_1, X_2, \ldots, X_n$ be independent and identically distributed (IID) random variables with a finite expected value μ and variance σ^2. Let

$$Z_n = \frac{X_1 + X_2 + \cdots + X_n - n\mu}{\sqrt{n}\sigma}. \tag{B.66}$$

In the limit as n approaches infinity, the distribution of Z_n approaches the standard normal distribution. ■

The central limit theorem shows why quasinormally distributed random variables appear so frequently in nature; the sum of numerous independent random variables produces an approximately normal random variable, regardless of the distribution of the underlying IID variables. In particular, this is one reason that measurement errors are often normally distributed. As we saw in Chapter 2, having normally distributed measurement errors leads us to consider least squares solutions to parameter estimation and inverse problems.

B.7 TESTING FOR NORMALITY

Many of the statistical procedures that we will use assume that data are normally distributed. Fortunately, the statistical techniques that we describe are generally robust in the face of small deviations from normality. Large deviations from the normal distribution can cause problems. Thus it is important to be able to examine a data set to see whether or not the distribution is approximately normal.

Plotting a histogram of the data provides a quick view of the distribution. The histogram should show a roughly "bell shaped" distribution, symmetrical around a single peak. If the histogram shows that the distribution is obviously skewed, then it would be unwise to assume that the data are normally distributed.

The **Q–Q plot** provides a more precise graphical test of whether a set of data could have come from a particular distribution. The data points

$$\mathbf{d} = [d_1, d_2, \ldots, d_n]^T \tag{B.67}$$

are first sorted in numerical order from smallest to largest into a vector $\mathbf{y}$, which is plotted versus

$$x_i = F^{-1}((i - 0.5)/n) \quad (i = 1, 2, \ldots, n) \tag{B.68}$$

where $F(x)$ is the CDF of the distribution against which we wish to compare our observations.

If we are testing to see if the elements of $\mathbf{d}$ could have come from the normal distribution, then $F(x)$ is the CDF for the standard normal distribution:

$$F_N(x) = \frac{1}{\sqrt{2\pi}} \int_{-\infty}^{x} e^{-\frac{1}{2}z^2} \, dz. \tag{B.69}$$

If the elements of $\mathbf{d}$ are normally distributed, the points (y_i, x_i) will follow a straight line with a slope and intercept determined by the standard deviation and expected value, respectively, of the normal distribution that produced the data.

■ **Example B.11** Figure B.8 shows the histogram from a set of 100 data points. The characteristic bell-shaped curve in the histogram makes it seem that these data might be normally distributed. The sample mean is 0.20 and the sample standard deviation is 1.81.

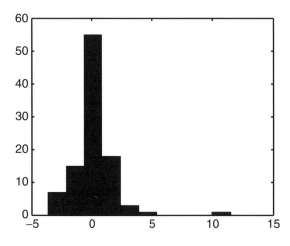

Figure B.8 Histogram of a sample data set.

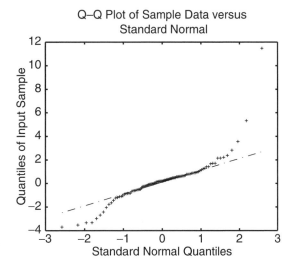

Figure B.9 $Q-Q$ plot for the sample data set.

Figure B.9 shows the $Q-Q$ plot for our sample data set. It is apparent that the data set contains more extreme values than the normal distribution would predict. In fact, these data were generated according to a t distribution with five degrees of freedom, which has broader tails than the normal distribution. See Figure B.6. Because of these deviations from normality, it would be wise not to treat these data as if they were normally distributed. ∎

There are a number of statistical tests for normality. These tests, including the Kolmogorov–Smirnov test, Anderson–Darling test, and Lilliefors test, each produce probabilistic measures called p-values. A small p-value indicates that the observed data would be unlikely if the distribution were in fact normal, while a larger p-value is consistent with normality.

B.8 ESTIMATING MEANS AND CONFIDENCE INTERVALS

Given a collection of noisy measurements $m_1, m_2, \ldots, m_n$ of some quantity of interest, how can we estimate the true value m, and how uncertain is our estimate? This is a classic problem in statistics.

We will assume first that the measurement errors are independent and normally distributed with expected value 0 and some unknown standard deviation σ. Equivalently, the measurements themselves are normally distributed with expected value m and standard deviation σ.

We begin by computing the measurement average

$$\bar{m} = \frac{m_1 + m_2 + \cdots + m_n}{n}. \tag{B.70}$$

This **sample mean** $\bar{m}$ will serve as our estimate of m. We will also compute an estimate s of the standard deviation

$$s = \sqrt{\frac{\sum_{i=1}^{n}(m_i - \bar{m})^2}{n-1}}. \tag{B.71}$$

The key to our approach to estimating m is the following theorem.

■ **Theorem B.8** (The Sampling Theorem) Under the assumption that measurements are independent and normally distributed with expected value m and standard deviation σ, the random quantity

$$t = \frac{m - \bar{m}}{s/\sqrt{n}} \tag{B.72}$$

has a **Student's t distribution** with $n-1$ degrees of freedom.   ■

If we had the true standard deviation σ instead of the estimate s, then t would in fact be normally distributed with expected value 0 and standard deviation 1. This does not quite work out because we have used an estimate s of the standard deviation. For smaller values of n, the estimate s is less accurate, and the t distribution therefore has fatter tails than the standard normal distribution. As n goes to infinity, s becomes a better estimate of σ and it can be shown that the t distribution converges to a standard normal distribution [40].

Let $t_{n-1,0.975}$ be the 97.5%-tile of the t distribution and let $t_{n-1,0.025}$ be the 2.5%-tile of the t distribution. Then

$$P\left(t_{n-1,0.025} \leq \frac{m - \bar{m}}{s/\sqrt{n}} \leq t_{n-1,0.975}\right) = 0.95. \tag{B.73}$$

This can be rewritten as

$$P\left(\left(t_{n-1,0.025}s/\sqrt{n}\right) \leq (m - \bar{m}) \leq \left(t_{n-1,0.975}s/\sqrt{n}\right)\right) = 0.95. \tag{B.74}$$

We can construct the 95% **confidence interval** for m as the interval from $\bar{m} + t_{n-1,0.025}s/\sqrt{n}$ to $\bar{m} + t_{n-1,0.975}s/\sqrt{n}$. Because the t distribution is symmetric, this can also be written as $\bar{m} - t_{n-1,0.975}s/\sqrt{n}$ to $\bar{m} + t_{n-1,0.975}s/\sqrt{n}$.

As we have seen, there is a 95% probability that when we construct the confidence interval, that interval will contain the true mean, m. Note that we have not said that, given a particular set of data and the resulting confidence interval, there is a 95% probability that m is in the confidence interval. The semantic difficulty here is that m is not a random variable, but is rather some true fixed quantity that we are estimating; the measurements $m_1, m_2, \ldots, m_n$, and the calculated $\bar{m}$, s and confidence interval are the random quantities.

■ **Example B.12** Suppose that we want to estimate the mass of an object and obtain the following ten measurements of the mass (in grams):

$$\begin{matrix} 9.98 & 10.07 & 9.94 & 10.22 & 9.98 \\ 10.01 & 10.11 & 10.01 & 9.99 & 9.92 \end{matrix} \cdot \tag{B.75}$$

The sample mean is $\bar{m} = 10.02$ g. The sample standard deviation is $s = 0.0883$. The 97.5%-tile of the t distribution with $n - 1 = 9$ degrees of freedom is (from a t-table or function) 2.262. Thus our 95% confidence interval for the mean is

$$\left[\bar{m} - 2.262s/\sqrt{n}, \bar{m} + 2.262s/\sqrt{n} \right] \text{ g.} \tag{B.76}$$

Substituting the values for $\bar{m}$, s, and n, we get an interval of

$$\left[10.02 - 2.262 \times 0.0883/\sqrt{10}, \ 10.02 + 2.262 \times 0.0883/\sqrt{10} \right] \text{ g} \tag{B.77}$$

or

$$[9.96, 10.08] \text{ g.} \tag{B.78}$$

■

The foregoing procedure for constructing a confidence interval for the mean using the t distribution was based on the assumption that the measurements were normally distributed. In situations where the data are not normally distributed this procedure can fail in a very dramatic fashion. See Exercise B.8. However, it may be safe to generate an approximate confidence interval using this procedure if (1) the number n of data is large (50 or more) or (2) the distribution of the data is not strongly skewed and n is at least 15.

B.9 HYPOTHESIS TESTS

In some situations we want to test whether or not a set of normally distributed data could reasonably have come from a normal distribution with expected value μ_0. Applying the sampling theorem, we see that if our data did come from a normal distribution with expected value μ_0, then there would be a 95% probability that

$$t_{\text{obs}} = \frac{\mu_0 - \bar{m}}{s/\sqrt{n}} \tag{B.79}$$

would lie in the interval

$$[F_t^{-1}(0.025), \ F_t^{-1}(0.975)] = [t_{n-1,0.025}, t_{n-1,0.975}] \tag{B.80}$$

and only a 5% probability that t would lie outside this interval. Equivalently, there is only a 5% probability that $|t_{\text{obs}}| \geq t_{n-1,0.975}$.

This leads to the ***t*–test**: If $|t_{obs}| \geq t_{n-1,0.975}$, then we reject the hypothesis that $\mu = \mu_0$. On the other hand, if $|t| < t_{n-1,0.975}$, then we cannot reject the hypothesis that $\mu = \mu_0$. Although the 95% confidence level is traditional, we can also perform the *t*–test at a 99% or some other confidence level. In general, if we want a confidence level of $1 - \alpha$, then we compare $|t|$ to $t_{n-1,1-\alpha/2}$.

In addition to reporting whether or not a set of data passes a *t*–test it is good practice to report the associated ***t*–test *p*–value** . The *p*-value associated with a *t*–test is the largest value of α for which the data passes the *t*–test. Equivalently, it is the probability that we could have gotten a greater *t* value than we have observed, given that all of our assumptions are correct.

■ **Example B.13** Consider the following data:

$$
\begin{array}{ccccc}
1.2944 & -0.3362 & 1.7143 & 2.6236 & 0.3082 \\
1.8580 & 2.2540 & -0.5937 & -0.4410 & 1.5711
\end{array}
\tag{B.81}
$$

These appear to be roughly normally distributed, with a mean that seems to be larger than 0. We will test the hypothesis $\mu = 0$. The *t* statistic is

$$
t_{obs} = \frac{\mu_0 - \bar{m}}{s\sqrt{n}},
\tag{B.82}
$$

which for this data set is

$$
t_{obs} = \frac{0 - 1.0253}{1.1895/\sqrt{10}} \approx -2.725.
\tag{B.83}
$$

Because $|t_{obs}|$ is larger than $t_{9,0.975} = 2.262$, we reject the hypothesis that these data came from a normal distribution with expected value 0 at the 95% confidence level.   ■

The *t*–test (or any other statistical test) can fail in two ways. First, it could be that the hypothesis that $\mu = \mu_0$ is true, but our particular data set contained some unlikely values and failed the *t*–test. Rejecting the hypothesis when it is in fact true is called a **type I error** . We can control the probability of a type I error by decreasing α.

The second way in which the *t*–test can fail is more difficult to control. It could be that the hypothesis $\mu = \mu_0$ was false, but the sample mean was close enough to μ_0 to pass the *t*–test. In this case, we have a **type II error**. The probability of a type II error depends very much on how close the true mean is to μ_0. If the true mean $\mu = \mu_1$ is very close to μ_0, then a type II error is quite likely. If the true mean $\mu = \mu_1$ is very far from μ_0, then a type II error will be less likely. Given a particular alternative hypothesis, $\mu = \mu_1$, we call the probability of a type II error $\beta(\mu_1)$ and call the probability of not making a type II error $(1 - \beta(\mu_1))$ the **power** of the test. We can estimate $\beta(\mu_1)$ by repeatedly generating sets of *n* random numbers with $\mu = \mu_1$ and performing the hypothesis test on the sets of random numbers. See Exercise B.9.

The results of a hypothesis test should always be reported with care. It is important to discuss and justify any assumptions (such as the normality assumption made in the *t*–test) underlying the test. The *p*-value should always be reported along with whether or not the hypothesis was rejected. If the hypothesis was not rejected and some particular alternative hypothesis is available, it is good practice to estimate the power of the hypothesis test against this alternative hypothesis. Confidence intervals for the mean should be reported along with the results of a hypothesis test.

It is important to distinguish between the statistical significance of a hypothesis test and the actual magnitude of any difference between the observed mean and the hypothesized mean. For example, with very large n it is nearly always possible to achieve statistical significance at the 95% confidence level, even though the observed mean may differ from the hypothesis by only 1% or less.

B.10 EXERCISES

B.1 Compute the expected value and variance of a uniform random variable in terms of the parameters a and b.

B.2 Compute the CDF of an exponential random variable with parameter λ.

B.3 Show that

$$\text{Cov}(aX, Y) = a\text{Cov}(X, Y) \tag{B.84}$$

and that

$$\text{Cov}(X + Y, Z) = \text{Cov}(X, Z) + \text{Cov}(Y, Z). \tag{B.85}$$

B.4 Show that the PDF for the sum of two independent uniform random variables on $[a, b] = [0, 1]$ is

$$f(x) = \begin{cases} 0 & (x \le 0) \\ x & (0 \le x \le 1) \\ 2 - x & (1 \le x \le 2) \\ 0 & (x \ge 0) \end{cases}. \tag{B.86}$$

B.5 Suppose that X and Y are independent random variables. Use conditioning to find a formula for the CDF of $X + Y$ in terms of the PDF's and CDF's of X and Y.

B.6 Suppose that $\mathbf{x} = (X_1, X_2)^T$ is a vector composed of two random variables with a multivariate normal distribution with expected value $\boldsymbol{\mu}$ and covariance matrix $\mathbf{C}$, and that $\mathbf{A}$ is a 2 by 2 matrix. Use properties of expected value and covariance to show that $\mathbf{y} = \mathbf{A}\mathbf{x}$ has expected value $\mathbf{A}\boldsymbol{\mu}$ and covariance $\mathbf{A}\mathbf{C}\mathbf{A}^T$.

B.7 Consider the following data, which we will assume are drawn from a normal distribution:

$$\begin{array}{ccccc} -0.4326 & -1.6656 & 0.1253 & 0.2877 & -1.1465 \\ 1.1909 & 1.1892 & -0.0376 & 0.3273 & 0.1746 \end{array}.$$

Find the sample mean and standard deviation. Use these to construct a 95% confidence interval for the mean. Test the hypothesis $H_0 : \mu = 0$ at the 95% confidence level. What do you conclude? What was the corresponding p-value?

B.8   Using MATLAB, repeat the following experiment 1000 times. Use the Statistics Tool-box function **exprnd()** to generate five exponentially distributed random numbers with $\mu = 10$. Use these five random numbers to generate a 95% confidence interval for the mean. How many times out of the 1000 experiments did the 95% confidence interval cover the expected value of 10? What happens if you instead generate 50 exponentially distributed random numbers at a time? Discuss your results.

B.9   Using MATLAB, repeat the following experiment 1000 times. Use the **randn** function to generate a set of 10 normally distributed random numbers with expected value 10.5 and standard deviation 1. Perform a t–test of the hypothesis $\mu = 10$ at the 95% confidence level. How many type II errors were committed? What is the approximate power of the t–test with $n = 10$ against the alternative hypothesis $\mu = 10.5$? Discuss your results.

B.10   Using MATLAB, repeat the following experiment 1000 times. Using the **exprnd()** function of the Statistics Toolbox, generate five exponentially distributed random numbers with expected value 10. Take the average of the five random numbers. Plot a histogram and a probability plot of the 1000 averages that you computed. Are the averages approximately normally distributed? Explain why or why not. What would you expect to happen if you took averages of 50 exponentially distributed random numbers at a time? Try it and discuss the results.

B.11 NOTES AND FURTHER READING

Most of the material in this appendix can be found in virtually any introductory textbook in probability and statistics. Some recent textbooks include [5, 22]. The multivariate normal distribution is a somewhat more advanced topic that is often ignored in introductory courses. [142] has a good discussion of the multivariate normal distribution and its properties. Numerical methods for probability and statistics are a specialized topic. Two standard references include [84, 161].

Appendix C

REVIEW OF VECTOR CALCULUS

Synopsis: A review is given of key vector calculus topics, including the gradient, Hessian, Jacobian, Taylor's theorem, and Lagrange multipliers.

C.1 THE GRADIENT, HESSIAN, AND JACOBIAN

In vector calculus, the familiar first and second derivatives of a single-variable function are generalized to operate on vectors.

■ **Definition C.1** Given a scalar-valued function with a vector argument, $f(\mathbf{x})$, the **gradient** of f is

$$\nabla f(\mathbf{x}) = \begin{bmatrix} \dfrac{\partial f}{\partial x_1} \\[6pt] \dfrac{\partial f}{\partial x_2} \\[6pt] \vdots \\[6pt] \dfrac{\partial f}{\partial x_n} \end{bmatrix}. \tag{C.1}$$

■

$\nabla f(\mathbf{x}^0)$ has an important geometric interpretation in that it points in the direction in which $f(\mathbf{x})$ increases most rapidly at the point $\mathbf{x}^0$.

Recall from single-variable calculus that if a function f is continuously differentiable, then a point x^* can only be a minimum or maximum point of f if $f'(x)|_{x=x^*} = 0$. Similarly in vector calculus, if $f(\mathbf{x})$ is continuously differentiable, then a point $\mathbf{x}^*$ can only be a minimum or maximum point if $\nabla f(\mathbf{x}^*) = \mathbf{0}$. Such a point $\mathbf{x}^*$ is called a **critical point**.

■ **Definition C.2** Given a scalar-valued function of a vector, $f(\mathbf{x})$, the **Hessian** of f is

$$\nabla^2 f(\mathbf{x}) = \begin{bmatrix} \dfrac{\partial^2 f}{\partial x_1 \partial x_1} & \dfrac{\partial^2 f}{\partial x_1 \partial x_2} & \cdots & \dfrac{\partial^2 f}{\partial x_1 \partial x_n} \\[2mm] \dfrac{\partial^2 f}{\partial x_2 \partial x_1} & \dfrac{\partial^2 f}{\partial x_2 \partial x_2} & \cdots & \dfrac{\partial^2 f}{\partial x_2 \partial x_n} \\[2mm] \vdots & \vdots & \ddots & \vdots \\[2mm] \dfrac{\partial^2 f}{\partial x_n \partial x_1} & \dfrac{\partial^2 f}{\partial x_n \partial x_2} & \cdots & \dfrac{\partial^2 f}{\partial x_n \partial x_n} \end{bmatrix}. \tag{C.2}$$

■

If f is twice continuously differentiable, the Hessian is symmetric. The Hessian is analogous to the second derivative of a function of a single variable. We have used the symbol ∇^2 here to denote the Hessian, but beware that this symbol is used by some authors to denote an entirely different differential operator, the Laplacian.

■ **Theorem C.1** If $f(\mathbf{x})$ is a twice continuously differentiable function, and $\nabla^2 f(\mathbf{x}^0)$ is a positive semidefinite matrix, then $f(\mathbf{x})$ is a **convex function** at $\mathbf{x}^0$. If $\nabla^2 f(\mathbf{x}^0)$ is positive definite, then $f(\mathbf{x})$ is **strictly convex** at $\mathbf{x}^0$.   ■

This theorem can be used to check whether a critical point is a minimum of f. If $\mathbf{x}^*$ is a critical point of f and $\nabla^2 f(\mathbf{x}^*)$ is positive definite, then f is convex at $\mathbf{x}^*$, and $\mathbf{x}^*$ is thus a local minimum of f.

It will be necessary to compute derivatives of quadratic forms.

■ **Theorem C.2** Let $f(\mathbf{x}) = \mathbf{x}^T \mathbf{A}\mathbf{x}$ where $\mathbf{A}$ is an n by n symmetric matrix. Then

$$\nabla f(\mathbf{x}) = 2\mathbf{A}\mathbf{x} \tag{C.3}$$

and

$$\nabla^2 f(\mathbf{x}) = 2\mathbf{A}. \tag{C.4}$$

■

■ **Definition C.3** Given a vector-valued function of a vector, $\mathbf{F}(\mathbf{x})$, where

$$\mathbf{F}(\mathbf{x}) = \begin{bmatrix} f_1(\mathbf{x}) \\ f_2(\mathbf{x}) \\ \vdots \\ f_m(\mathbf{x}) \end{bmatrix}, \tag{C.5}$$

the **Jacobian** of **F** is

$$
\mathbf{J}(\mathbf{x}) =
\begin{bmatrix}
\dfrac{\partial f_1}{\partial x_1} & \dfrac{\partial f_1}{\partial x_2} & \cdots & \dfrac{\partial f_1}{\partial x_n} \\[2mm]
\dfrac{\partial f_2}{\partial x_1} & \dfrac{\partial f_2}{\partial x_2} & \cdots & \dfrac{\partial f_2}{\partial x_n} \\[2mm]
\vdots & \vdots & \ddots & \vdots \\[2mm]
\dfrac{\partial f_m}{\partial x_1} & \dfrac{\partial f_m}{\partial x_2} & \cdots & \dfrac{\partial f_m}{\partial x_n}
\end{bmatrix}.
\tag{C.6}
$$

∎

Some authors use the notation $\nabla \mathbf{F}(\mathbf{x})$ for the Jacobian. Notice that the rows of $\mathbf{J}(\mathbf{x})$ are the gradients (C.1) of the functions $f_1(\mathbf{x}), f_2(\mathbf{x}), \ldots, f_m(\mathbf{x})$.

C.2 TAYLOR'S THEOREM

In the calculus of single-variable functions, Taylor's theorem produces an infinite series for $f(x + \Delta x)$ in terms of $f(x)$ and its derivatives. Taylor's theorem can be extended to a function of a vector $f(\mathbf{x})$, but in practice, derivatives of order higher than two are extremely inconvenient. The following form of Taylor's theorem is often used in optimization theory.

∎ **Theorem C.3** Suppose that $f(\mathbf{x})$ and its first and second partial derivatives are continuous. For any vectors $\mathbf{x}$ and $\Delta\mathbf{x}$ there is a vector $\mathbf{c}$, with $\mathbf{c}$ on the line between $\mathbf{x}$ and $\mathbf{x} + \Delta\mathbf{x}$, such that

$$
f(\mathbf{x} + \Delta\mathbf{x}) = f(\mathbf{x}) + \nabla f(\mathbf{x})^T \Delta\mathbf{x} + \frac{1}{2}\Delta\mathbf{x}^T \nabla^2 f(\mathbf{c}) \Delta\mathbf{x}.
\tag{C.7}
$$

∎

This form of **Taylor's theorem with remainder term** is useful in many proofs. However, in computational work there is no way to determine $\mathbf{c}$. For that reason, when $\Delta\mathbf{x}$ is a small perturbation, we often make use of the approximation

$$
f(\mathbf{x} + \Delta\mathbf{x}) \approx f(\mathbf{x}) + \nabla f(\mathbf{x})^T \Delta\mathbf{x} + \frac{1}{2}\Delta\mathbf{x}^T \nabla^2 f(\mathbf{x}) \Delta\mathbf{x}.
\tag{C.8}
$$

An even simpler version of Taylor's theorem, called the **mean value theorem**, uses only the first derivative.

∎ **Theorem C.4** Suppose that $f(\mathbf{x})$ and its first partial derivatives are continuous. For any vectors $\mathbf{x}$ and $\Delta\mathbf{x}$ there is a vector $\mathbf{c}$, with $\mathbf{c}$ on the line between $\mathbf{x}$ and $\mathbf{x} + \Delta\mathbf{x}$ such that

$$
f(\mathbf{x} + \Delta\mathbf{x}) = f(\mathbf{x}) + \nabla f(\mathbf{c})^T \Delta\mathbf{x}.
\tag{C.9}
$$

∎

We will make use of a truncated version of (C.8),

$$f(\mathbf{x} + \Delta\mathbf{x}) \approx f(\mathbf{x}) + \nabla f(\mathbf{x})^T \Delta\mathbf{x}. \tag{C.10}$$

By applying (C.10) to each of the functions $f_1(\mathbf{x}), f_2(\mathbf{x}), \ldots, f_m(\mathbf{x})$, we obtain the approximation

$$\mathbf{F}(\mathbf{x} + \Delta\mathbf{x}) \approx \mathbf{F}(\mathbf{x}) + \mathbf{J}(\mathbf{x})\Delta\mathbf{x}. \tag{C.11}$$

C.3 LAGRANGE MULTIPLIERS

The method of **Lagrange multipliers** is an important technique for solving optimization problems of the form

$$\begin{aligned} \min \quad & f(\mathbf{x}) \\ & g(\mathbf{x}) = 0 \end{aligned} \tag{C.12}$$

where the scalar-valued function of a vector argument, $f(\mathbf{x})$, is called the **objective function**.

Figure C.1 shows a typical situation. The curve represents the set of points (a contour) where $g(\mathbf{x}) = 0$. At a particular point $\mathbf{x}_0$ on this curve, the gradient of $g(\mathbf{x}_0)$ must be perpendicular to the curve because the function is constant along the curve. Moving counterclockwise, we can trace out a curve $\mathbf{x}(t)$, parameterized by the variable $t \geq 0$, with $\mathbf{x}(0) = \mathbf{x}_0$ and $g(\mathbf{x}(t)) = 0$. By the chain rule,

$$f'(\mathbf{x}(t)) = \mathbf{x}'(t)^T \nabla f(\mathbf{x}(t)). \tag{C.13}$$

Here $\mathbf{x}'(t)$ is the tangent to the curve. Since $\nabla f(\mathbf{x}_0)$ and $\mathbf{x}'(0)$ are at an acute angle, their dot product, $f'(0)$, is positive. Thus $f(\mathbf{x})$ is increasing as we move counterclockwise around the

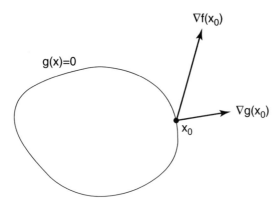

Figure C.1 The situation at a point which is not a minimum of (C.12).

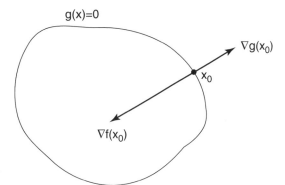

Figure C.2 The situation at a minimum point of (C.12).

curve $g(\mathbf{x}) = 0$ starting at $\mathbf{x}_0$. Similarly, by moving clockwise around the curve, we will have $f'(0) < 0$. Thus $\mathbf{x}_0$ cannot be a minimum point.

In Figure C.2, $\nabla f(\mathbf{x}_0)$ is perpendicular to the curve $g(\mathbf{x}) = 0$. In this case, $f'(0) = 0$, and $\nabla f(\mathbf{x}_0)$ provides no information on what will happen to the function value as we move away from $\mathbf{x}_0$ along the curve $g(\mathbf{x}) = 0$. The point $\mathbf{x}_0$ might be a minimum point, or it might not.

In general, a point $\mathbf{x}_0$ on the curve $g(\mathbf{x}) = 0$ can only be a minimum or maximum point if $\nabla g(\mathbf{x}_0)$ and $\nabla f(\mathbf{x}_0)$ are parallel, or where

$$\nabla f(\mathbf{x}_0) = \lambda \nabla g(\mathbf{x}_0) \tag{C.14}$$

where λ is scalar. A point $\mathbf{x}_0$ satisfying (C.14) is called a **stationary point**, and λ is called a **Lagrange multiplier**.

The Lagrange multiplier condition (C.14) is necessary but not sufficient for finding a minimum of $f(\mathbf{x})$ along the contour $g(\mathbf{x}) = 0$, because a stationary point may be a minimum, maximum, or saddle point. Furthermore, a problem may have several local minima. Thus it is necessary to examine the behavior of $f(\mathbf{x})$ at all stationary points to find a global minimum.

■ **Theorem C.5** A minimum of (C.12) can occur only at a point $\mathbf{x}_0$ where

$$\nabla f(\mathbf{x}_0) = \lambda \nabla g(\mathbf{x}_0) \tag{C.15}$$

for some λ.   ■

The Lagrange multiplier condition can be extended to problems of the form

$$\begin{aligned} \min \quad & f(\mathbf{x}) \\ & g(\mathbf{x}) \leq 0. \end{aligned} \tag{C.16}$$

Since points along the curve $g(\mathbf{x}) = 0$ are still feasible in (C.16), the previous necessary condition must still hold true. However, there is an additional restriction. Suppose that $\nabla g(\mathbf{x}_0)$ and $\nabla f(\mathbf{x}_0)$ both point in the outward direction. In that case, we could move in the opposite direction, into the feasible region, and decrease the function value. Thus a point $\mathbf{x}_0$ cannot be a minimum point of (C.16) unless the two gradients point in opposite directions.

■ **Theorem C.6** A minimum point of (C.16) can occur only at a point $\mathbf{x}_0$ where

$$\nabla f(\mathbf{x}_0) + \lambda \nabla g(\mathbf{x}_0) = \mathbf{0} \tag{C.17}$$

for some Lagrange multiplier $\lambda \geq 0$.   ■

■ **Example C.1** Consider the problem

$$\min \quad x_1 + x_2 \\ x_1^2 + x_2^2 - 1 \leq 0. \tag{C.18}$$

The Lagrange multiplier condition is

$$\begin{bmatrix} 1 \\ 1 \end{bmatrix} + \lambda \begin{bmatrix} 2x_1 \\ 2x_2 \end{bmatrix} = \mathbf{0} \tag{C.19}$$

with $\lambda \geq 0$. One solution to this nonlinear system of equations is $x_1 = 0.7071$, $x_2 = 0.7071$, with $\lambda = -0.7071$. Since $\lambda < 0$, this point does not satisfy the Lagrange multiplier condition. In fact, it is the maximum of $f(\mathbf{x})$ subject to $g(\mathbf{x}) \leq 0$. The second solution to the Lagrange multiplier equations is $x_1 = -0.7071$, $x_2 = -0.7071$, with $\lambda = 0.7071$. Since this is the only solution with $\lambda \geq 0$, this point solves the minimization problem. ■

Note that (C.17) is (except for the condition $\lambda > 0$) the necessary condition for a minimum point of the unconstrained minimization problem

$$\min \quad f(\mathbf{x}) + \lambda g(\mathbf{x}). \tag{C.20}$$

Here the parameter λ can be adjusted so that, for the optimal solution, $\mathbf{x}^*$, $g(\mathbf{x}^*) \leq 0$. We will make frequent use of this technique to convert constrained optimization problems into unconstrained optimization problems.

C.4 EXERCISES

C.1 Let

$$f(\mathbf{x}) = x_1^2 x_2^2 - 2x_1 x_2^2 + x_2^2 - 3x_1^2 x_2 + 12x_1 x_2 - 12x_2 + 6. \tag{C.21}$$

Find the gradient, $\nabla f(\mathbf{x})$, and Hessian, $\nabla^2 f(\mathbf{x})$. What are the critical points of f? Which of these are minima and maxima of f?

C.2 Find a Taylor's series approximation for

$$f(\mathbf{x}) = e^{-(x_1+x_2)^2} \tag{C.22}$$

near the point

$$\mathbf{x} = \begin{bmatrix} 2 \\ 3 \end{bmatrix}. \tag{C.23}$$

C.3 Use the method of Lagrange multipliers to solve the problem

$$\begin{aligned} \min \quad & 2x_1 + x_2 \\ & 4x_1^2 + 3x_2^2 - 5 \leq 0. \end{aligned} \tag{C.24}$$

C.4 Derive the formula (A.91) for the 2-norm of a matrix. Begin with the maximization problem

$$\max_{\|\mathbf{x}\|_2=1} \|\mathbf{A}\mathbf{x}\|_2^2. \tag{C.25}$$

Note that we have squared $\|\mathbf{A}\mathbf{x}\|_2$. We will take the square root at the end of the problem.

(a) Using the formula $\|\mathbf{x}\|_2 = \sqrt{\mathbf{x}^T\mathbf{x}}$, rewrite the above maximization problem without norms.
(b) Use the Lagrange multiplier method to find a system of equations that must be satisfied by any stationary point of the maximization problem.
(c) Explain how the eigenvalues and eigenvectors of $\mathbf{A}^T\mathbf{A}$ are related to this system of equations. Express the solution to the maximization problem in terms of the eigenvalues and eigenvectors of $\mathbf{A}^T\mathbf{A}$.
(d) Use this solution to get $\|\mathbf{A}\|_2$.

C.5 Derive the normal equations (2.3) using vector calculus, by letting

$$f(\mathbf{m}) = \|\mathbf{G}\mathbf{m} - \mathbf{d}\|_2^2 \tag{C.26}$$

and minimizing $f(\mathbf{m})$. Note that in problems with many least squares solutions, all of the least squares solutions will satisfy the normal equations.

(a) Rewrite $f(\mathbf{m})$ as a dot product and then expand the expression.
(b) Find $\nabla f(\mathbf{m})$.
(c) Set $\nabla f(\mathbf{m}) = \mathbf{0}$, and obtain the normal equations.

C.5 NOTES AND FURTHER READING

Basic material on vector calculus can be found in many calculus textbooks. However, more advanced topics, such as Taylor's theorem for functions of a vector, are often skipped in basic texts. The material in this chapter is particularly important in optimization and can often be found in associated references [98, 108, 115].

Appendix D

GLOSSARY OF NOTATION

- α, β, γ, ... : Scalars.
- a, b, c, ... : Scalars or functions.
- $\mathbf{a}$, $\mathbf{b}$, $\mathbf{c}$, ... : Column vectors.
- a_i: ith element of vector $\mathbf{a}$.
- A, B, C, ... : Scalar valued functions or random variables.
- $\mathcal{A}$, $\mathcal{B}$, $\mathcal{C}$, ... : Fourier transforms.
- $\mathbf{A}$, $\mathbf{B}$, $\mathbf{C}$, ... : Vector valued functions or matrices.
- $\mathbf{A_{i,.}}$: ith row of matrix $\mathbf{A}$.
- $\mathbf{A_{.,i}}$: ith column of matrix $\mathbf{A}$.
- $A_{i,j}$: (i, j)th element of matrix $\mathbf{A}$.
- $\mathbf{A}^{-1}$: The inverse of the matrix $\mathbf{A}$.
- $\mathbf{A}^T$: The transpose of the matrix $\mathbf{A}$.
- $\mathbf{A}^{-T}$: The transpose of the matrix $\mathbf{A}^{-1}$.
- R^n: The space of n–dimensional real vectors.
- $N(\mathbf{A})$: Null space of the matrix $\mathbf{A}$.
- $R(\mathbf{A})$: Range of the matrix $\mathbf{A}$.
- rank($\mathbf{A}$): Rank of the matrix $\mathbf{A}$.
- s_i: Singular values.
- λ_i: Eigenvalues.
- $\|\mathbf{x}\|$: Norm of a vector $\mathbf{x}$. A subscript is used to specify the 1-norm, 2-norm, or ∞-norm.
- $\|\mathbf{A}\|$: Norm of a matrix $\mathbf{A}$. A subscript is used to specify the 1-norm, 2-norm, or ∞-norm.
- $\mathbf{G}^\dagger$: Generalized inverse of the matrix $\mathbf{G}$.
- $\mathbf{m}_\dagger$: Generalized inverse solution $\mathbf{m}_\dagger = \mathbf{G}^\dagger \mathbf{d}$.
- $\mathbf{G}^\sharp$: A regularized generalized inverse of the matrix $\mathbf{G}$.
- $\mathbf{m}_{\alpha,L}$: Tikhonov regularized solution with regularization parameter α and regularization matrix $\mathbf{L}$.
- $\mathbf{R_m}$: Model resolution matrix.

- $\mathbf{R}_d$: Data resolution matrix.
- $E[X]$: Expected value of the random variable X.
- $\bar{\mathbf{a}}$: Mean value of the elements in vector $\mathbf{a}$.
- $N(\mu,\ \sigma^2)$: Normal probability distribution with expected value μ and variance σ^2.
- $\text{Cov}(X,\ Y)$: Covariance of the random variables X and Y.
- $\text{Cov}(\mathbf{x})$: Matrix of covariances of elements of the vector $\mathbf{x}$.
- $\rho(X,\ Y)$: Correlation between the random variables X and Y.
- $\text{Var}(X)$: Variance of the random variable X.
- $f(\mathbf{d}|\mathbf{m})$: Conditional probability density for $\mathbf{d}$, conditioned on a particular model $\mathbf{m}$.
- $L(\mathbf{m}|\mathbf{d})$: Likelihood function for a model $\mathbf{m}$ given a particular data vector $\mathbf{d}$.
- σ: Standard deviation.
- σ^2: Variance.
- $t_{v,p}$: p-percentile of the t distribution with v degrees of freedom.
- χ^2_{obs}: Observed value of the χ^2 statistic.
- $\chi^2_{v,p}$: p-percentile of the χ^2 distribution with v degrees of freedom.
- $\nabla f(\mathbf{x})$: Gradient of the function $f(\mathbf{x})$.
- $\nabla^2 f(\mathbf{x})$: Hessian of the function $f(\mathbf{x})$.

BIBLIOGRAPHY

[1] U. Amato and W. Hughes. Maximum entropy regularization of Fredholm integral equations of the 1st kind. *Inverse Problems*, 7(6):793–808, 1991.

[2] R. C. Aster. On projecting error ellipsoids. *Bulletin of the Seismological Society of America*, 78(3):1373–1374, 1988.

[3] G. Backus and F. Gilbert. Uniqueness in the inversion of inaccurate gross earth data. *Philosophical Transactions of the Royal Society A*, 266:123–192, 1970.

[4] Z. Bai and J. W. Demmel. Computing the generalized singular value decomposition. *SIAM Journal on Scientific Computing*, 14:1464–1486, 1993.

[5] L. J. Bain and M. Englehardt. *Introduction to Probability and Mathematical Statistics,* 2nd ed. Brooks/Cole, Pacific Grove, CA, 2000.

[6] R. Barrett, M. Berry, T. F. Chan, J. Demmel, J. Donato, V. Eijkhout, R. Pozo, C. Romine, and H. van der Vorst. *Templates for the Solution of Linear Systems: Building Blocks for Iterative Methods,* 2nd ed. SIAM, Philadelphia, 1994.

[7] I. Barrowdale and F. D. K. Roberts. Solution of an overdetermined system of equations in the l_1 norm. *Communications of the ACM*, 17(6):319–326, 1974.

[8] D. M. Bates and D. G. Watts. *Nonlinear Regression Analysis and Its Applications*. Wiley, New York, 1988.

[9] J. Baumeister. *Stable Solution of Inverse Problems*. Vieweg, Braunschweig, 1987.

[10] A. Ben-Israel and T. N. E. Greville. *Generalized Inverses,* 2nd ed. Springer-Verlag, New York, 2003.

[11] A. Ben-Tal and A. Nemirovski. *Lectures on Modern Convex Optimization: Analysis, Algorithms, and Engineering Applications*. SIAM, Philadelphia, 2001.

[12] J. G. Berryman. Analysis of approximate inverses in tomography I. Resolution analysis. *Optimization and Engineering*, 1(1):87–115, 2000.

[13] J. G. Berryman. Analysis of approximate inverses in tomography II. Iterative inverses. *Optimization and Engineering*, 1(4):437–473, 2000.

[14] M. Bertero and P. Boccacci. *Introduction to Inverse Problems in Imaging*. Institute of Physics, London, 1998.

[15] J. T. Betts. *Practical Methods for Optimal Control Using Nonlinear Programming*. SIAM, Philadelphia, 2001.

[16] Å. Björck. *Numerical Methods for Least Squares Problems*. SIAM, Philadelphia, 1996.

[17] P. T. Boggs, J. R. Donaldson, R. H. Byrd, and R. B. Schnabel. ODRPACK software for weighted orthogonal distance regression. *ACM Transactions on Mathematical Software*, 15(4):348–364, 1989. The software is available at http://www.netlib.org/odrpack/.

[18] R. Bracewell. *The Fourier Transform and its Applications,* 3rd ed. McGraw-Hill, Boston, 2000.

[19] D. S. Briggs. *High Fidelity Deconvolution of Moderately Resolved Sources*. Ph.D. thesis, New Mexico Institute of Mining and Technology, 1995.

[20] S. L. Campbell and C. D. Meyer, Jr. *Generalized Inverses of Linear Transformations*. Dover, Mineola, New York, 1991.

[21] P. Carrion. *Inverse Problems and Tomography in Acoustics and Seismology*. Penn Publishing Company, Atlanta, 1987.

[22] G. Casella and R. L. Berger. *Statistical Inference,* 2nd ed. Duxbury, Pacific Grove, CA, 2002.

[23] Y. Censor and S. A. Zenios. *Parallel Optimization: Theory, Algorithms, and Applications.* Oxford University Press, New York, 1997.

[24] B. G. Clark. An efficient implementation of the algorithm "CLEAN." *Astronomy and Astrophysics*, 89(3):377–378, 1980.

[25] T. F. Coleman and Y. Li. A globally and quadratically convergent method for linear l_1 problems. *Mathematical Programming*, 56:189–222, 1992.

[26] S. C. Constable, R. L. Parker, and C. G. Constable. Occam's inversion: A practical algorithm for generating smooth models from electromagnetic sounding data. *Geophysics*, 52(3):289–300, 1987.

[27] G. Corliss, C. Faure, A. Griewank, and L. Hascoet. *Automatic Differentiation of Algorithms.* Springer-Verlag, Berlin, 2000.

[28] T. J. Cornwell and K. F. Evans. A simple maximum entropy deconvolution algorithm. *Astronomy and Astrophysics*, 143(1):77–83, 1985.

[29] R. T. Cox. *Algebra of Probable Inference.* The Johns Hopkins University Press, Baltimore, 2002.

[30] P. Craven and G. Wahba. Smoothing noisy data with spline functions: Estimating the correct degree of smoothing by the method of generalized cross-validation. *Numerische Mathematik*, 31:377–403, 1979.

[31] J. W. Demmel. *Applied Numerical Linear Algebra.* SIAM, Philadelphia, 1997.

[32] J. E. Dennis, Jr. and R. B. Schnabel. *Numerical Methods for Unconstrained Optimization and Nonlinear Equations.* SIAM, Philadelphia, 1996.

[33] N. R. Draper and H. Smith. *Applied Regression Analysis,* 3rd ed. Wiley, New York, 1998.

[34] C. Eckart and G. Young. A principal axis transformation for non-Hermitian matrices. *Bulletin of the American Mathematical Society*, 45:118–121, 1939.

[35] A. W. F. Edwards. *Likelihood.* The Johns Hopkins University Press, Baltimore, 1992.

[36] L. Eldèn. Solving quadratically constrained least squares problems using a differential-geometric approach. *BIT*, 42(2):323–335, 2002.

[37] H. W. Engl. Regularization methods for the stable solution of inverse problems. *Surveys on Mathematics for Industry*, 3:71–143, 1993.

[38] H. W. Engl, M. Hanke, and A. Neubauer. *Regularization of Inverse Problems.* Kluwer Academic Publishers, Boston, 1996.

[39] R. M. Errico. What is an adjoint model. *Bulletin of the American Meteorological Society*, 78(11):2577–2591, 1997.

[40] M. Evans, N. Hasting, and B. Peacock. *Statistical Distributions,* 2nd ed. John Wiley & Sons, New York, 1993.

[41] S. N. Evans and P. B. Stark. Inverse problems as statistics. *Inverse Problems*, 18:R1–R43, 2002.

[42] J. G. Ferris and D. B. Knowles. The slug-injection test for estimating the coefficient of transmissibility of an aquifer. In R. Bentall, editor, *Methods of Determining Permeability, Transmissibility and Drawdown*, pages 299–304. U.S. Geological Survey, 1963.

[43] A. Frommer and P. Maass. Fast CG-based methods for Tikhonov–Phillips regularization. *SIAM Journal on Scientific Computing*, 20(5):1831–1850, 1999.

[44] I. M. Gelfand and S. V. Fomin. *Calculus of Variations.* Dover, Mineola, New York, 2000.

[45] A. Gelman, J. B. Carlin, H. S. Stern, and D. B. Rubin. *Bayesian Data Analysis,* 2nd ed. Chapman & Hall/CRC, Boca Raton, FL, 2003.

[46] L. El Ghaoui and H. Lebret. Robust solutions to least-squares problems with uncertain data. *SIAM Journal on Matrix Analysis and Applications*, 18(4):1035–1064, 1997.

[47] W. R. Gilks, S. Richardson, and D. J. Spiegelhalter. *Markov Chain Monte Carlo in Practice.* Chapman & Hall, London, 1996.

[48] G. H. Golub and D. P. O'Leary. Some history of the conjugate gradient and Lanczos methods. *SIAM Review*, 31(1):50–102, 1989.

[49] G. H. Golub and C. F. Van Loan. *Matrix Computations,* 3rd ed. The Johns Hopkins University Press, Baltimore, 1996.

[50] G. H. Golub and U. von Matt. Generalized cross-validation for large-scale problems. *Journal of Computational and Graphical Statistics*, 6(1):1–34, 1997.

[51] W. P. Gouveia and J. A. Scales. Resolution of seismic waveform inversion: Bayes versus Occam. *Inverse Problems*, 13(2):323–349, 1997.

[52] A. Griewank. *Evaluating Derivatives: Principles and Techniques of Algorithmic Differentiation*. SIAM, Philadelphia, 2000.

[53] C. W. Groetsch. *Inverse Problems in the Mathematical Sciences*. Vieweg, Braunschweig, 1993.

[54] D. Gubbins. *Time Series Analysis and Inverse Theory for Geophysicists*. Cambridge University Press, Cambridge, U.K., 2004.

[55] C. Guus, E. Boender, and H. Edwin Romeijn. Stochastic methods. In R. Horst and P. M. Pardalos, editors, *Handbook of Global Optimization*, pages 829–869. Kluwer Academic Publishers, Dordrecht, 1995.

[56] P. C. Hansen. Relations between SVD and GSVD of discrete regularization problems in standard and general form. *Linear Algebra and Its Applications*, 141:165–176, 1990.

[57] P. C. Hansen. Analysis of discrete ill-posed problems by means of the L-curve. *SIAM Review*, 34(4):561–580, 1992.

[58] P. C. Hansen. Regularization tools: A MATLAB package for analysis and solution of discrete ill-posed problems. *Numerical Algorithms*, 6(I–II):1–35, 1994. The software is available at http://www.imm.dtu.dk/documents/users/pch/Regutools/regutools.html.

[59] P. C. Hansen. *Rank-Deficient and Discrete Ill-Posed Problems: Numerical Aspects of Linear Inversion*. SIAM, Philadelphia, 1998.

[60] P. C. Hansen. Deconvolution and regularization with Toeplitz matrices. *Numerical Algorithms*, 29:323–378, 2002.

[61] P. C. Hansen and K. Mosegaard. Piecewise polynomial solutions without a priori break points. *Numerical Linear Algebra with Applications*, 3(6):513–524, 1996.

[62] J. M. H. Hendrickx. Bosque del Apache soil data. Personal communication, 2003.

[63] J. M. H. Hendrickx, B. Borchers, J. D. Rhoades, D. L. Corwin, S. M. Lesch, A. C. Hilgendorf, and J. Schlue. Inversion of soil conductivity profiles from electromagnetic induction measurements; theory and experimental verification. *Soil Science Society of America Journal*, 66(3):673–685, 2002.

[64] G. T. Herman. *Image Reconstruction from Projections*. Academic Press, San Francisco, 1980.

[65] M. R. Hestenes. Conjugacy and gradients. In S. G. Nash, editor, *A History of Scientific Computing*, pages 167–179. ACM Press, New York, 1990.

[66] M. R. Hestenes and E. Stiefel. Methods of conjugate gradients for solving linear systems. *Journal of Research, National Bureau of Standards*, 49:409–436, 1952.

[67] J. A. Hildebrand, J. M. Stevenson, P. T. C. Hammer, M. A. Zumberge, R. L. Parker, C. J. Fox, and P. J. Meis. A sea-floor and sea-surface gravity survey of Axial volcano. *Journal of Geophysical Research*, 95(B8):12751–12763, 1990.

[68] J. A. Högbom. Aperture synthesis with a non-regular distribution of interferometer baselines. *Astronomy and Astrophysics Supplement*, 15:417–426, 1974.

[69] R. A. Horn and C. R. Johnson. *Matrix Analysis*. Cambridge University Press, Cambridge, 1985.

[70] R. Horst and P. M. Pardalos. *Handbook of Global Optimization*. Kluwer Academic Publishers, Dordrecht, 1995.

[71] R. Horst, P. M. Pardalos, and N. V. Thoai. *Introduction to Global Optimization,* 2nd ed. Kluwer Academic Publishers, Dordrecht, 2001.

[72] P. J. Huber. *Robust Statistical Procedures,* 2nd ed. SIAM, Philadelphia, 1996.

[73] S. Van Huffel and J. Vandewalle. *The Total Least Squares Problem: Computational Aspects and Analysis*. SIAM, Philadelphia, 1991.

[74] H. Iyer and K. Hirahara, editors. *Seismic Tomography*. Chapman and Hall, New York, 1993.

[75] E. T. Jaynes. *Probability Theory: The Logic of Science*. Cambridge University Press, Cambridge, 2003.

[76] H. Jeffreys. *Theory of Probability,* 3rd ed. Oxford University Press, New York, 1998.

[77] W. H. Jeffreys, M. J. Fitzpatrick, and B. E. McArthur. Gaussfit—A system for least squares and robust estimation. *Celestial Mechanics*, 41(1–4):39–49, 1987. The software is available at http://clyde.as.utexas.edu/Gaussfit.html.

[78] A. C. Kak and M. Slaney. *Principles of Computerized Tomographic Imaging*. SIAM, Philadelphia, 2001.

[79] J. N. Kapur and H. K. Kesavan. *Entropy Optimization Principles with Applications*. Academic Press, Boston, 1992.

[80] L. Kaufman and A. Neumaier. PET regularization by envelope guided conjugate gradients. *IEEE Transactions on Medical Imaging*, 15(3):385–389, 1996.

[81] L. Kaufman and A. Neumaier. Regularization of ill-posed problems by envelope guided conjugate gradients. *Journal of Computational and Graphical Statistics*, 6(4):451–463, 1997.

[82] C. T. Kelley. *Iterative Methods for Solving Linear and Nonlinear Equations*. SIAM, Philadelphia, 1995.

[83] C. T. Kelley. *Solving Nonlinear Equations with Newton's Method*. SIAM, Philadelphia, 2003.

[84] W. J. Kennedy, Jr. and J. E. Gentle. *Statistical Computing*. Marcel Dekker, New York, 1980.

[85] A. Kirsch. *An Introduction to the Mathematical Theory of Inverse Problems*. Springer-Verlag, New York, 1996.

[86] F. J. Klopping, G. Peter, D. S. Robertson, K. A. Berstis, R. E. Moose, and W. E. Carter. Improvements in absolute gravity observations. *Journal of Geophysical Research*, 96(B5):8295–8303, 1991.

[87] S. Kullback. *Information Theory and Statistics,* 2nd ed. Dover, Mineola, New York, 1997.

[88] C. Lanczos. Solutions of systems of linear equations by minimized iterations. *Journal of Research, National Bureau of Standards*, 49:33–53, 1952.

[89] C. Lanczos. *Linear Differential Operators*. Dover, Mineola, New York, 1997.

[90] C. L. Lawson and R. J. Hanson. *Solving Least Squares Problems*. SIAM, Philadelphia, 1995.

[91] D. C. Lay. *Linear Algebra and its Applications,* 3rd ed. Addison-Wesley, Boston, 2003.

[92] T. Lay and T. Wallace. *Modern Global Seismology*. Academic Press, San Diego, 1995.

[93] J. J. Leveque, L. Rivera, and G. Wittlinger. On the use of the checker-board test to assess the resolution of tomographic inversions. *Geophysical Journal International*, 115(1):313–318, 1993.

[94] Z. P. Liang and P. C. Lauterbur. *Principles of Magnetic Resonance Imaging: A Signal Processing Perspective*. IEEE Press, New York, 2000.

[95] L. R. Lines, editor. *Inversion of Geophysical Data*. Society of Exploration Geophysicists, Tulsa, OK, 1988.

[96] T. W. Lo and P. Inderwiesen. *Fundamentals of Seismic Tomography*. Society of Exploration Geophysicists, Tulsa, OK, 1994.

[97] D. G. Luenberger. *Optimization by Vector Space Methods*. John Wiley & Sons, New York, 1969.

[98] W. H. Marlow. *Mathematics for Operations Research*. Dover, Mineola, New York, 1993.

[99] P. J. McCarthy. Direct analytic model of the L-curve for Tikhonov regularization parameter selection. *Inverse Problems*, 19:643–663, 2003.

[100] W. Menke. *Geophysical Data Analysis: Discrete Inverse Theory,* revised edition, volume 45 of *International Geophysics Series*. Academic Press, San Diego, 1989.

[101] C. D. Meyer. *Matrix Analysis and Applied Linear Algebra*. SIAM, Philadelphia, 2000.

[102] E. H. Moore. On the reciprocal of the general algebraic matrix. *Bulletin of the American Mathematical Society*, 26:394–395, 1920.

[103] J. J. More, B. S. Garbow, and K. E. Hillstrom. User guide for MINPACK-1. Technical Report ANL-80-74, Argonne National Laboratory, 1980.

[104] V. A. Morozov. *Methods for Solving Incorrectly Posed Problems*. Springer-Verlag, New York, 1984.

[105] K. Mosegaard and M. Sambridge. Monte Carlo analysis of inverse problems. *Inverse Problems*, 18(2):R29–R54, 2002.

[106] R. H. Myers. *Classical and Modern Regression with Applications,* 2nd ed. PWS Kent, Boston, 1990.

[107] R. Narayan and R. Nityananda. Maximum entropy image restoration in astronomy. *Annual Review of Astronomy and Astrophysics*, 24:127–170, 1986.

[108] S. G. Nash and A. Sofer. *Linear and Nonlinear Programming*. McGraw-Hill, New York, 1996.

[109] F. Natterer. *The Mathematics of Computerized Tomography*. SIAM, Philadelphia, 2001.

[110] F. Natterer and F. Wübbeling. *Mathematical Methods in Image Reconstruction*. SIAM, Philadelphia, 2001.

[111] A. Neumaier. Solving ill-conditioned and singular linear systems: A tutorial on regularization. *SIAM Review*, 40(3):636–666, 1998.

[112] R. Neupauer and B. Borchers. A MATLAB implementation of the minimum relative entropy method for linear inverse problems. *Computers & Geosciences*, 27(7):757–762, 2001.

[113] R. Neupauer, B. Borchers, and J. L. Wilson. Comparison of inverse methods for reconstructing the release history of a groundwater contamination source. *Water Resources Research*, 36(9):2469–2475, 2000.

[114] I. Newton. *The Principia, Mathematical Principles of Natural Philosophy (A new translation by I. B. Cohen and A. Whitman)*. University of California Press, Berkeley, 1999.

[115] J. Nocedal and S. J. Wright. *Numerical Optimization*. Springer-Verlag, New York, 1999.

[116] G. Nolet. Solving or resolving inadequate and noisy tomographic systems. *Journal of Computational Physics*, 61(3):463–482, 1985.

[117] G. Nolet, editor. *Seismic Tomography with Applications in Global Seismology and Exploration Geophysics*. D. Reidel, Boston, 1987.

[118] S. J. Osher and R. P. Fedkiw. *Level Set Methods and Dynamic Implicit Surfaces*. Springer-Verlag, New York, 2002.

[119] C. C. Paige and M. A. Saunders. Algorithm 583 LSQR: Sparse linear equations and least-squares problems. *ACM Transactions on Mathematical Software*, 8(2):195–209, 1982.

[120] C. C. Paige and M. A. Saunders. LSQR: An algorithm for sparse linear equations and sparse least squares. *ACM Transactions on Mathematical Software*, 8(1):43–71, 1982.

[121] R. L. Parker. A theory of ideal bodies for seamount magnetism. *Journal of Geophysical Research*, 96(B10):16101–16112, 1991.

[122] R. L. Parker. *Geophysical Inverse Theory*. Princeton University Press, Princeton, NJ, 1994.

[123] R. L. Parker and M. K. McNutt. Statistics for the one-norm misfit measure. *Journal of Geophysical Research*, 85:4429–4430, 1980.

[124] R. L. Parker and M. A. Zumberge. An analysis of geophysical experiments to test Newton's law of gravity. *Nature*, 342:29–32, 1989.

[125] R. Penrose. A generalized inverse for matrices. *Proceedings of the Cambridge Philosophical Society*, 51:406–413, 1955.

[126] S. Portnoy and R. Koenker. The Gaussian hare and the Laplacian tortoise: Computability of squared-error versus absolute-error estimators. *Statistical Science*, 12:279–296, 1997.

[127] M. B. Priestley. *Spectral Analysis and Time Series*. Academic Press, London, 1983.

[128] F. Rendl and H. Wolkowicz. A semidefinite framework for trust region subproblems with applications to large scale minimization. *Mathematical Programming*, 77(2):273–299, 1997.

[129] W. Rison, R. J. Thomas, P. R. Krehbiel, T. Hamlin, and J. Harlin. A GPS-based three-dimensional lightning mapping system: Initial observations in central New Mexico. *Geophysical Research Letters*, 26(23):3573–3576, 1999.

[130] C. P. Robert and G. Cassella. *Monte Carlo Statistical Methods*. Springer-Verlag, New York, 1999.

[131] C. D. Rodgers. *Inverse Methods for Atmospheric Sounding: Theory and Practice*. World Scientific Publishing, Singapore, 2000.

[132] M. Rojas, S. A. Santos, and D. C. Sorensen. A new matrix-free algorithm for the large-scale trust-region subproblem. *SIAM Journal on Optimization*, 11(3):611–646, 2000.

[133] M. Rojas and D. C. Sorensen. A trust-region approach to the regularization of large-scale discrete forms of ill-posed problems. *SIAM Journal on Scientific Computing*, 23(6):1843–1861, 2002.

[134] C. A. Rowe, R. C. Aster, B. Borchers, and C. J. Young. An automatic, adaptive algorithm for refining phase picks in large seismic data sets. *Bulletin of the Seismological Society of America*, 92:1660–1674, 2002.

[135] W. Rudin. *Real and Complex Analysis,* 3rd ed. McGraw-Hill, New York, 1987.

[136] Y. Saad. *Iterative Methods for Sparse Linear Systems,* 2nd ed. SIAM, Philadelphia, 2003.

[137] M. Sambridge and K. Mosegaard. Monte Carlo methods in geophysical inverse problems. *Reviews of Geophysics*, 40(3):1–29, 2002.

[138] R. J. Sault. A modification of the Cornwell and Evans maximum entropy algorithm. *Astrophysical Journal*, 354(2):L61–L63, 1990.

[139] L. J. Savage. *The Foundation of Statistics,* 2nd ed. Dover, Mineola, New York, 1972.

[140] J. Scales and M. Smith. DRAFT: Geophysical inverse theory. http://landau.Mines.EDU/~samizdat/inverse_theory/, 1997.

[141] J. A. Scales, A. Gersztenkorn, and S. Treitel. Fast lp solution of large, sparse, linear systems: Application to seismic travel time tomography. *Journal of Computational Physics*, 75(2): 314–333, 1988.

[142] S. R. Searle. *Matrix Algebra Useful for Statistics*. Wiley, New York, 1982.

[143] M. K. Sen and P. L. Stoffa. *Global Optimization Methods in Geophysical Inversion*. Number 4 in *Advances in Exploration Geophysics*. Elsevier, New York, 1995.

[144] C. B. Shaw, Jr. Improvement of the resolution of an instrument by numerical solution of an integral equation. *Journal of Mathematical Analysis and Applications*, 37:83–112, 1972.

[145] J. R. Shewchuk. An introduction to the conjugate gradient method without the agonizing pain, edition 1–1/4. Technical report, School of Computer Science, Carnegie Mellon University, August 1994. http://www.cs.cmu.edu/~jrs/jrspapers.html.

[146] D. S. Sivia. *Data Analysis, A Bayesian Tutorial*. Oxford University Press, New York, 1996.

[147] T. H. Skaggs and Z. J. Kabala. Recovering the release history of a groundwater contaminant. *Water Resources Research*, 30(1):71–79, 1994.

[148] T. H. Skaggs and Z. J. Kabala. Recovering the history of a groundwater contaminant plume: Method of quasi-reversibility. *Water Resources Research*, 31:2669–2673, 1995.

[149] J. Skilling and R. K. Bryan. Maximum entropy image reconstruction: General algorithm. *Monthly Notices of the Royal Astronomical Society*, 211:111–124, 1984.

[150] W. Spakman and G. Nolet. Imaging algorithms, accuracy and resolution in delay time tomography. In N. J. Vlaar, G. Nolet, M. J. R. Wortel, and S. A. P. L. Cloetingh, editors, *Mathematical Geophysics*, pages 155–187. D. Reidel, Dordrecht, 1988.

[151] P. B. Stark and R. L. Parker. Velocity bounds from statistical estimates of $\tau(p)$ and $x(p)$. *Journal of Geophysical Research*, 92(B3):2713–2719, 1987.

[152] P. B. Stark and R. L. Parker. Correction to "velocity bounds from statistical estimates of $\tau(p)$ and $x(p)$". *Journal of Geophysical Research*, 93:13821–13822, 1988.

[153] P. B. Stark and R. L. Parker. Bounded-variable least-squares: An algorithm and applications. *Computational Statistics*, 10(2):129–141, 1995.

[154] G. W. Stewart. On the early history of the singular value decomposition. *SIAM Review*, 35: 551–566, 1993.

[155] G. Strang. *Linear Algebra and Its Applications,* 3rd ed. Harcourt Brace Jovanovich Inc., San Diego, 1988.

[156] G. Strang and K. Borre. *Linear Algebra, Geodesy, and GPS*. Wellesley-Cambridge Press, Wellesley, MA, 1997.

[157] N. Z. Sun. *Inverse Problems in Groundwater Modeling*. Kluwer Academic Publishers, Boston, 1984.

[158] A. Tarantola. *Inverse Problem Theory: Methods for Data Fitting and Model Parameter Estimation*. Elsevier, New York, 1987.

[159] A. Tarantola and B. Valette. Inverse problems = quest for information. *Journal of Geophysics*, 50(3):159–170, 1982.

[160] G. B. Taylor, C. L. Carilli, and R. A. Perley, editors. *Synthesis Imaging in Radio Astronomy II*. Astronomical Society of the Pacific, San Francisco, 1999.

[161] R. A. Thisted. *Elements of Statistical Computing*. Chapman and Hall, New York, 1988.

[162] C. Thurber. Hypocenter-velocity structure coupling in local earthquake tomography. *Physics of the Earth and Planetary Interiors*, 75(1–3):55–62, 1992.

[163] C. Thurber and K. Aki. Three-dimensional seismic imaging. *Annual Review of Earth and Planetary Sciences*, 15:115–139, 1987.

[164] A. N. Tikhonov and V. Y. Arsenin. *Solutions of Ill-Posed Problems*. Halsted Press, New York, 1977.

[165] A. N. Tikhonov and A. V. Goncharsky, editors. *Ill-Posed Problems in the Natural Sciences*. MIR Publishers, Moscow, 1987.

[166] J. Trampert and J. J. Leveque. Simultaneous iterative reconstruction technique: Physical interpretation based on the generalized least squares solution. *Journal of Geophysical Research*, 95(B8):12553–12559, 1990.

[167] L. N. Trefethen and D. Bau. *Numerical Linear Algebra*. SIAM, Philadelphia, 1997.

[168] S. Twomey. *Introduction to the Mathematics of Inversion in Remote Sensing and Indirect Measurements*. Dover, Mineola, New York, 1996.

[169] T. Ulrych, A. Bassrei, and M. Lane. Minimum relative entropy inversion of 1-d data with applications. *Geophysical Prospecting*, 38(5):465–487, 1990.

[170] J. Um and C. Thurber. A fast algorithm for two-point seismic ray tracing. *Bulletin of the Seismological Society of America*, 77(3):972–986, 1987.

[171] A. van der Sluis and H. A. van der Vorst. Numerical solution of large, sparse linear algebraic systems arising from tomographic problems. In G. Nolet, editor, *Seismic Tomography with Applications in Global Seismology and Exploration Geophysics*, chapter 3, pages 49–83. D. Reidel, 1987.

[172] M. Th. van Genuchten. A closed-form equation for predicting the hydraulic conductivity of unsaturated soils. *Soil Science Society of America Journal*, 44:892–898, 1980.

[173] C. F. Van Loan. Generalizing the singular value decomposition. *SIAM Journal on Numerical Analysis*, 13:76–83, 1976.

[174] C. R. Vogel. Non-convergence of the L-curve regularization parameter selection method. *Inverse Problems*, 12:535–547, 1996.

[175] C. R. Vogel. *Computational Methods for Inverse Problems*. SIAM, Philadelphia, 2002.

[176] G. Wahba. *Spline Models for Observational Data*. SIAM, Philadelphia, 1990.

[177] G. A. Watson. Approximation in normed linear spaces. *Journal of Computational and Applied Mathematics*, 121(1–2):1–36, 2000.

[178] G. M. Wing. *A Primer on Integral Equations of the First Kind: The Problem of Deconvolution and Unfolding*. SIAM, Philadelphia, 1991.

[179] A. D. Woodbury and T. J. Ulrych. Minimum relative entropy inversion: Theory and application to recovering the release history of a ground water contaminant. *Water Resources Research*, 32(9):2671–2681, 1996.

[180] Z. Zhang and Y. Huang. A projection method for least squares problems with a quadratic equality constraint. *SIAM Journal on Matrix Analysis and Applications*, 25(1):188–212, 2003.

INDEX

International Geophysics Series

EDITED BY

RENATA DMOWSKA

Division of Applied Science
Harvard University
Cambridge, Massachusetts

JAMES R. HOLTON†

Department of Atmospheric Sciences
University of Washington
Seattle, Washington

H. THOMAS ROSSBY

Graduate School of Oceanography
University of Rhode Island
Narragansett, Rhode Island

---

* Out of print

Volume 10   MICHELE CAPUTO. The Gravity Field of the Earth from Classical and Modern Methods. 1967*

Volume 11   S. MATSUSHITA AND WALLACE H. CAMPBELL (eds.). Physics of Geomagnetic Phenomena (In two volumes). 1967*

Volume 12   K. YA KONDRATYEV. Radiation in the Atmosphere. 1969*

Volume 13   E. PALMÅN AND C. W. NEWTON. Atmospheric Circulation Systems: Their Structure and Physical Interpretation. 1969*

Volume 14   HENRY RISHBETH AND OWEN K. GARRIOTT. Introduction to Ionospheric Physics. 1969*

Volume 15   C. S. RAMAGE. Monsoon Meteorology. 1971*

Volume 16   JAMES R. HOLTON. An Introduction to Dynamic Meteorology. 1972*

Volume 17   K. C. YEH AND C. H. LIU . Theory of Ionospheric Waves. 1972*

Volume 18   M. I. BUDYKO. Climate and Life. 1974*

Volume 19   MELVIN E. STERN. Ocean Circulation Physics. 1975

Volume 20   J. A. JACOBS. The Earth's Core. 1975*

Volume 21   DAVID H. MILLER. Water at the Surface of the Earth: An Introduction to Ecosystem Hydrodynamics. 1977

Volume 22   JOSEPH W. CHAMBERLAIN. Theory of Planetary Atmospheres: An Introduction to Their Physics and Chemistry. 1978*

Volume 23   JAMES R. HOLTON. An Introduction to Dynamic Meteorology, Second Edition. 1979*

Volume 24   ARNETT S. DENNIS. Weather Modification by Cloud Seeding. 1980*

Volume 25   ROBERT G. FLEAGLE and JOOST A. BUSINGER. An Introduction to Atmospheric Physics, Second Edition. 1980*

Volume 26   KUO-NAN LIOU. An Introduction to Atmospheric Radiation. 1980*

Volume 27   DAVID H. MILLER. Energy at the Surface of the Earth: An Introduction to the Energetics of Ecosystems. 1981

Volume 28   HELMUT G. LANDSBERG. The Urban Climate. 1991

Volume 29   M. I. BUDKYO. The Earth's Climate: Past and Future. 1982*

Volume 30   ADRIAN E. GILL. Atmosphere-Ocean Dynamics. 1982

Volume 31   PAOLO LANZANO. Deformations of an Elastic Earth. 1982*

Volume 32   RONALD T. MERRILL AND MICHAEL W. MCELHINNY. The Earth's Magnetic Field: Its History, Origin, and Planetary Perspective. 1983*

Volume 33   JOHN S. LEWIS AND RONALD G. PRINN. Planets and Their Atmospheres: Origin and Evolution. 1983

About the CD-ROM

The CD-ROM accompanying this book contains MATLAB programs and data files for all MATLAB-based examples in the book, along with Per Hansen's MATLAB package, Regularization Tools. The disk also includes data files required in the exercises. Throughout the book, you will find a CD icon indicating that the example or exercise has associated material on the CD. Instructors and students may find these materials to be especially useful both as general learning aids, and as useful examples of basic MATLAB programming. These programs can be run on many platforms including Windows, Linux, Mac OS X, and Solaris. The programs have been tested under versions 6.5.1 and 7.0 of MATLAB.

The CD has three directories:

- Regutools: This directory contains Hansen's Regularization Tools.
- Homework: This directory contains files referred to in various exercises, organized by chapter.
- Examples: This directory contains the MATLAB codes for examples, organized by chapter and example number.

A companion Web site containing supplemental materials for this book can be found at books.elsevier.com/companions/0120656043.